Studies in Computational Intelligence

Volume 655

Series editor

Janusz Kacprzyk, Polish Academy of Sciences, Warsaw, Poland
e-mail: kacprzyk@ibspan.waw.pl

About this Series

The series "Studies in Computational Intelligence" (SCI) publishes new developments and advances in the various areas of computational intelligence—quickly and with a high quality. The intent is to cover the theory, applications, and design methods of computational intelligence, as embedded in the fields of engineering, computer science, physics and life sciences, as well as the methodologies behind them. The series contains monographs, lecture notes and edited volumes in computational intelligence spanning the areas of neural networks, connectionist systems, genetic algorithms, evolutionary computation, artificial intelligence, cellular automata, self-organizing systems, soft computing, fuzzy systems, and hybrid intelligent systems. Of particular value to both the contributors and the readership are the short publication timeframe and the worldwide distribution, which enable both wide and rapid dissemination of research output.

More information about this series at http://www.springer.com/series/7092

Stefka Fidanova
Editor

Recent Advances in Computational Optimization

Results of the Workshop on Computational Optimization WCO 2015

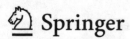 Springer

Editor
Stefka Fidanova
Department of Parallel Algorithms
Institute of Information and Communication
   Technologies
Bulgarian Academy of Sciences
Sofia
Bulgaria

ISSN 1860-949X   ISSN 1860-9503   (electronic)
Studies in Computational Intelligence
ISBN 978-3-319-82037-8   ISBN 978-3-319-40132-4   (eBook)
DOI 10.1007/978-3-319-40132-4

Printed on acid-free paper

This Springer imprint is published by Springer Nature
The registered company is Springer International Publishing AG Switzerland

Preface

Many real-world problems arising in engineering, economics, medicine, and other domains can be formulated as optimization tasks. Every day we solve optimization problems. Optimization occurs in minimizing time and cost or maximizing profit, quality, and efficiency. Such problems are frequently characterized by nonconvex, nondifferentiable, discontinuous, noisy or dynamic objective functions, and constraints which ask for adequate computational methods.

This volume is a result of vivid and fruitful discussions held during the workshop on computational optimization. The participants have agreed that the relevance of the conference topic and the quality of the contributions have clearly suggested that a more comprehensive collection of extended contributions devoted to the area would be very welcome and would certainly contribute to a wider exposure and proliferation of the field and ideas.

This volume includes important real problems such as parameter settings for controlling processes in bioreactor, control of ethanol production, minimal convex hill with application in routing algorithms, graph coloring, flow design in photonic data transport system, predicting indoor temperature, crisis control center monitoring, fuel consumption of helicopters, portfolio selection, GPS surveying, and so on. Some of them can be solved applying traditional numerical methods, but others need huge amount of computational resources. Therefore it is more appropriate to develop an algorithms based on some metaheuristic method like evolutionary computation, ant colony optimization, constrain programming, etc., for them.

Sofia, Bulgaria
April 2016

Stefka Fidanova
Co-Chair, WCO 2015

Organization Committee

Workshop on Computational Optimization (WCO 2015) is organized in the framework of Federated Conference on Computer Science and Information Systems (FedCSIS)—2015

Conference Co-chairs

Stefka Fidanova, IICT, Bulgarian Academy of Sciences, Bulgaria
Antonio Mucherino, IRISA, Rennes, France
Daniela Zaharie, West University of Timisoara, Romania

Program Committee

David Bartl, University of Ostrava, Czech Republic
Tibérius Bonates, Universidade Federal do Ceará, Brazil
Mihaela Breaban, University of Iasi, Romania
Camelia Chira, Technical University of Cluj-Napoca, Romania
Douglas Gonçalves, Universidade Federal de Santa Catarina, Brazil
Stefano Gualandi, University of Pavia, Italy
Hiroshi Hosobe, National Institute of Informatics, Japan
Hideaki Iiduka, Kyushu Institute of Technology, Japan
Nathan Krislock, Northern Illinois University, USA
Carlile Lavor, IMECC-UNICAMP, Campinas, Brazil
Pencho Marinov, Bulgarian Academy of Science, Bulgaria
Stelian Mihalas, West University of Timisoara, Romania
Ionel Muscalagiu, Politehnica University Timisoara, Romania
Giacomo Nannicini, University of Technology and Design, Singapore
Jordan Ninin, ENSTA-Bretagne, France
Konstantinos Parsopoulos, University of Patras, Greece

Contents

Fast Output-Sensitive Approach for Minimum Convex Hulls Formation

Artem Potebnia and Sergiy Pogorilyy

Abstract The paper presents an output-sensitive approach for the formation of the minimum convex hulls. The high speed and close to the linear complexity of this method are achieved by means of the input vertices distribution into the set of homogenous units and their filtration. The proposed algorithm uses special auxiliary matrices to control the process of computation. Algorithm has a property of the massive parallelism, since the calculations for the selected units are independent, which contributes to their implementation by using the graphics processors. In order to demonstrate its suitability for processing of the large-scale problems, the paper contains a number of experimental studies for the input datasets prepared according to the uniform, normal, log-normal and Laplace distributions.

1 Introduction

Finding the minimum convex hull (MCH) of the graph's vertices is a fundamental problem in many areas of modern research [9]. A set of nodes V in an affine space E is convex if $c \in V$ for any point $c = \sigma a + (1 - \sigma)b$, where $a, b \in V$ and $\sigma \in [0, 1]$ [9]. Formation of the convex hull for any given subset S of E requires calculation of the minimum convex set containing S (Fig. 1a). It is known that MCH is a common tool in computer-aided design and computer graphics packages [23].

In computational geometry convex hull is just as essential as the "sorted sequence" for a collection of numbers. For example, Bezier's curves used in *Adobe Photoshop*, *GIMP* and *CorelDraw* for modeling smooth lines fully lie in the convex hull of their control nodes (Fig. 1b). This feature greatly simplifies finding the points of intersection between curves and allows their transformation (moving, scaling, rotating,

A. Potebnia (✉) · S. Pogorilyy
Kyiv National Taras Shevchenko University, Kyiv, Ukraine
e-mail: potebnia@mail.ua

S. Pogorilyy
e-mail: sdp@univ.net.ua

© Springer International Publishing Switzerland 2016
S. Fidanova (ed.), *Recent Advances in Computational Optimization*,
Studies in Computational Intelligence 655, DOI 10.1007/978-3-319-40132-4_1

(a)   **(b)**

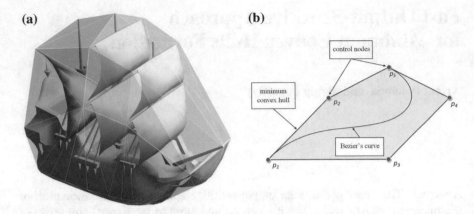

Fig. 1 Examples of the minimum convex hulls

etc.) by appropriate control nodes [26]. The formation of some fonts and animation effects in the *Adobe Flash* package also uses splines composed of quadratic Bezier's curves [10].

Convex hulls are commonly used in *Geographical Information Systems* and routing algorithms in determining the optimal ways for avoiding obstacles. The papers [1] offer the methods for solving complex optimization problems using them as the basic data structures. For example, the process of the set diameter calculation can be accelerated by means of the preliminary MCH computation. This approach is finished by application of the rotating calipers method to obtained hull, and its expediency is based on the reduction of the problem dimensionality.

MCHs are also used for simplifying the problem of classification by implementing the similar ideas. Let's consider the case of the binary classification which requires the finding of the hyperplane that separates two given sets of points and determines the maximum possible margin between them. Acceleration of the corresponding algorithms is associated with the analyzing of only those points that belong to the convex hulls of the initial sets.

Last decades are associated with rapid data volume growth in research processed by the information systems [22]. According to IBM, about 15 petabytes of new information are created daily in the world. Therefore, in modern science, there is a separate area called *Big Data* related to the study of large data sets [25]. However, most of the known algorithms for MCH construction have time complexity $O(n \log n)$, making them useless when forming solutions for large-scale graphs. Therefore, there is a need to develop efficient algorithms with the complexity close to linear $O(n)$.

It is known that Wolfram Mathematica is one of the most powerful mathematical tools for the high performance computing. Features of this package encapsulate a number of algorithms and, depending on the input parameters of the problem, select the most productive ones [21]. Therefore, Wolfram Mathematica 9.0 is used to track the performance of the algorithm proposed in this article.

In recent years, CPU+GPU hybrid systems (GPGPU technology) allowing for a significant acceleration of computations have become widespread. Unlike CPU, consisting of several cores, the graphics processor is a multicore structure and the number of its components is measured in hundreds [17]. In this case, the sequential steps of algorithm are executed on the CPU, while its parallel parts are implemented on the GPU [21]. For example, the latest generation of NVIDIA Fermi GPUs contains 512 computing cores, allowing for the introduction of new algorithms with large-scale parallelism [11]. Thus, the usage of NVIDIA GPU ensures the conversion of standard workstations to powerful supercomputers with cluster performance [19].

This paper is organized as follows. Section 2 contains the analysis of the problem complexity and provides the theoretical background to the development of new method. A short review of the existing traditional methods and their improvements is given in Sect. 3. Sections 4 and 5 are devoted to the description of the proposed algorithm. Section 6 presents the experimental data and their discussion for the uniform distribution of initial nodes. In this case, the time complexity of the proposed method is close to linear. However, effective processing of the input graphs in the case of the low entropy distributions requires the investigation of the algorithm execution for such datasets. Section 7 contains these experiments and their discussion.

2 Complexity of the Problem

Determination of similarities and differences between the computational problems is a powerful tool for the efficient algorithms development. In particular, the reduction method is now widely used for providing an estimation of problems complexity and defining the basic principles for the formation of classification [4].

The polynomial-time reduction of the combinatorial optimization problem A to another problem B is presented by two transformations f and h, which have the polynomial time complexity. Herewith, the algorithm f determines the mapping of any instance I for the original problem A to the sample $f(I)$ for the problem B. At the same time, the algorithm h implements the transformation of the global solution S for the obtained instance $f(I)$ to the solution $h(S)$ for the original sample I. On the basis of these considerations, any algorithm for solving the problem B can be applied for calculation of the problem A solutions by including the special operations f and h, as shown in Fig. 2.

The reduction described above is denoted by $A \leq_{poly} B$. The establishment of such transformation indicates that the problem B is at least as complex as the problem A [21]. Therefore, the presence of the polynomial time algorithm for B leads to the possibility of its development for the original problem A.

Determination of a lower bound on the complexity for the problem of convex hull computing requires establishing of the reduction from sorting (SORT) to MCH. In this case, the initial instances I of the original problem are represented by collections $X = \langle x_1, x_2, \ldots, x_n \rangle$, while the samples $f(I)$ form the planar point set P. The function f provides the formation of the set P by the introduction of the individual vertices

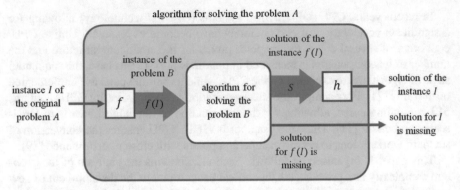

Fig. 2 Diagram illustrating the reduction of the computational problem A to the problem B

$p_i = (x_i, x_i{}^2)$ for items x_i of the input collection. Therefore, the solutions S are represented by the convex hulls of the points $p_i \in P$, arranged on a parabola. The mapping of solutions h requires only traversing of the obtained hull starting from the leftmost point. This relation $SORT \leq_{poly} MCH$ describes the case in which the hull contains all given points. But, in general, the number of nodes on MCH, denoted by h, may be less than n.

It is known that the reduction relation is not symmetric. However, for the considered problems the reverse transformation $MCH \leq_{poly} SORT$ is also established, and the *Graham's* algorithm [14] demonstrates the example of its usage for the formation of the convex hulls. This relation is satisfied for an arbitrary number of vertices in the hull. Therefore, the complexity of the MCH construction problem is described by the following system of reductions:

$$SORT \leq_{poly} MCH, \text{ where } h = n;$$
$$MCH \leq_{poly} SORT, \text{ where } h \leq n.$$

According to the first relation, the lower bound on the complexity of the convex hull computation for the case $h = n$ is limited by its value for the sorting problem and equals $O(n \log n)$. However, the second reduction shows that in the case $h < n$, there is a potential to develop the output-sensitive algorithms that overcome this limitation. The purpose of this paper is to design an approach for solving the MCH construction problem, which can realize this potential.

In order to form the classification of the combinatorial optimization problems, they are grouped into separate classes from the perspective of their complexity. Class P is represented by a set of problems whose solutions can be obtained in a polynomial time using a deterministic Turing machine. However, the problems belonging to the class P have different suitability for the application of the parallel algorithms. Fundamentally sequential problems which have no natural parallelism are considered as P-complete. An example of such problem is the calculation of the maximum flow. Per contra, the problems which have the ability to efficient parallel

Fig. 3 Internal structure of the complexity class P

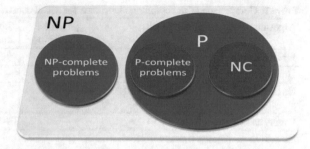

implementation are combined by the class $NC \subseteq P$. The problem of determining the exact relationship between the sets NC and P is still open, but the assumption $NC \subset P$ is the most common, as shown in Fig. 3.

A formal condition for the problem inclusion to the class NC is determined as the achievement of the time complexity $O(\log^k n)$ using $O(n^c)$ parallel processors, where k and c are constants, and n is the dimensionality of the input parameters [12]. Both sorting and MCH construction problems belong to the class NC. Therefore, the computation of the convex hulls has a high suitability for parallel execution, which should be realized by the effective algorithms.

3 A Review of Algorithms for Finding the Minimum Convex Hulls

Despite intensive research, which lasted for the past 40 years, the problem of developing efficient algorithms for MCH formation is still open. The main achievement is the development of numerous methods based on the extreme points determination of the original graph and the link establishment among them [8]. These techniques include the *Jarvis's march* [16], *Graham's Scan* [14], *QuickHull* [5], *Divide and Conquer* algorithm, and many others. The main features of their practical usage are given in Table 1.

For parallelization the *Divide and Conquer* algorithm is the most suitable. It provides a random division of the original vertex set into subsets, formation of partial solutions and their connection to the general hull [23]. Although the hull connection phase has linear complexity, it leads to a significant slowdown of the algorithm, and as a result, to the unsuitability of its application in the hull processing for large-scale graphs.

Chan's algorithm, which is a combination of slower algorithms, has the lowest time complexity $O(n \log h)$. However, it can work by the known number of vertices contained in the hull [3]. Therefore, currently, its usage in practice is limited [6].

Study [2] gives a variety of acceleration tools for known MCH formation algorithms by cutting off the graph's vertices falling inside an octagon or rectangle and appropriate reducing the dimensionality of the original problem. The paper [15]

Table 1 Comparison of the common algorithms for MCH construction

Algorithm	Complexity	Parallel versions	Ability of generalization for the multidimensional cases
Jarvis's march	$O(nh)$	+	+
Graham's Scan	$O(n \log n)$	–	–
QuickHull	$O(n \log n)$, in the worst case $-O(n^2)$	+	+
Divide and Conquer	$O(n \log n)$	+	+

suggests numerous methods of convex hull approximate formation, which have linear complexity. Such algorithms are widely used for tasks where speed is a critical parameter. But linearithmic time complexity of the fastest exact algorithms demonstrates the need for the introduction of new high-speed methods of convex hulls formation for large-scale graphs.

4   Overview of the Proposed Algorithm

We shall consider non-oriented planar graph $G = (V, E)$. The proposed algorithm provides a division of the original graph's vertex set into a set of output units $U = \langle U_1, U_2, \ldots, U_n \rangle$, $U_i \subseteq V$. However, unlike the *Divide and Conquer* method, this division is not random, but it is based on the spatial distribution of vertices. All nodes of the graph should be distributed by the formed subsets, i.e. $\bigcup_{i=1}^{n} U_i = V$. This allows the presence of empty units, which don't contain vertices. Additionally, the condition of orthogonality division is met, i.e. one vertex cannot be a part of the different blocks: $U_i \cap U_j = \varnothing, \forall i \neq j$. In this study, all allocated units are homogeneous and have the same geometrical sizes.

The next stage of the proposed algorithm involves the formation of an auxiliary matrix based on the distribution of nodes by units. The purpose of this procedure is the primary filtration of the graph's vertices, which provides a significant decrease in the original problem dimensionality. In addition, the following matrices define the sets of blocks for the calculation in the subsequent stages of the algorithm and the sequence of their connection to the overall result. An auxiliary matrix formation involves the following operations:

1. Each block of the original graph $U_{i,j}$ must be mapped to one cell $c_{i,j}$ of the supporting matrix. Accordingly, the dimensionality of this matrix is $n \times m$, where n and m are the numbers of blocks allocated by the relevant directions.
2. The following operations provide the necessary coding of matrix's cells. Thus, the value of cell $c_{i,j}$ is zero if the corresponding block $U_{i,j}$ of original graph contains no vertices. Coding of blocks that contain extreme nodes (which have the lowest

and largest values of both plane coordinates) of a given set is important for the algorithm. In particular, units, which contain the highest, rightmost, lowest and leftmost points of the given data set are respectively coded with 2, 3, 4 and 5 in the matrix representation. Other units that are filled, and contain no extreme peaks, shall be coded with ones in auxiliary matrix.

3. Further, primary filtration of allocated blocks is carried out using the filled matrix. Empty subsets thus shall be excluded from consideration. Blocks containing extreme vertices shall determine the graph division into parts (called northwest, southwest, southeast, and northeast) for which the filtration procedure is applied. We shall consider the example of the block selection for the northwest section limited with cells 2–3. If $c_{i,j} = 2$, then the next non-zero cell is searched by successive increasing of j. In their absence, the next matrix's row $i + 1$ is reviewed. Selection of blocks is completed, if the value of the next chosen cell is $c_{i,j} = 3$ (Fig. 4). Processing of the southwest, southeast, and northeast parts is based on a similar principle. These operations can be interpreted as solving of the problem at the macro level.

At the next step partial solutions are formed for selected blocks. Such operations require the formation of fragments rather than full-scale hulls that provides the secondary filtration of the graph's vertices. The last step of the algorithm involves the connection of partial solutions to the overall result. Thus, the sequential merging of local fragments is done on a principle similar to *Jarvis's march*. It should be noted that at this stage filtration mechanism leads to a significant reduction in the dimensionality

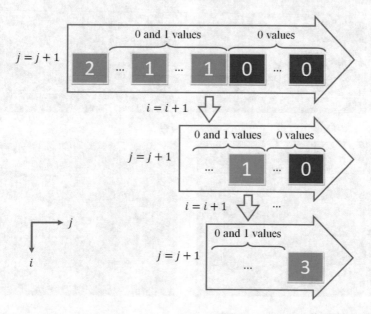

Fig. 4 Diagram demonstrating the traversing of the northwest matrix section limited with cells 2–3

of the original problem. Therefore, when processing the hulls for large graphs, the combination operations constitute about 0.1 % of the algorithm total operation time.

We shall consider the example of this algorithm execution. Let the set of the original graph's vertices have undergone division into 30 blocks (Fig. 5a). Auxiliary matrix calculated for this case is given in Fig. 5b. After application of the primary filtration, only 57 % of the graph's nodes were selected for investigation at the following stages of the algorithm (Fig. 5c). The next operations require the establishment of

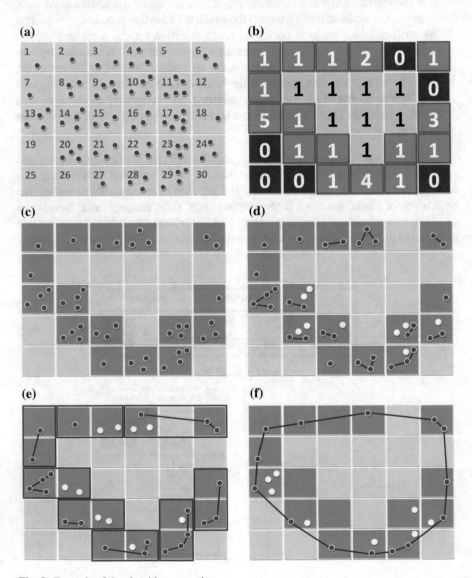

Fig. 5 Example of the algorithm execution

the local hulls (Fig. 5d) and their aggregations are given in Fig. 5e. After performing of the pairwise connections, this operation is applied repeatedly until a global convex hull is obtained (Fig. 5f).

5   The Development of Hybrid CPU–GPU Algorithm

It is known that the video cards have much greater processing power compared to the central processing elements. GPU computing cores work simultaneously, enabling to use them to solve problems with the large volume of data. CUDA (*Compute Unified Device Architecture*), the technology created by NVIDIA, is designed to increase the productivity of conventional computers through the usage of video processors computing power [24].

CUDA architecture is based on SIMD (*Single Instruction Multiple Data*) concept, which provides the possibility to process the given set of data via one function. Programming model provides for consolidation of threads into blocks, and blocks— into a grid, which is performed simultaneously. Accordingly, the key to effective usage of GPU hardware capabilities is algorithm parallelization into hundreds of blocks performing independent calculations on the video card [27].

It is known that GPU consists of several clusters. Each of them has a texture unit and two streaming multiprocessors, each containing 8 computing devices and 2 superfunctional units [13]. In addition, multiprocessors have their own distributed memory resources (16 KB) that can be used as a programmable cache to reduce delays in data accessing by computing units [24]. From these features of CUDA architecture, it may be concluded that it is necessary to implement massively-parallel parts of the algorithm on the video cards, while sequential instructions must be executed on the CPU. Accordingly, the stage of partial solutions formation is suitable for implementation on the GPU since the operations for each of the numerous blocks are carried out independently.

It is known that function designed for executing on the GPU is called a kernel. The kernel of the proposed algorithm contains a set of instructions to create a local hull of any selected subset. In this case, distinguishing between the individual subproblems is realized only by means of the current thread's number. Thus, the hybrid algorithm (Fig. 6) has the following execution stages:

1. Auxiliary matrix is calculated on the CPU. The program sends cells' indexes that have passed the primary filtration procedure and corresponding sets of vertices to the video card.
2. Based on the received information, particular solutions are formed on the GPU, recorded to its global memory and sent to the CPU.
3. Further, the procedure of their merging is carried out and the overall result is obtained.

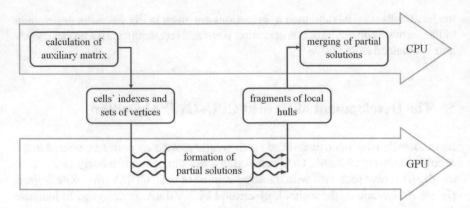

Fig. 6 Diagram illustrating a strategy of the parallel algorithm formation

It should be noted that an important drawback of hybrid algorithms is the need to copy data from the CPU to the GPU and vice versa, which leads to significant time delays [17, 20]. Communication costs are considerably reduced by means of the filtration procedure.

When developing high-performance algorithms for the GPU it is important to organize the correct usage of the memory resources. It is known that data storage in the global video memory is associated with significant delays in several hundred GPU cycles. Therefore, in the developed algorithm, the global memory is used only as a means of communication between the processor and video card. The results of intermediate calculations for each of the threads are recorded in the shared memory, access speed of which is significantly higher and is equal to 2–4 cycles.

6 Experimental Studies of the Proposed Algorithm for Uniformly Distributed Datasets

In this section, both coordinates of the input vertices have uniform distribution $U[a, b]$, where a and b are the minimum and maximum values of the distribution's support. The probability density of this distribution is constant in the specified interval $[a, b]$. Figure 11a shows the example of such dataset with the grid of units and obtained global convex hull for $a = 0$ and $b = 10$.

For the respective datasets, the number of allocated homogeneous units, which have the fixed average size, increases linearly with enhancing of the processed graphs dimensionality. The complexity of calculating the relevant auxiliary matrices grows by the same principle, and partial problems have the constant average dimensionality. The stages of multi-step filtration and local hulls construction provide a significant simplification of the final connection procedure. Therefore, its contribution to the total running time is insignificant. Thus, the complexity of the developed algorithm is close to linear $O(n)$ for uniformly distributed data.

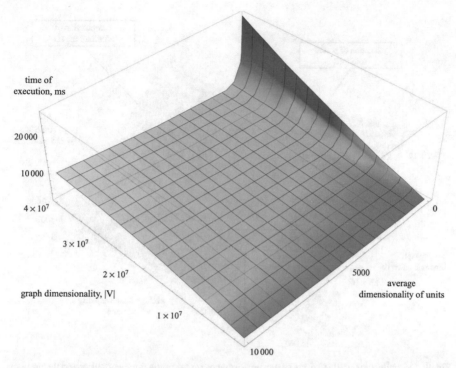

Fig. 7 Dependence of the developed algorithm performance on the graph dimensionality and the number of vertices in the selected units for uniformly distributed datasets

MCH instances composed of all graph's vertices are the worst for investigation. In this case, the filtration operations don't provide the required acceleration and the lower bound on the complexity of the algorithm is determined by the reduction $SORT \leq_{poly} MCH$ and equals $O(n \log n)$.

In the current survey, experimental tests were run on a computer system with an Intel Core i7-3610QM processor (2.3 GHz), 8 GB RAM and DDR3-1600 NVIDIA GeForce GT 630M video card (2GB VRAM). This graphics accelerator contains 96 CUDA kernels, and its clock frequency is 800 MHz.

Figure 7 shows the dependence of the proposed algorithm execution time on the graph dimensionality and the number of vertices in the selected blocks. These results confirm the linear complexity of the proposed method for uniformly distributed data. In addition, it is important to set the optimal dimensionality of the subsets allocated in the original graph. A selection of smaller blocks (up to 1000 nodes) leads to a dramatic increase in the algorithm operation time.

This phenomenon is caused by the significant enhancing of the auxiliary matrices dimensionality, making it difficult to control the computing process (Fig. 8). Per contra, the allocation of large blocks (over 5000 vertices) is associated with the elimination of the massive parallel properties, enhancing of the partial problems dimensionality, and as a consequence, increasing of the algorithm execution time.

A. Potebnia and S. Pogorilyy

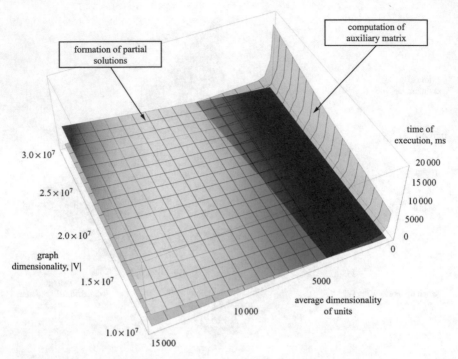

Fig. 8 Dependencies of the various stages performance on the graph dimensionality and the number of vertices in the selected units

Thus, the highest velocity of the proposed method is observed for intermediate values of the blocks dimensionality (1000–5000 vertices). In this case, auxiliary matrices are relatively small, and the second stage of the algorithm preserves the properties of massive parallelism.

One of the most important means to ensure the algorithm's high performance is the multi-step filtration of the graph's vertices. Figure 9a shows the dependence of the primary selection quality on the dimensionality of the original problem and allocated subsets. These results show that such filtration is the most efficient with the proviso that the graph's vertices are distributed into small blocks. Furthermore, the number of selected units increases with the raising of the problem's size, providing rapid solutions to graphs of extra large dimensionality. By virtue of a riddance from the discarded blocks, the next operations of the developed algorithm are applied only to 1–3 % of the initial graph's vertices.

However, the results of the secondary filtration (Fig. 9b) are the opposite. In this case, the highest quality of the selection is obtained on the assumption that the original vertices are grouped into large subsets. Withal, the secondary filtration is much slower than the primary procedure, so the most effective selection occurs at intermediate values of the blocks dimensionality. As a result of these efforts, only 0.05–0.07 % of the initial graph's vertices are involved in the final operations of the proposed algorithm.

(a)

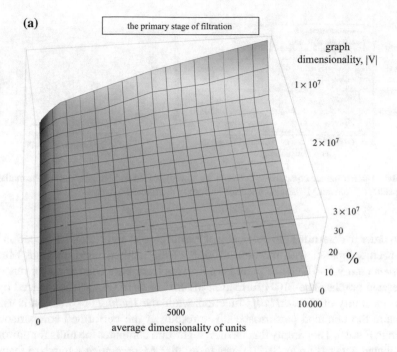

(b)

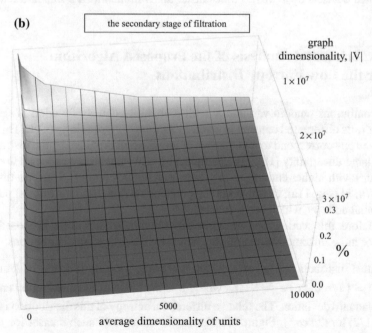

Fig. 9 The influence of the primary and secondary filtration procedures over the reduction in the problem size

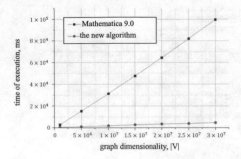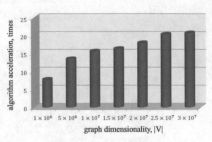

Fig. 10 A performance comparison between the new algorithm and built-in tools of the mathematical package Wolfram Mathematica 9.0 for uniformly distributed datasets

In order to determine the efficiency of the developed algorithm, its execution time has been compared with the built-in tools of the mathematical package *Wolfram Mathematica 9.0*. All choice paired comparison tests were conducted for randomly generated graphs. The MCH formation in *Mathematica* package is realized by the instrumentality of *ConvexHull[ ]* function, while the *Timing[ ]* expression is used to measure the obtained performance. The results of the performed comparison are given in Fig. 10. They imply that the new algorithm computes the hulls for uniformly distributed datasets up to 10–20 times faster than *Mathematica's* standard features.

7 Experimental Analysis of the Proposed Algorithm for the Low Entropy Distributions

For a continuous random variable X with probability density function $p(x)$ in the interval I, its differential entropy is given by $h(X) = -\int_I p(x) \log p(x) dx$. High values of entropy correspond to less amount of information provided by the distribution and its large uncertainty [18]. For example, physical systems are expected to evolve into states with higher entropy as they approach equilibrium [7]. Uniform distributions $U[a, b]$ (Fig. 11a), examined in the previous section, have the highest possible differential entropy, which value equals $\log(b - a)$.

Therefore, this section focuses on the experimental studies of the proposed algorithm for more informative datasets presented by the following distributions:

1. Normal distribution $N(\mu, \sigma^2)$, which probability density function is defined as $p(x) = \left(1/\sigma\sqrt{2\pi}\right) e^{-\frac{1}{2\sigma^2}(x-\mu)^2}$, where μ is the mean of the distribution and σ is its standard deviation. The relative differential entropy of this distribution is equal to $(1/2)\log(2\pi e\sigma^2)$. Figure 11b shows the example of such dataset for $\mu = 5$ and $\sigma = 1$.

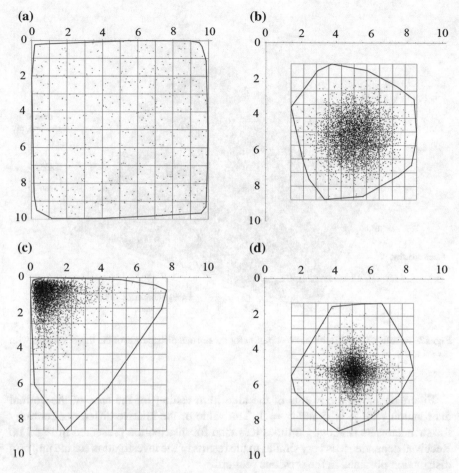

Fig. 11 Examples of the distributions used for the algorithm investigation with the structures of allocated units and received hulls

2. Log-normal distribution, whose logarithm is normally distributed. In contrast to the previous distribution, it is single-tailed with a semi-infinite range and the random variable takes on only positive values. Its differential entropy is equal to $\log(2\pi\sigma^2 e^{\mu+1/2})$. Example of this distribution for $\mu = 0$ and $\sigma = 0.6$ is shown in Fig. 11c.
3. Laplace (or double exponential) distribution, which has the probability density function $p(x) = (1/2b)\,e^{-\frac{|x-\mu|}{b}}$ that consists of two exponential functions, where μ and b are the location and scale parameters. The entropy of this distribution is equal to $\log(2be)$. Laplace distribution for $\mu = 5$ and $b = 0.5$ is illustrated in Fig. 11d.

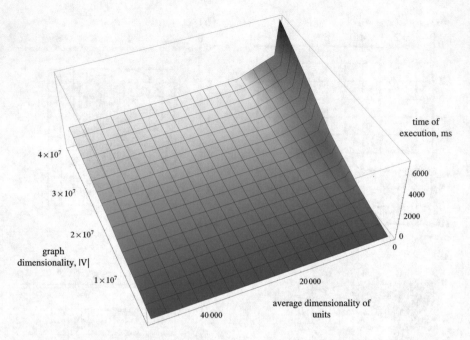

Fig. 12 Results of the algorithm investigation for the normal distribution of the initial vertices with $\mu = 5$ and $\sigma = 1$

Figure 12 shows the results of the algorithm testing for the case of the normal distribution with $\mu = 5$ and $\sigma = 1$. The value of the relative differential entropy for such datasets is about 1.6 times less than for distribution presented in Fig. 11a. Received dependence is very similar to the results of the investigation for the uniform distribution obtained in the previous section.

However, the speed of such datasets processing is 2–4 times larger due to the significant increasing in the primary filtration efficiency. And the reason for this is that the density of the normal distribution is maximum for the central units, which are discarded after formation of the auxiliary matrix. For example, if $|V| = 10^6$ and the average block dimensionality $U_{av} = 1000$, only 0.02 % of the initial vertices are passing the primary filtration procedure, while for the uniform distribution, this value is about 12 %.

In addition, the running time of the algorithm for the normal distribution doesn't increase if the large subsets for which $U_{av} = 10000$ are allocated. Execution time even slowly decreases with a further enhancing of their dimensionality. This is associated with the reduction in the complexity of the auxiliary matrices calculation, which is accompanied by the sufficiently slow degradation of the primary filtration procedure. For example, if $|V| = 4 \times 10^7$, then the running time is increasing only when allocated blocks contain more than 10^6 vertices in average.

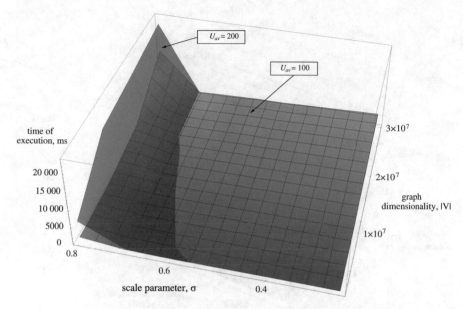

Fig. 13 Dependencies of the algorithm performance on the graph dimensionality, average dimensionality of units U_{av} and value of σ for the case of log-normal distribution with $\mu = 0$

The results obtained for the log-normal distribution (with $\mu = 0$) are the most difficult to analyze. The relevant dependencies of the algorithm performance are given in Fig. 13 for different σ values of the distribution and two values of the allocated units dimensionality. It is worth noting that the parameter σ determines the skewness of the distribution. In particular, the increasing of the σ value results in the enhancing of the tolerance interval and distancing of the mode (the global maximum of the probability density function) from the median.

As a consequence, selection of the high σ values leads to the increasing in the geometric sizes of the homogeneous units and accumulation of the input vertices near the extreme blocks of the grid. When reaching a certain critical value σ_c, the main congestion of the nodes appears in the extreme blocks. This effect leads to a rapid degradation of the primary filtration procedure and is accompanied by a drastic jump in the dependence of the algorithm performance. However, the selection of the small units allows to increase the critical value σ_c and minimize the magnitude of such jump. Therefore, unlike the cases of the uniform and normal distributions, the lognormal distribution requires the allocation of the small units, which dimensionality is less than 100 nodes.

The results obtained for the Laplace distribution with $\mu = 5$ and $b = 1$ (Fig. 14) are reminiscent of the dependence received for the normal distribution. However, in the case of the Laplace distribution, the input vertices are even more concentrated in the central units of the grid. As a consequence, the primary filtration procedure is more efficient and the time required for the processing of the corresponding datasets is approximately 30 % less than the results obtained for the normal distribution.

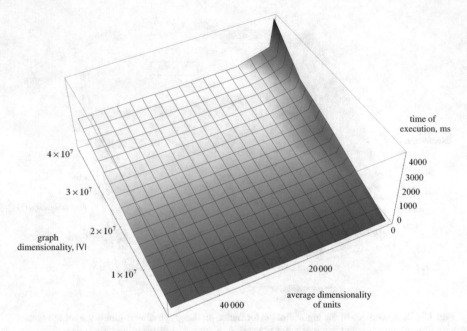

Fig. 14 Results of the algorithm execution for the Laplace distribution of the initial vertices with $\mu = 5$ and $b = 1$

8 Conclusion

The paper suggests an output-sensitive approach for computation of the minimum convex hulls, which is based on the division of the graph's vertex set into a set of output units and usage of the special auxiliary matrices for solving of the problem at the macro level. By means of the primary and secondary filtration procedures, the proposed algorithm is adapted to fast processing of the large-scale problems and, therefore, is suitable for using with respect to *Big Data* direction.

The problem of the convex hull formation belongs to the class *NC*, and the developed algorithm has a property of the massive parallelism. In particular, the calculations of the partial hulls are carried out independently, which contributes to their implementation by using the graphics processors. The property of output sensitivity has a key value for the effective processing of the input datasets prepared according to the uniform, normal and Laplace distributions, as shown in Sects. 6 and 7. Therefore, the algorithm implements a potential which is represented by the reduction $MCH \leq_{poly} SORT$ for the case $h \leq n$. The effective application of the developed algorithm for processing of the datasets, which have the log-normal distribution, requires the allocation of the sufficiently small units to ensure the condition $\sigma < \sigma_c$.

Another advantage of the proposed algorithm is its ability to the dynamic adjustment of the convex hulls. When adding new vertices to the initial set, calculations are executed only for units that have been modified. These operations require only local updating of the convex hulls because the results for intact parts of the original graph are invariable. In addition, the algorithm concept envisages the possibility of its generalization for the multidimensional problem instances. In these cases, the allocated units are represented by the n-dimensional cubes to which operations of the developed method are applied.

References

1. Aardal, K., Van Hoesel, S.: Polyhedral techniques in combinatorial optimization I: theory. Statistica Neerlandica **50**, 3–26 (1995). doi:10.1111/j.1467-9574.1996.tb01478.x
2. Akl, S.G., Toussaint, G.T.: A fast convex hull algorithm. Inf. Process. Lett. **7**, 219–222 (1978)
3. Allison, D.C.S., Noga, M.T.: Some performance tests of convex hull algorithms. BIT **24**(1), 2–13 (1984). doi:10.1007/BF01934510
4. Arora, S., Barak, B.: Computational Complexity: A Modern Approach. Cambridge University Press, New York (2009)
5. Barber, C.B., Dobkin, D.P., Huhdanpaa, H.: The quickhull algorithm for convex hulls. ACM Trans. Math. Softw. **22**(4), 469–483 (1996). doi:10.1145/235815.235821
6. Chan, T.M.: Optimal output-sensitive convex hull algorithms in two and three dimensions. Discret. Comput. Geom. **16**, 361–368 (1996). doi:10.1007/BF02712873
7. Conrad, K.: Probability Distributions and Maximum Entropy. Available via DIALOG. http://www.math.uconn.edu/~kconrad/blurbs/analysis/entropypost.pdf (2005). Accessed 7 Feb 2016
8. Cormen, T.H., Leiserson, C.E., Rivest, R.L., Stein, C.: Introduction to Algorithms, 2nd edn. MIT Press, Cambridge (2001)
9. De Berg, M., Cheong, O., van Kreveld, M., Overmars, M.: Computational Geometry: Algorithms and Applications. Springer, Heidelberg (2008)
10. Duncan, M.: Applied Geometry for Computer Graphics and CAD. Springer, London (2005). doi:10.1007/b138823
11. Farber, R.: CUDA Application Design and Development. Elsevier, Waltham (2011)
12. Fujiwara, A., Inoue, M., Masuzawa, T.: Parallelizability of some P-complete problems. In: International Parallel and Distributed Processing Symposium, pp. 116–122. Springer, Cancun (2000)
13. Govindaraju, N.K., Larsen, S., Gray, J., Manocha, D.: A memory model for scientific algorithms on graphics processors. In: Proceedings of the ACM/IEEE Conference on Supercomputing. IEEE Press, Tampa (2006). doi:10.1109/SC.2006.2
14. Graham, R.L.: An efficient algorithm for determining the convex hull of a finite planar set. Info. Proc. Lett. **1**(1), 132–133 (1972). doi:10.1016/0020-0190(72)90045-2
15. Hossain, M., Amin, M.: On constructing approximate convex hull. Am. J. Comput. Math. **3**(1A), 11–17 (2013). doi:10.4236/ajcm.2013.31A003
16. Jarvis, R.A.: On the identification of the convex hull of a finite set of points in the plane. Info. Proc. Lett. **2**(1), 18–21 (1973). doi:10.1016/0020-0190(73)90020-3
17. Lee, C., Ro, W.W., Gaudiot, J.-L.: Boosting CUDA applications with CPU-GPU hybrid computing. Int. J. Parallel Program. **42**(2), 384–404 (2014). doi:10.1007/s10766-013-0252-y
18. Michalowicz, J.V., Nichols, J.M., Bucholtz, F.: Handbook of Differential Entropy. Taylor & Francis Group, Boca Raton (2014)
19. Nickolls, J., Dally, W.: The GPU computing era. IEEE Micro **30**(2), 56–69 (2010). doi:10.1109/MM.2010.41
20. Novakovic, V., Singer, S.: A GPU-based hyperbolic SVD algorithm. BIT **51**(4), 1009–1030 (2011). doi:10.1007/s10543-011-0333-5
21. Potebnia, A.V., Pogorilyy, S.D.: Exploration of data coding methods in wireless computer networks. In: Proceedings of the Fourth International Conference on Theoretical and Applied Aspects of Cybernetics, pp. 17–31. Bukrek, Kyiv, (2014). doi:10.13140/RG.2.1.3186.3844

22. Potebnia, A.V., Pogorilyy, S.D.: Innovative GPU accelerated algorithm for fast minimum convex hulls computation. In: Proceedings of the Federated Conference on Computer Science and Information Systems, pp. 555–561. IEEE Press, Lodz, (2015). doi:10.15439/2015F305
23. Preparata, F.P., Shamos, M.I.: Computational Geometry: An Introduction. Springer, New York (1985). doi:10.1007/978-1-4612-1098-6
24. Sanders, J., Kandrot, E.: CUDA by Example: An Introduction to General-Purpose GPU Programming. Addison-Wesley Professional, Boston (2010)
25. Schroeck, M., Shockley, R., Smart, J., Romero-Morales, D., Tufano, P.: Analytics: The real-world use of big data. How innovative enterprises extract value from uncertain data. IBM Institute for Business Value (2012)
26. Sederberg, T.W.: In: Computer aided geometric design course notes. Available via DIALOG. http://tom.cs.byu.edu/557/text/cagd.pdf (2011). Accessed 7 Feb 2016
27. Sklodowski, P., Zorski, W.: Movement tracking in terrain conditions accelerated with CUDA. In: Proceedings of the Federated Conference on Computer Science and Information Systems, pp. 709–717. IEEE Press, Warsaw (2014). doi:10.15439/2014F282

Local Search Algorithms for Portfolio Selection: Search Space and Correlation Analysis

Giacomo di Tollo and Andrea Roli

Abstract Modern Portfolio Theory dates back from the fifties, and quantitative approaches to solve optimization problems stemming from this field have been proposed ever since. We propose a metaheuristic approach for the Portfolio Selection Problem that combines local search and Quadratic Programming, and we compare our approach with an exact solver. Search space and correlation analysis are performed to analyse the algorithm's performance, showing that metaheuristics can be efficiently used to determine optimal portfolio allocation.

1 Introduction

Modern Portfolio Theory dates back to the 1950s and concerns wealth allocation over assets: the investor has to decide which asset to invest in and by how much. Many optimization problem have been formulated to express this principle, and the main example is to minimize a risk measure for a given minimum required target return. Variance of portfolio's return was used as risk measure in the seminal work by Markowitz [25] and is still the most used, even though there exists a wide literature about risk measures to be implemented.

Portfolio Selection Problem (PSP) can be viewed as an optimisation problem, defined in terms of three objects: variables, objective, and constraints. Every object has to be instantiated by a choice in a set of possible choices, the combination of which induces a specific formulation (model) of the problem, and different optimisation results. For instance, as observed by di Tollo and Roli [8], two main choices are

G. di Tollo (✉)
Dipartimento di Economia, Universitá Ca' Foscari, Cannaregio 873,
30121 Venice, Italy
e-mail: giacomo.ditollo@unive.it

A. Roli
Dipartimento di Informatica e Ingegneria,
Alma Mater Studiorum – Universitá di Bologna,
Campus of Cesena, Via Venezia 52, 47521 Cesena, Italy
e-mail: andrea.roli@unibo.it

© Springer International Publishing Switzerland 2016
S. Fidanova (ed.), *Recent Advances in Computational Optimization*,
Studies in Computational Intelligence 655, DOI 10.1007/978-3-319-40132-4_2

21

possible for variable domains: continuous [15, 28, 29, 31] and integer [22, 30]. Choosing continuous variables is a very 'natural' option and leads to a representation independent of the actual budget, while integer values (ranging between zero and the maximum available budget, or equal to the number of 'rounds') makes it possible to add constraints taking into account actual budget, minimum lots and to tackle other objective functions to capture specific features of the problem at hand. As for the different results, the integer formulation is more suitable to explain the behaviour of rational operators such small investors, whose activity is strongly influenced by integer constraint [23].

In addition, the same representation can be modelled by means of different formulations, e.g., by adding auxiliary variables [21], symmetry breaking [27] or redundant [32] constraints, which may provide beneficial effects on, or on the contrary harm, the efficiency of the search algorithms yet preserving the possibility of finding an optimal solution.

In this work we investigate how the use of different formulations for the very same problem can lead to different behaviours of the algorithm used. We address this question by solving the PSP by means of metaheuristic techniques [4, 8], which are general problem-solving strategies conceived as high level strategies that coordinate the behaviour of lower level heuristics. Although most metaheuristics can not return a proof of optimality of the solution found, they represent a good compromise between solution quality and computational effort. Through the use of metaheuristic, and using the paradigm of separation between model and algorithm [17], we show that different formulations affect algorithm performance and we study the reasons of this phenomenon.

The paper will start by recalling Portfolio Theory in Sect. 2, before introducing the concept of metaheuristics in Sect. 3. Then we will introduce a metaheuristic approach for the Portfolio Selection Problem in Sect. 4. In Sect. 5 will briefly present the principles of the search space analysis we perform. Search Space Analysis is applied to instances of PSP and results are discussed in Sect. 6. Finally, we conclude in Sect. 7.

2 Portfolio Selection Basis

We associate to each asset belonging to a set A of n assets ($A = \{a_1, \ldots, a_n\}$) a real-valued *expected return* r_i, and the corresponding return variance σ_i. We furthermore associate, to each pair of assets $\langle a_i, a_j \rangle$, a real-valued return *covariance* σ_{ij}. We are furthermore given a value r_e representing the minimum required return.

In this context, a portfolio is defined as the n-sized real vector $X = \{x_1, \ldots, x_n\}$ in which x_i represents the relative amount invested in asset a_i. For each portfolio we can define its variance as $\sum_{i=1}^{n} \sum_{j=1}^{n} \sigma_{ij} x_i x_j$ and its return as $\sum_{i=1}^{n} r_i x_i$. In the original formulation [25], PSP is formulated as the minimization of portfolio variance, imposing that the portfolio's return must be not smaller than r_e, leading to the following optimisation problem:

$$\min \sum_{i=1}^{n} \sum_{j=1}^{n} \sigma_{ij} x_i x_j, \tag{1}$$

$$s.t. \sum_{i=1}^{n} r_i x_i \geq r_e, \tag{2}$$

$$\sum_{i=1}^{n} x_i = 1, \tag{3}$$

$$x_i \geq 0 \quad (i = 1, \ldots, n). \tag{4}$$

The aforecited return constrained is introduced in constraint (2); constraint (3) is referred to as *budget constraint*, meaning that all the capital must be invested; constraint (4) imposes that variables have to be non-negative (i.e., short sales are not allowed).

If we define a finite set of values for r_e and solve the problem for all defined r_e values, we obtain the *Unconstrained Efficient Frontier* (UEF), in which the minimum risk value is associated to each r_e.

This formulation may be improved to grasp financial market features, by introducing a binary variable Z for each asset ($z_i = 1$ if asset i is on the portfolio, 0 otherwise). Additional constraints which can be added to the basic formulation are:

- **Cardinality constraint**, used either to impose an upper bound k to the cardinality of assets in the portfolio

$$\sum_{i=1}^{n} z_i \leq k, \tag{5}$$

or to force the resulting portfolio to contain exactly k assets:

$$\sum_{i=1}^{n} z_i = k_{max}. \tag{6}$$

This constraint is important for practitioners in order to reduce the portfolio management costs.

- **Floor and ceiling constraints**, used to set, for each asset, the minimum (ε_i) and maximum (δ_i) quantity allowed to be held in the portfolio

$$\varepsilon_i z_i \leq x_i \leq \delta_i z_i. \tag{7}$$

Those constraints are used to ensure diversification and to avoid tiny portions of assets in the portfolios, which would make their management difficult and lead to unnecessary transaction costs.

- **Preassignments**. This constraint is used to express subjective preferences: we want certain specific assets to be held in the portfolio, by determining a n-sized binary vector P (i.e., $p_i = 1$ if a_i has to be held in the portfolio) and imposing the following:

$$z_i \geq p_i \quad (i = 1, \ldots, n). \tag{8}$$

3 Metaheuristics

As stated in the Introduction, in this work we are solving the PSP by using meta-heuristics [4], which can be defined as high-level strategies that coordinate the action of low-level algorithms (heuristics) in order to find near-optimal solutions for combinatorial optimization problem. They are used when it is impossible to find the certified optimum solution in a reasonable amount of time, and their features can be outlined as follows:

- They are used to explore the search space and to determine principles to guide the action of subordinated heuristics.
- Their level of complexity ranges from a simple escape-mechanism to complex populations procedures.
- They are stochastic, hence escape and restart procedures have to be devised in the experimental phase.
- The concepts they are built upon allow an abstract descriptions, that is useful to design hybrid procedures.
- They are not problem-specific, but additional components may be used to exploit the structure of the problem or knowledge acquired during the search process.
- They may make use of problem-specific knowledge in the form of heuristics that are controlled by the upper level strategy.

The main paradigm metaheuristics are build upon is the *intensification-diversification* paradigm, meaning that they should incorporate a mechanism to balance the exploration of promising regions of the search landscape (intensification) and the identification of new areas in the search landscape (diversification). The way of implementing this balance is different depending on the specific metaheuristic used. A completed description is out of the scope of this paper, and we forward the interested reader to Hoos and Stuetzle [18].

4 Our Approach for Portfolio Choice

We are using the solver introduced by di Tollo et al. [7, 9] to tackle a constrained PSP, in which the Markowitz' variance minimisation in a continuous formulation is enhanced by adding constraints (4), (6) and (7), leading to the following formulation:

$$\min \sum_{i=1}^{n} \sum_{j=1}^{n} \sigma_{ij} x_i x_j, \tag{9}$$

subject to

$$\sum_{i=1}^{n} r_i x_i \geq r_e, \tag{10}$$

$$\sum_{i=1}^{n} x_i = 1, \tag{11}$$

$$x_i \geq 0 \quad i = 1 \ldots n, \tag{12}$$

$$k_{min} \leq \sum_{i=1}^{n} z_i \leq k_{max}, \tag{13}$$

$$\varepsilon_i z_i \leq x_i \leq \delta_i z_i, \tag{14}$$

$$x_i \leq z_i \quad i = 1 \ldots n. \tag{15}$$

where k_{min} and k_{max} are respectively lower and upper bounds on cardinality. This problem formulation contains two classes of decision variables: integer (i.e., Z) and continuous (i.e., X). Hence, it is possible to devise an hybrid procedure in which each variable class is tackled by a different component. Starting from this principle, we have devised a master–slave decomposition, in which a metaheuristic procedure is used in order to determine, for each search step, assets contained in the portfolio (Z). Once the assets contained in the portfolio are decided, the corresponding continuous X values can be determined with proof of optimality. Hence at each step, after having selected which assets to be taken into account, we are resorting to a the Goldfarb–Idnani algorithm for quadratic programming (QP) [16] to determine their optimum value. The stopping criterion and escape mechanism depend on the metaheuristic used, which will be detailed in what follows.

As explained in Sect. 6, this master–slave decomposition has a dramatic impact on the metaheuristic performance due to the different structure determined by this formulation, in which the basin of attraction are greater than the ones determined by a monolithic approach based on the same metaheuristic approaches. In what follows we are outlining the components of our metaheuristic approach.

- **Search space** Since the *master* metaheuristic component takes into account the Z variables only, the search space S is composed of the 2^n portfolios that are feasible w.r.t cardinality and pre-assignment constraints, while other constraints are directly ensured by the *slave* QP procedure. If the QP procedure does not succeed in finding a feasible portfolio, a greedy procedure is used to find the portfolio with maximum return and minimum constraint violations.

- **Cost function** In our approach the cost function corresponds to the objective function of the problem σ^2, and is computed, at each step of the search process, by the *slave* QP procedure.
- **Neighborhood relations** As in di Tollo et al. [9], we are using three neighborhood relations in which the neighbor portfolio are generated by *adding*, *deleting* or *replacing* one asset: the neighbor is created by defining the asset pair $\langle i, j \rangle (i \neq j)$, inserting asset i, and deleting asset j. Addition is implemented by setting $j = 0$; deletion is implemented by $i = 0$.
- **Initial solution** The initial solution must be generated to create a configuration of Z. Since the we aim to generate an approximation of the unconstrained efficient frontier, we are devising three different procedures for generating the starting port-folio, which are used w.r.t. different r_e values: MaxReturn (in which the starting portfolio corresponds to the maximum return portfolio, without constraints on the risk); RandomCard (in which cardinality and assets are randomly generated); and WarmRestart (in which the starting portfolio corresponds to the optimal solution found for the previous r_e value). MaxReturn is used when setting the highest r_e value (i.e., first computed value); for all other r_e values both RandomCard and WarmRestart have been used.

4.1 Solution Techniques

As specific metaheuristics for the *master* procedure, we have used Steepest Descent (SD), First Descent (FD) and Tabu Search (TS). SD and FD are considered as the most simple metaheuristic strategies, since they accept the candidate solution only when its cost function is better than the current one, otherwise the search stops. They differ to each other in the neighborhood exploration, since in SD all neighbors are generated and the best one is compared to the current solution, while in FD the first better solution found is selected as current one. TS enhances this schema by selecting, as the new current solution, the best one amongst the neighborhood, and using an additional memory (Tabu list) in which forbidden states (i.e., former solutions) are stored, so that they cannot be generated as neighbors. In our implementation, we have used a dynamic-sized tabu list, in which solutions are put in the Tabu list for a randomly generated period of time. The length range of the Tabu list has been determined by using F-Race [3], and has been set to [3, 10].

The three metaheuristics components have been coded in C++ by Luca Di Gaspero and Andrea Schaerf and are available upon request.

As for the *slave* Quadratic programming procedure, we have used the Goldfarb and Idnani dual set method [16] to determine the optimal X values corresponding to Z values computed by the *master* metaheuristic component. This method has been coded in C++ by Luca Di Gaspero: it is available upon request, and has achieved good performances when matrices at hand are dense.

To sum up, the *master* metaheuristic component determines the actual configuration of Z variables (i.e., point of the search space), the *slave* QP procedure computes the cost of the determined configuration, which is accepted (or not) depending on the mechanism embedded in FD, SD or TS.

4.2 Benchmark Instances

We have used instances from the repository ORlib (http://people.brunel.ac.uk/~mastjjb/jeb/info.html) and instances used in Crama and Schyns [6], which have been kindly provided to us by the authors. The UEF for the ORlib instances is provided in the aforementioned website; the UEF for instances from Crama and Schyns [6] has been generated by us by using our *slave* QP procedure. In both cases, the resulting UEF consists of 100 portfolios corresponding to 100 equally distributed r_e values. Benchmarks' main features are highlighted in Table 1.

By measuring the distance of the obtained frontier (CEF) from the UEF we obtain the *average percentage loss*, which is an indicator of the solution quality and which is defined as:

$$\text{apl} = \frac{100}{p} \sum_{l=1}^{p} (V(r_e) - V_U(r_e))/V_U(r_e) \tag{16}$$

in which r_e is the minimum required return, p is the frontier cardinality, $V(r_e)$ and $V_U(r_e)$ are the values of the function F returned by the solver and the risk on the UEF.

4.3 Experimental Analysis

Our experiments have been run on a computer equipped with a Pentium 4 (3.2 GHz), and in what follows we are showing results obtained on both instance classes. In

Table 1 Our instances

ORlib dataset				Crama and Schyns dataset			
ID	Country	Assets	AVG(UEF)risk	ID	Country	Assets	AVG(UEF)risk
1	Hong Kong (Hang Seng)	31	1.55936×10^{-3}	S1	USA (DataStream)	20	4.812528
2	Germany (DAX 100)	85	0.412213×10^{-3}	S2	USA (DataStream)	30	8.892189
3	UK (FTSE 100)	89	0.454259×10^{-3}	S3	USA (DataStream)	151	8.64933
4	USA (S&P 100)	98	0.502038×10^{-3}				
5	Japan (NIKKEI)	225	0.458285×10^{-3}				

Table 2 Results over ORlib instances

Inst.	FD+QP		SD+QP		TS+QP		TS [29]		GA+QP [26]	
	Min apl	Time	Min apl	Time	Min apl	Time	Min apl	Time	Min apl	Time
1	0.00366	1.5	0.00321	3.1	0.00321	29.1	0.00409	251	0.00321	415.1
2	2.66104	9.6	2.53139	14.1	2.53139	100.9	2.53617	531	2.53180	552.7
3	2.00146	10.1	1.92146	16.1	1.92133	114.4	1.92597	583	1.92150	886.3
4	4.77157	11.2	4.69371	18.8	4.69371	130.5	4.69816	713	4.69507	1163.7
5	0.24176	25.3	0.20219	45.9	0.20210	361.8	0.20258	1603	0.20198	1465.8

Table 3 Results over Crama and Schyns instances

Inst.	FD+QP			SD+QP			TS+QP			SA [6]		
	apl		Time	apl		Time	apl		Time	apl		Time
S1	0.72	0.094	0.3	0.35	0.0	1.4	0.35	0.0	4.6	1.13	0.13	3.2
S2	1.79	0.22	0.5	1.48	0.0	3.1	1.48	0.0	8.5	3.46	0.17	5.4
S3	10.50	0.51	10.2	8.87	0.003	53.3	8.87	0.0003	124.3	16.12	0.43	30.1

order to assess the quality of our approach, in the following tables we also report results obtained by other works tackling the same instances. Table 2 reports results over ORlibinstances, showing that our approach outperforms the metaheuristic approach by Schaerf [29], and compares favourably with Moral-Escudero et al. [26].

Table 3 compares our results with the one by Crama and Schyns [6]: solutions found by our hybrid approach have better quality than the ones found by SA [6], but running times are higher, due to our QP procedure and to our complete neighbourhood exploration, which are not implemented by Crama and Schyns.

We have also compared our approach with Mixed Integer Non-linear Programming (MINLP) solvers, by encoding the problem in AMPL [14] and solving it using CPLEX 11.0.1 and MOSEK 5. We have run the MINLP solvers over ORLib instances, and compared their results with SD+QP (10 runs), obtaining the same solutions in the three approaches, hence showing that our approach is able to find the optimal solution in a low computational time. Computational times for SD+QP and for the MINLP solvers are reported in Table 4 and in Fig. 1. We can notice that for big-sized instances exact solvers require higher computation time to generate points in which cardinality constraints are binding (i.e., left part of the frontier). Our approach instead scales very well w.r.t. size and provides results which are comparable.

We can conclude this section by observing that SD+QP provides as satisfactory results as the more complex TS+QP. Since Tabu Search is conceived to better explore the search space, this can be considered rather surprising. The next sections will enlighten us about this phenomenon.

Table 4 Computational times over ORLib instances 1–4, SD+QP and MINLP

Instance	Avg(SD+QP) (s)	CPLEX 11 (s)	MOSEK 5 (s)
1	3.1	2.1	15.8
2	14.7	397.1	5.0
3	18.0	890.7	1,903.3
4	20.9	169,461.0	239,178.4

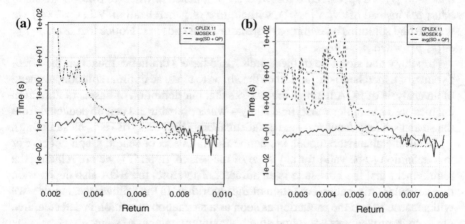

Fig. 1 Computational time: comparison between SD+QP and MINLP approaches over ORLib Instances. **a** Instance 2. **b** Instance 3

5 Search Space Analysis

The search process executed by a metaheuristic method can be viewed as a probabilistic walk over a discrete space, which in turn can be modelled as a graph: the vertices (usually named 'nodes' in this case) of the graph correspond to candidate solutions to the problem, while edges denote the possibility of locally transforming a solution into the other by means of the application of a local move. Therefore, algorithm behaviour depends heavily on the properties of this search space. A principled and detailed illustration of the most relevant techniques for search space analysis can be found in the book by Hoos and Stützle [18].

In this work we focus on a specific and informative feature of the search space, the *basin of attraction* (BOA), defined in the following.

Definition Given a deterministic algorithm $\mathscr{A}$, the basin of attraction $\mathscr{B}(\mathscr{A}|s)$ of a point s, is defined as the set of states that, taken as initial states, give origin to trajectories that include point s.

Let S^* be the set of global optima: for each $s \in S^*$ there exist a basin of attraction, and their union $I^* = \bigcup_{i \in S^*} \mathscr{B}(\mathscr{A}|i)$ contains the states that, taken as a starting solution, would have the search provide a certified global optimum. Hence, if we use a randomly chosen state as a starting solution, the ratio $|I^*|/|S|$ would measure the probability to find an optimal solution. As a generalization, we are defining a probabilistic basin of attraction as follows:

Definition Given a stochastic algorithm $\mathscr{A}$, the basin of attraction $\mathscr{B}(\mathscr{A}\,|s;\,p^*)$ of a point s, is defined as the set of states that, taken as initial states, give origin to trajectories that include point s *with probability* $p \geq p^*$. Accordingly, the union of the BOA of global optima is defined as $I^*(p) = \bigcup_{i \in S^*} \mathscr{B}(\mathscr{A}\,|i;\,p)$. It is clear that that $\mathscr{B}(\mathscr{A}\,|s)$ is a special case for $\mathscr{B}(\mathscr{A}\,|s;\,p^*)$, hence in what follows we are using $\mathscr{B}(s;\,p^*)$ instead of $\mathscr{B}(\mathscr{A}\,|s;\,p^*)$, without loss of generalization. When $p^* = 1$ we want to find solutions belonging to trajectories that ends in s. Notice that $\mathscr{B}(s;\,p_1) \subseteq \mathscr{B}(s;\,p_2)$ when $p_1 > p_2$.

Topology and structure of the search space have a dramatic impact on the effectiveness of a metaheuristic, and since the aim is to reach an optimal solution, the need of an analysis of BOA features arises.Note that our definition of basins of attraction enables both a complete/analytical study—when probabilities can be deducted from the search strategy features—and a statistical/empirical analysis (e.g., by sampling).

In our metaheuristic model, we define BOAs as sets of search graph nodes. For this definition to be valid for any state of the search graph [2], we are relaxing the requirement that the goal state is an attractor. Therefore, the BOA also depends on the particular termination condition of the algorithm. In the following examples, we will suppose to end the execution as soon as a stagnation condition is detected, i.e., when no improvements are found after a maximum number of steps.

6 Search Space Analysis for Portfolio Selection Problem

When solving an optimisation problem, a sound modelling and development phase should be based on the separation between the model and the algorithm: this stems from constraint programming, and several tools foster this approach (i.e., Comet [17]). In this way, it is possible to draw information about the structure of the optimisation problem, and this knowledge can be used, for instance, for the choice of the algorithm to be used. Up to the author's knowledge, literature about portfolio selection by metaheuristics has hardly dealt with this aspect, though some attempts have been made to study the problem structure. For instance, Maringer and Winker [24] draw some conclusion about the objective function landscape by using a memetic algorithm which embeds, in turn, Simulated Annealing (SA) [20] and Threshold Acceptance (TA) [11]. They compare the use of SA and TA inside the memetic algorithm dealing with different objective functions: Value-at-Risk(*Var*) and Expected Shortfall (*ES*) [8]. Their results indicates that TA is suitable when using *VaR*, while SA performs best when using *ES*. An analysis of the search space is made to understand this phenomenon.

Other works compare different algorithms on the same instance to understand which algorithm perform best, and in what portion of the frontier. Amongst them, Crama and Schyns [6] introduce three different Simulated Annealing strategies, showing that there is no clear dominance among them. Armañanzas and Lozano [1] introduces Ant Colony Optimisation (ACO) [10], refining solutions with a greedy search, comparing results with Simulated Annealing and Iterative Improvement, and showing that ACO and SA performances greatly depends on the expected return (see Sect. 2). A common way of tackling this analysis is to run the different algorithms, and then to pool the obtained solutions. After this phase, the dominated solutions are deleted and it is possible to understand which algorithm performs best w.r.t. a given part of the frontier [5, 13].

The main shortcoming of these approaches is that they identify which algorithm performs well in a given portion of the frontier, without explaining the motivation beneath this behaviour. Hence, an additional effort has to be made to understand the model and how it can affect the algorithm performance. In this section, we are aimed in comparing different formulations for the PSP and in understanding how the structure of the problem affects the algorithm's performances through Search Space Analysis.

When using a metaheuristic, search space analysis represents an effective tool to assess the algorithm performances and the instance hardness. In what follows we are discussing results obtained over real instances and over hard-handmade instances in order to outline the connections between search space analysis and algorithm performances.

Analysis for Real Instances

We define five equally distributed r_e values, referred to as R_i ($i = 1 \ldots 5$) and we analyse the search space corresponding to each r_i over the five ORlib instances in order to assess the local minima distribution, that is an indicator of the search space ruggedness. This concept is important since it has been shown that there exists a negative correlation between ruggedness and metaheuristic performances [18]. We have implemented and run a deterministic version of SD (referred to as SD_{det}) to estimate the number of minima of an instance of the problem discussed in Sect. 4, which combines continuous variables x with integer variables z. As for the constraints, we have set both a minimum (k_{min}) or a maximum (k_{max}) bound on cardinality in order to understand the differences arising when using a maximum or strict cardinality constraint. As for determining the initial states, we have resorted either to complete enumeration (if the instance at hand is small) or to uniform sampling.

Results are shown in Table 5, where we report the number of the different local minima found by 30 runs of SD_{det}. Dashed entries mean that no feasible solution exists.

Results indicate that instances at hand show a small number of local minima and only one global minimum. This clearly indicates a situation in which the search landscape is rather smooth, and explains why different strategies such TS and FD/SD lead to similar optimization results: since local optimum are few and far between, there is no need of using complex strategies or escape mechanisms, since the probability of meeting a trajectory leading to one of the optima are quite high. We recall that those values have been found by using a deterministic version of SD, and their inverse

Table 5 Instance 4, number of minima found

k_{min}, k_{max}	$R_1 = 0.00912$	$R_2 = 0.00738$	$R_3 = 0.00556$	$R_4 = 0.00375$	$R_5 = 0.00193$
1,3	1	1	1	1	1
1,6	1	1	1	5	1
1,10	1	1	1	1	3
3,3	1	1	3	5	3
6,6	–	1	1	2	1
10,10	–	1	1	3	2

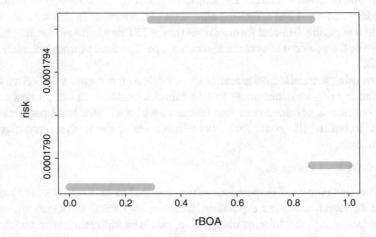

Fig. 2 Instance 4: BOA analysis. $k_{min} = 1$, $k_{max} = 10$, $R = 0.00375$

represents an upper bound on the probability to reach the certified optimum when using the stochastic SD and TS defined in Sect. 4.1.

We conclude that when using our formulation, global minima have a quite large BOA. A pictorial view of an example of this is provided in Fig. 2, where segments length corresponds to *rBOA* (i.e., ratio between size of *BOA(s)* and search space size) and their y-value corresponds to the minimum found: global minima *rBOA* ranges from 30 to 60 %.

Search space autocorrelation

A further analysis of the search space with respect to the study of BOA is the estimation of the *autocorrelation* of the search landscape [19]. This measure estimate the extent to which a local move from a solution leads to a destination state with similar objective function value. Smooth landscape, where it is easy to move towards better solutions, are characterized by a high autocorrelation value; conversely, low autocorrelation values are typical of landscape in which the search if often trapped in local optima or anyway very loosely guided towards good solution just by exploring the

Table 6 Autocorrelation of lag $k = 1, \ldots 10$ estimated for the three instances

Lag	Instance 2	Instance 3	Instance 4
1	0.99	0.95	0.91
2	0.98	0.90	0.90
3	0.97	0.86	0.87
4	0.97	0.81	0.86
5	0.96	0.76	0.77
6	0.95	0.70	0.75
7	0.94	0.65	0.73
8	0.93	0.61	0.72
9	0.93	0.55	0.69
10	0.92	0.50	0.57

neighbor of incumbent solutions. The autocorrelation of the search space may help elucidating the differences among algorithm performance across different instances, providing further bits of information besides the BOA analysis.

The autocorrelation of a series $G = (g_1, \ldots, g_m)$ of objective function values is computed as

$$r = \frac{\sum_{k=1}^{m-1} (g_k - \overline{g}) \cdot (g_{k+1} - \overline{g})}{\sum_{k=1}^{m} (g_k - \overline{g})^2},$$

where $\overline{g}$ is the average value of the series. This definition refers to the autocorrelation of length one, i.e., that corresponding to series generated by sampling neighbouring states at distance 1 in the search space. In general, we can consider the autocorrelation of lag k.

We performed a random walk of 1000 steps over the search space in the case of the three instances considered in this study and computed the autocorrelation of lag $k = 1, 2, \ldots 10$. In this way we can estimate with more precision the hardness of each instance. Results are shown in Table 6. As we can observe, the autocorrelation value for $k = 1$ is quite high for all the three instances, confirming the fact that in general the instances are quite easy for this combination of model and algorithm, as already observed in the case of BOA analysis. However, some differences can be observed considering the autocorrelation decay when the lag increases: instance 2 is by far the one with the highest autocorrelation, while we observe that instance 3 has a faster decrease with respect to the other two instances, making search slightly more difficult in case that the initial solution is not in the BOA of the optimal solution.

In conclusion, we can confirm that the three instances are considerably easy for local search. In the next paragraph we will show that the same problem, modeled in a different way, leads to different basin of attractions.

Monolithic Search Basin of Attraction

In the previous paragraph we have shown that, when using our problem formulation, the BOAs of local optima are quite big, making the search landscape smooth and the problem easy to be tackled by our hybrid solver. BOAs depend on the search strategy used and on the problem formulation, and this can be shown by running a different strategy, i.e., a monolithic one, on the same problem instances. We have used a SD based on a variant of Threshold Accepting [12], in which only a variable class is considered, i.e., w variables corresponding to actual asset weights. The desired outcome of this problem is the same as the previously introduced one, but they are represented in a different way. In the following we explain the main features of this metaheuristic approach:

- **Search Space** The *master–slave* decomposition is not used anymore, and a state is represented by a sequence $W = w_1 \ldots w_n$ such that w_b corresponds to the relative amount invested in asset b. Furthermore, the portfolio has to be feasible w.r.t. cardinality, budget, floor and ceiling constraints.
- **Neighborhood relations** A given amount ($step$) is transferred from asset a to another b, no matter if b is already in the portfolio or not. If this leads one asset value to be smaller than ε_i, its value is set to ε_i. If the move consists in decreasing the value of an asset being set to ε_i, its value is set to 0.
- **Initial solution** The initial solution has to be feasible w.r.t. cardinality, budget, floor and ceiling constraints and is always created from scratch.
- **Cost Function** As for the cost function we are using a penalty approach, hence it is given by adding the degree of constraints violations to the portfolio risk.
- **Local Search Strategies** SD that explores the space of w variables.

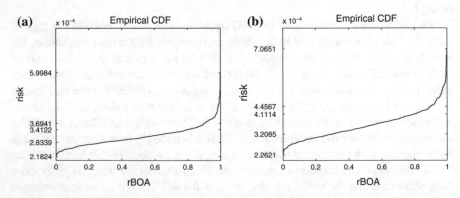

Fig. 3 Two ORlib instances: Monolithic BOA analysis with different constraints. **a** Instance 4: $k_{min} = 1$, $k_{max} = 10$, $R = 0.00375$. **b** Instance 4: $k_{min} = 1$, $k_{max} = 6$, $R = 0.00193$

$$\sigma = \begin{pmatrix}
1 & -1 & 0 & 0 & 0 & & \cdots & & & 0 \\
-1 & 1 & 0 & 0 & 0 & & \cdots & & & 0 \\
0 & 0 & 1 & -0.9 & 0 & 0 & 0 & \cdots & & \vdots \\
0 & 0 & -0.9 & 1 & 0 & 0 & 0 & \cdots & & \vdots \\
\vdots & & 0 & 0 & 1 & -0.9 & 0 & \cdots & & \vdots \\
\vdots & & 0 & 0 & -0.9 & 1 & 0 & \cdots & & \vdots \\
\vdots & & \ddots & & 0 & 0 & \ddots & & & \vdots \\
\vdots & & & \ddots & \vdots & \vdots & & \ddots & & \vdots \\
\vdots & & & \cdots & & & & 1 & 0 & 0 \\
\vdots & & & \cdots & & & & 0 & 1 & -0.9 \\
0 & & & \cdots & & & & 0 & -0.9 & 1
\end{pmatrix} \qquad (6)$$

Results about BOAs analysis for this approach are shown in Fig. 3. Even from visual inspection only, it turns out that the number of local minima is dramatically higher than the one corresponding to the *master–slave* approach; furthermore basin of attraction are tiny, and the certified optimum has not been found.

Analysis for artificial instances

In the previous paragraph we have shown that, for the PSP we are solving, instances at hand are easy to solve, since our *master–slave* decomposition leads to search spaces with a small number of local optima with huge BOAs. Hence, there is no need for complex approaches and escape mechanisms, and this explains why simple metaheuristics performances are comparable with more sophisticated one such TS. Furthermore, preliminary analysis have suggested us that this is a common feature in financial market related instances: this could be considered as a good point for practitioners, but makes impossible to test the robustness of our approach, in which we have developed TS+QP in order to tackle more difficult instances. Hence, we have designed an artificial hand-made instance featuring a huge number of minima with tiny BOAs, containing an even number n of assets i, in which $r_i = 1 \forall i$ and whose covariance matrix is depicted here above.

It is easy to see that for every r_e the best portfolio contains the first two assets only, but also that portfolios consisting of assets i (odd) and $i + 1$ only are local optima, since all their neighbors feature higher risk.

It can be shown show that it is necessary to visit a portfolio s having $z_1 = 1$ or $z_2 = 1$ to reach the global optimum s^*. Furthermore, portfolios containing an odd asset i ($i > 1$) whose $z_i = 1$ and $z_{i+1} = 1$ will never entry in a trajectory in which this couple would be removed. Hence, $\mathcal{B}(s^*)$ contains all portfolios featuring $z_1 = 1$ or $z_2 = 1$, and in which there is no i odd and > 1 such that $z_i = 1$ and $z_{i+1} = 1$. In this case, $rBOA(s^*)$ is inversely proportional to n.

By running our master–slave approach over this instance ($\varepsilon_i = 0.01$ and $\delta_i = 1$ for $i = 1 \ldots n$) we have remarked that TS+QP easily find a solution comparable to that provided by CPLEX, while SD and FD performances are greatly affected by the starting solution (and anyhow much poorer than TS+QP).

It has to be noticed that such an instance could be hardly found over real markets, even its presence is not forbidden by structural properties, but when tackling it the need of larger neighborhoods arises. Anyhow, no matter the neighborhood size, it is always possible to devise artificial instances whose minima are composed by subsets that have to be moved jointly.

From the Search Space Analysis conducted in this section, we may conclude that different formulations (hybrid vs continuous only) lead to different Basin of Attraction analysis on the instances at hand. This turns into different algorithm behaviours. The formulation that leads to a smooth search landscape (hybrid) can be tackled by algorithms with weak diversification capabilities (i.e., SD in the proposed hybrid formulation), whilst these algorithms are to be replaced by more sophisticated ones when the search landscape becomes rugged (see the behaviour of SD in the monolithic version). The artificial instance places itself in the middle of these phenomena, as it provides room for the use of more complex strategies (i.e., TS) in the hybrid case, due to the neighbor moves used which make the search to get stuck in the first local optimum found, but when embedded in the continuous only formulation doesn't provide different performances from the real instances.

7 Conclusion

In this work we have used a metaheuristic approach to study the impact of different formulations on the Portfolio Selection algorithm's behaviour, and we have devised a methodology to understand the root of the different behaviours (search space analysis through BOA analysis). To this aim we have compared an approach based on a master–slave decomposition with a monolithic approach. Results have shown that the search space defined by the monolithic approach is quite rugged and need an algorithm featuring an escape mechanism to be solved efficiently, whilst the hybrid approach leads to a smoother search landscape to be explored efficiently also by simpler algorithms such SD.

References

1. Armañanzas, R., Lozano, J.A.: A multiobjective approach to the portfolio optimization problem. In: Proceedings of the 2005 IEEE Congress on Evolutionary Computation, vol. 2, pp. 1388–1395 (2005)
2. Roli, A.: A note on a model of local search. Technical Report TR/IRIDIA/2004/23.01, IRIDIA, ULB, Belgium (2004)
3. Birattari, M., Stützle, T., Paquete, L., Varrentrapp, K.: A racing algorithm for configuring metaheuristics. In: Proceedings of the Genetic and Evolutionary Computation Conference (GECCO 2002), pp. 11–18. Morgan Kaufmann Publishers (2002)

4. Blum, C., Roli, A.: Metaheuristics in combinatorial optimization: overview and conceptual comparison. ACM Comput. Surv. **35**(3), 268–308 (2003)
5. Chang, T.J., Meade, N., Beasley, J.E., Sharaiha, Y.M.: Heuristics for cardinality constrained portfolio optimisation. Comput. Oper. Res. **27**(13), 1271–1302 (2000)
6. Crama, Y., Schyns, M.: Simulated annealing for complex portfolio selection problems. Eur. J. Oper. Res. **150**, 546–571 (2003)
7. Di Gaspero, L., di Tollo, G., Roli, A., Schaerf, A.: Hybrid local search for constrained financial portfolio selection problems. In: Proceedings of Integration of AI and OR Techniques in Constraint Programming for Combinatorial Optimization Problems, pp, 44–58 (2007)
8. di Tollo, G., Roli, A.: Metaheuristics for the portfolio selection problem. Int. J. Oper. Res. **5**(1), 443–458 (2008)
9. di Tollo, G., Stützle, T., Birattari, M.: A metaheuristic multi-criteria optimisation approach to portfolio selection. J. Appl. Oper. Res. **6**(4), 222–242 (2014)
10. Dorigo, M., Gambardella, L.M., Middendorf, M., Stützle, T. (eds.): Special Section on Ant Colony Optimization. IEEE Trans. Evol. Comput. **6**(4), 317–365 (2002)
11. Dueck, G., Scheuer, T.: Threshold accepting: a general purpose optimization algorithm appearing superior to simulated annealing. J. Comput. Phys. **90**(1), 161–175 (1990)
12. Dueck, G., Winker, P.: New concepts and algorithms for portfolio choice. Appl. Stoch. Models Data Anal. **8**, 159–178 (1992)
13. Fernandez, A., Gomez, S.: Portfolio selection using neural networks. Comput. Oper. Res. **34**, 1177–1191 (2007)
14. Fourer, R., Gay, D.M., Kernighan, B.W.: AMPL: A Modeling Language for Mathematical Programming. Duxbury Press/Brooks/Cole Publishing Company, Pacific Grove (2002)
15. Di Gaspero, L., di Tollo, G., Roli, A., Schaerf, A.: Hybrid metaheuristics for constrained portfolio selection problems. Quant. Financ. **11**(10), 1473–1487 (2011)
16. Goldfarb, D., Idnani, A.: A numerically stable dual method for solving strictly convex quadratic programs. Math. Program. **27**, 1–33 (1983)
17. Van Hentenryck, P., Michel, L.: Constraint-Based Local Search. The MIT Press, Cambridge (2005)
18. Hoos, H., Stützle, T.: Stochastic Local Search Foundations and Applications. Morgan Kaufmann Publishers, Burlington (2005)
19. Hordijk, W.: A measure of landscapes. Evol. Comput. **4**, 335–360 (1996)
20. Kirkpatrick, S., Gelatt, C.D., Vecchi, M.P.: Optimization by simulated annealing. Science **220**(4598), 671–680 (1983)
21. Mansini, R., Ogryczak, W., Speranza, M.G.: LP solvable models for portfolio optimization: a classification and computational comparison. IMA J. Manag. Math. **14**(3), 187–220 (2003)
22. Mansini, R., Speranza, M.G.: Heuristic algorithms for the portfolio selection problem with minimum transaction lots. Eur. J. Oper. Res. **114**(2), 219–233 (1999)
23. Maringer, D.: Portfolio Management with Heuristic Optimization. Springer, Heidelberg (2005)
24. Maringer, D., Winker, P.: Portfolio optimization under different risk constraints with modified memetic algorithms. Technical Report 2003-005E, University of Erfurt, Faculty of Economics, Law and Social Sciences (2003)
25. Markowitz, H.: Portfolio selection. J. Financ. **7**(1), 77–91 (1952)
26. Moral-Escudero, R., Ruiz-Torrubiano, R., Suárez, A.: Selection of optimal investment with cardinality constraints. In: Proceedings of the IEEE World Congress on Evolutionary Computation, pp. 2382–2388 (2006)
27. Prestwich, S., Roli, A.: Symmetry breaking and local search spaces. In: Proceedings of the 2nd International Conference on Integration of AI and OR Techniques in Constraint Programming for Combinatorial Optimization Problems, pp. 273–287 (2005)
28. Rolland, E.: A tabu search method for constrained real number search:applications to portfolio selection. Technical report, Department of Accounting and Management Information Systems, Ohio State University, Columbus. U.S.A. (1997)
29. Schaerf, A.: Local search techniques for constrained portfolio selection problems. Comput. Econ. **20**(3), 177–190 (2002)

30. Speranza, M.G.: A heuristic algorithm for a portfolio optimization model applied to the Milan stock market. Comput. Oper. Res. **23**(5), 433–441 (1996)
31. Streichert, F., Ulmer, H., Zell, A.: Comparing discrete and continuous genotypes on the constrained portfolio selection problem. In: Proceedings of Genetic and Evolutionary Computation Conference. LNCS, vol. 3103, pp. 1239–1250 (2004)
32. Yokoo, M.: Why adding more constraints makes a problem easier for hill-climbing algorithms: Analyzing landscapes of CSPs. In: Proceedings of the Third Conference on Principles and Practice of Constraint Programming, pp. 356–370 (1997)

Optimization of Fuel Consumption in Firefighting Water Capsule Flights of a Helicopter

Jacek M. Czerniak, Dawid Ewald, Grzegorz Śmigielski,
Wojciech T. Dobrosielski and Łukasz Apiecionek

Abstract This article presents a possible use of ABC method for optimization of fuel consumption of helicopter which transport water capsule. There are lot of attibutes of flight e.g. the mass of the capsule, velocity, altitude, aerodynamic coefficients of the capsule, and horizontal and vertical winds. The presented method is focused on mentioned attributes. The problem is very important because helicopters are used in real hughe fire and it will be quite good if the fueal needed for this will be as low as possible. That is why authors presents theoretical models of flight of a bag filled with water which is then dropped from an aircraft moving horizontally. The resulet achived by numerical computations are later compared with the results measured for the trajectory of a capsule of water dropped from a helicopter. There ia also a discussion of the experimental and numerical results achieved in mentioned experiment.

Keywords ABC · Water capsule flight · Bee optimization

J.M. Czerniak (✉) · D. Ewald · W.T. Dobrosielski · Ł. Apiecionek
Institute of Technology, Casimir the Great University in Bydgoszcz, ul. Chodkiewicza 30,
85-064 Bydgoszcz, Poland
e-mail: jczerniak@ukw.edu.pl

D. Ewald
e-mail: dawidewald@ukw.edu.pl

W.T. Dobrosielski
e-mail: wdobrosielski@ukw.edu.pl

Ł. Apiecionek
e-mail: lapiecionek@ukw.edu.pl

G. Śmigielski
Institute of Mechanics and Applied Computer Science, Casimir the Great University
in Bydgoszcz, Bydgoszcz, Poland
e-mail: grzegorz.smigielski@ukw.edu.pl

© Springer International Publishing Switzerland 2016
S. Fidanova (ed.), *Recent Advances in Computational Optimization*,
Studies in Computational Intelligence 655, DOI 10.1007/978-3-319-40132-4_3

1 Introduction

Behavior of many animal species in nature is similar to the swarm behavior. Shoals of fish, flocks of birds and flocks of land animals are created as a result of the biological drive to live in a group. Specific individuals belonging to a flock or a shoal are characterized by higher survival probability because predators or raptors usually attack only one individual. Group movement is characteristic for flocks of birds and other animals as well as shoals of fish. Flocks of land animals react quickly to changes of movement direction and velocity of neighboring individuals. Herd behavior is also one of the main characteristic features of insects living in colonies (bees, wasps, ants, termites) Communication between individual insects of the swarm of social insects has already been thoroughly studied and is still subject of studies. The systems of communication between individual insects contribute to creation of "collective intelligence" of swarms of social insects [6, 14]. Thus the term "Swarm intelligence" emerged, meaning the above mentioned "collective intelligence" [8, 11, 12, 19, 20]. The swarm intelligence is part of the Artificial Intelligence as per examination of activities performed by separate individuals in decentralized systems [18]. The Artificial Bee Colony (ABC) metaheuristics has been introduced quite recently as a new trend in the Swarm intelligence domain [3, 16, 21]. Artificial bees represent agents solving complex combinatorial optimization problems. This article presents proposed optimization of water capsule flight using ABC method [11, 12]. Data obtained from real water capsule flights developed for firefighting was used here. A very efficient way of water spray formation is explosion method consisting in detonation of an explosive placed in a water container [23]. Water sufficiently eliminates undesired consequences of detonation, which provides potential possibility for applications of that method. Water spray can be used, e.g. to extinguish fire and to neutralize contaminated areas [7, 15]. Water capsule suspended under a helicopter or another aircraft enables fast transport of water to the area of airdrop. Described system allows automatic release of the water capsule at such a distance from the target so that, after some time of its free fall, it is located over the target at the specified altitude and then detonated to generate spray which covers specified area of the ground [7, 22].

2 Physical Methodology of the Water Capsule Flight Analysis

A simplified model of the water capsule flight was presented for the purpose of this paper. This is related to the fact that ABC algorithm does not directly use the data received from the model. However, to understand the problem, it is necessary to analyze information related to this issue. In principle the problem of delivering a water capsule to a given point on the ground is very similar to the problem of hitting a surface target by a bomber with an unguided bomb. There are, however, two problems

Fig. 1 Schematic view of
the procedure of delivering
water-capsule to a designed
point

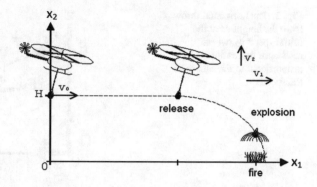

that make difficult a direct application of the procedures used by military aviation.
The first follows from the fact that such procedures, as majority of procedures used by
the military are either classified as a whole or comprise classified crucial components
[1, 9, 17]. The second problem is connected with much higher safety standards that
must be observed in the case of placing water-capsule "in target". It seems then more
reasonable to develop procedures from the very beginning than to try to adopt non-
classified components of similar military procedures. The ultimate objective consists
in developing a high precision system of delivering by an aircraft (presumably a
helicopter) a water-capsule to a defined point where it should be exploded in order to
generate cloud of water-spray playing role of the fire-extinguishing agent. A scheme
of such procedure is shown in Fig. 1.

By analysing the flight of water capsule dropped at a certain initial speed from
a given height we obtain the case of horizontal throw. This is a commonly known
physical phenomenon. The only difference in such an approach to this problem is
the fact that the resistance to motion and the effect of wind are taken into account.
The equations considering those corrections sufficiently describe the matter and a
proper analysis of the water capsule flight can be done on their basis. In the analysis
of a real horizontal throw one has to take into account air resistance during flight.
The resistance force is the resultant of:

- tangent forces (viscous drag),
- normal forces (pressure resistance),

relative to the surface of the body that is flown around by air—i.e. the water capsule.
When considering the horizontal throw of a body with a given mass m from the
height H at the initial speed v_0 (Fig. 2) it was assumed that the resistance to motion
Fresist is proportional to the square of the speed of the body (Bernoullis case). The
resultant force acting on the capsule is the vector sum of two components (Fig. 2).

Designing a suitable system must be based on theoretical models that can serve
as a foundation of numerical programs. The models are founded on the assumption
that the water-capsule moves in the air under the influence of a constant and vertical
gravitational force and of the Bernoulli drag (pressure drag) that acts against its
motion with respect to the air and is proportional to the square of the velocity of this

Fig. 2 The horizontal throw
from the height H at the
initial speed v_0, where
resistance forces are
proportional to the square of
the speed

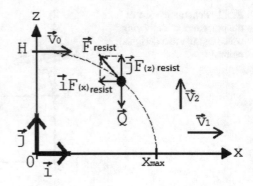

motion. After denoting the velocity by $\vec{v}$ one can write the following formula the
drag force.

$$\vec{O} = \frac{c\rho A}{2} v \vec{v}, \tag{1}$$

where $v = |\vec{v}| = \sqrt{v_1^2 + v_2^2}$,
c is the drag coefficient depending on the shape of the moving body, ρ denotes density
of the air, and A is the frontal cross-section of the body.

2.1 Equations Describing Flight of a Water Capsule

A water capsule dropped from a horizontally moving aircraft (e.g. helicopter) falls
down under composite action of the drag force that has both vertical and horizontal components and the gravitational force that acts all time vertically. Introducing
Cartesian coordinates: the horizontal one x_1 and the vertical one x_2, one can write
equations of motion in the form

$$\dot{v}_1 = -\frac{c_1 \rho A_1}{2M} \sqrt{v_1^2 + v_2^2} v_1, \quad \dot{v}_2 = -\frac{c_2 \rho A_2}{2M} \sqrt{v_1^2 + v_2^2} v_2 - g \tag{2}$$

where v_1 and v_2 are the horizontal and vertical coordinates of the capsule's velocity
respectively, M is its mass and g denotes gravitational acceleration. One has to do
with a set of ordinary, first order, nonlinear differential equation with respect to the
Cartesian coordinates of capsules velocity. Having these equations solved, one can
obtain coordinates of the capsule by simple integration coordinates of velocity with
respect to time. Unfortunately, the Eq. (2) cannot be solved analytically without far
going simplifications. It is so due to the coupling square root term. As such, one has
to apply numerical methods for solving the equations.

2.2 Numerical Solutions

In this case the standard fourth order Runge–Kutta method was used, and numerical computations were performed inside the MATLAB environment. In practice some additional work aimed, e.g., on optimization of the length of the step of integration, had to be done, but we will not go into technical details [18].

The solution is obtained for standard initial conditions given by the equations

$$v_1(0) = v_0, v_2(0) = v_0 \tag{3}$$

which corresponds to horizontal motion of the water-capsule at the moment of release. Provided the value of the drag coefficient c is known, one can obtain both components of capsule's velocity as functions of time. Since the main objective consists in computing trajectory of the capsule, one has to compute its horizontal and vertical component using integrals

$$x_1(t) = \int_0^t v_1(\tau)d\tau + x_1(0),$$

$$x_2(t) = \int_0^t v_2(\tau)d\tau + x_2(0) \tag{4}$$

that, in general, have to be computed numerically since the functional form of v_1 and v_2 with respect to time are not known. The numerical solution of equations for the components v_1 and v_2 of the capsule's velocity has one more advantage. After some modifications such a procedure can be applied to the problem of flight in the air moving with respect to the ground. In fact, Eq. (2) describe velocity of the capsule with respect to the ground under the assumption that the air is still. If, however, velocities of wind and that of ascending or descending current are considerable, the equations have to be modified

$$\dot{v}_1 = -\frac{c_1 \rho A_1}{2M} \sqrt{\tilde{v}_1^2 + \tilde{v}_2^2} \, \tilde{v}_1$$

$$\dot{v}_2 = -\frac{c_2 \rho A_2}{2M} \sqrt{\tilde{v}_1^2 + \tilde{v}_2^2} \, \tilde{v}_2 - g \tag{5}$$

where

$$\tilde{v}_i = \tilde{v}_i - \tilde{V}_i, i = 1, 2 \tag{6}$$

are coordinates of the capsule's velocity with respect to the air; V_1 denotes velocity of wind and V_2 velocity of vertical current (a further generalization, we will not discuss here, would be taking into account the fact that strong and random winds make the problem 3-dimensional instead of 2-dimensional planar problem of a flight in the still air).

Numerical solution of equation of motion requires inserting numerical data from the very beginning. Some of them like the mass M of the capsule or the density of the air are at hand, but the drag coefficients $k_1 = cA_1$ and $k_2 = cA_2$, appearing in (2) and (4) have to be determined from experimental data.

Otrzymane rozwizanie pozwala na.

3 ABC Application to Optimization of Fuel Consumption of a Helicopter

3.1 Numerical Solutions

Artificial bee colony (ABC) is a model proposed in 2005 by a Turkish scientist Dervis Karaboga [4, 11, 12]. Like other algorithms described herein, it is also based on herd behavior of honey bees. It differs from other algorithms in the application of higher number bee types in a swarm [2, 5, 13]. After the initialization phase, the algorithm consists of the following four stages repeated by iteration until the number of repetitions specified by the used is competed:

- Employed Bees stage,
- Onlooker Bees stage,
- Scout Bees stage,
- storage of the best solution so far.

The algorithm starts with initialization of the food source vectors x_m, where $m = 1 \ldots SN$, while SN, is the population size. Each of those vectors stores n values $x_m, i = 1 \ldots n$, that shall be optimized during execution of that method. The vectors are initialized using the following formula:

$$x_{mi} = l_i + rand(0, 1)x(u_i - l_i) \tag{7}$$

where:

l_i-lower limit of the searched range,

u_i-upper limit of the searched range,

Bees adapted to different tasks participate in each stage of the algorithm operation. In case of ABC, there are 3 types of objects involved in searching:

- Employed Bees—bees searching points near points already stored in memory,
- Onlooker Bees—objects responsible for searching neighborhood of points deemed the most attractive,
- Scout Bees—(also referred to as scouts) this kind of bees explores random points not related in any way to those discovered earlier.

Once initialization is completed, Employed Bees start their work. They are sent to places in the neighborhood of already known food sources to determine the amount of nectar available there. Results of the Employed Bees work are used by Onlooker Bees. Onlooker Bees randomly select a potential food source using the following relationship:

$$v_i = x_{mi} + \varphi_{mi}(x_{mi} + x_{ki}) \tag{8}$$

where:

v_i-vector of potential food sources,

x_k- randomly selected food source,

φ_{mi}- random number from the range [-a,a]. Once the vector is determined its fitting is calculated based on the formula dependent on the problem being solved and the fitting v_m is compared with x_m. If the new vector fits better than the former one, then the new replaces the old one. Another phase of the algorithm operation is the Onlooker Bees stage. Those bees are sent to food sources classified as the best ones and in those very points the amount of available nectar is determined. The probability of the x_m source selection is expressed with the formula:

$$p_m = \frac{fit_m(x_m)}{\sum_{k=1}^{SN} fit_k(x_k)} \tag{9}$$

where:

$fit_m(x_m)$- value of fitting functions for a given source.

Obviously, when onlooker bees gather information on the amount of nectar, such data is compared with results obtained so far and if the new food sources are better, they replace the old ones in the memory. The last phase of this algorithm operation is exploration by scouts. Bees of that type select random points from the search space and then check nectar volumes available there. If newly found volumes are higher than the volumes stored so far, they replace the old volumes. The activity of those bees makes it possible to explore the space unavailable for the remaining types of bees thus allowing to omit any extremes.

3.2 Application of ABC

The reach of the capsule flight is calculated so that the distance from the helicopter does not exceed 140 m and then the initial velocity V_0 and the altitude Z are optimized. The fuel consumption at the power of 2225 km–292 g/kmh (i.e. 292 g of fuel per horse power per hour) is assumed as the cost. At the moment the program estimates the results only approximately, but author believes that he shall be able to make the results more real in the near future.

Fig. 3 Diagram of the
optimization

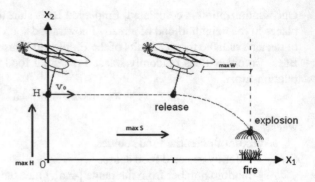

Random selection of the initial altitude and velocity of the helicopter;
REPEAT
The selected altitude is put into the water capsule flight reach formula;
The selected velocity is put into the water capsule flight reach formula;
Then we calculate the function of the cost of rising the
helicopter to the specified altitude and accelerating it to
the specified velocity so that the capsule is dropped not further than 140 m (max
W, max S, max H)
away from the target;
Sources;
The verified velocity and altitude are replaced by new values;
The best velocity and altitude are stored in the memory; UNTIL (the conditions
are met)

The main underlying idea of the optimization is to select such altitude and velocity of
the helicopter that shall enable the capsule to reach the target. The path covered by the
capsule falling from the helicopter to the vicinity of fire depends on the altitude from
which the capsule was dropped. It is obvious that increase of altitude or velocity
depends directly on helicopter rotor power. The power to be generated influences
specific fuel consumption. As illustrated in the graph, vertical climb of the helicopter
at zero horizontal velocity generates huge power demand. One can significantly
reduce power needed to climb the helicopter to the specified altitude by increasing its
horizontal velocity. This relationship results from the way of generating aerodynamic
lift by the helicopter [10]. However, too high horizontal velocity can significantly
increase power demand causing increased specific fuel consumption (Fig. 3).

4 Directions of Further Research

The results of the application using ABC algorithm achieved so far are based on theoretical model of the helicopter developed on the basis of the available literature. This
model describes the relationship between the height and the speed of the helicopter

Fig. 4 Specific PZL-10W engine

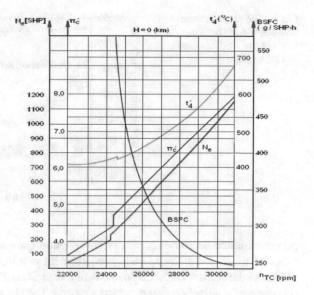

and the engine power demand. The model allows to estimate the fuel consumption which in turn gives direct picture of the flight economy. Knowing the engine power demand, the flight time and fuel cost we can estimate the cost of the fire extinguishing action using the water capsule.

As shown, through the optimization of the helicopter route, the operation costs can be significantly reduced. Therefore, the next step would be to change from the theoretical model of the helicopter to real unit commonly used in civil aviation.

The most commonly used helicopter in Poland is PZL Sokół driven by PZL-10W engine. Technical data of the PZL-10W engine:

- Take-off power: 662 kW (900 hp)
- Power—exceptional range 2,5 min: 846 kW (1150 hp)
- Specific fuel consumption : 268 g/hph

As we can see, the specific fuel consumption is less than that in the theoretical model and the aerodynamic properties which have an impact on the longitudinal speed also contribute to generation of an additional aerodynamic lift. All these factors significantly reduce power demand of the PZL Sokół engine (Fig. 4).

5 Conclusion

Obviously, the power required during forward flight will also be the function of GTOW (Gross Takeoff Weight). Representative results illustrating the effect of GTOW on the rotor power required are provided in Fig. 5 for a sample helicopter at sea-level (SL) conditions. It should be noted that with increasing GTOW, the excess

Fig. 5 The graph of fuel
consumption versus the
velocity and altitude of the
helicopter flight

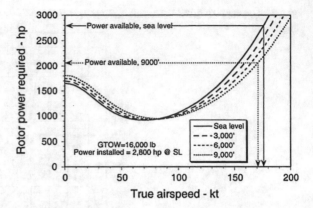

power available decreases gradually, this phenomenon applies in particular at lower
airspeed where the induced power requirement is a higher percentage of the total
power. In the subject case, the power available at SL is 2800 hp and for a gas turbine
this remains relatively constant versus airspeed. The airspeed value at the intersection
of the power required curve and the power available curve indicates the maximum
level flight speed. However, the maximum velocity is limited by probable onset of
rotor stall and compressibility effects before this point is reached. Multi-objective
optimization of a helicopter flight carrying a water capsule is a non-trivial problem.
ABC algorithm application enables efficient optimization of fuel costs. Due to high
complexity of that problem, one must bear in mind that optimization results may
deviate from real results. These can be caused by the fact that wind drag and direction were skipped. The distance at which the helicopter must approach the fire can
have significant impact on the final result while the air temperature can significantly
influence the fuel demand of the helicopter engine. There is also an issue of the angle
at which the capsule is dropped. That aspect can also be taken into account in further
studies on ABC application to fuel consumption optimization and as a consequence,
on reduction fire extinguishing cost using that method. Summing up, that problem
is very complex, which makes it a good example of ABC application.

References

1. Angryk, R.A., Czerniak, J.: Heuristic algorithm for interpretation of multi-valued attributes
   in similarity-based fuzzy relational databases. Int. J. Approx. Reason. **51**(8), 895–911 (2010).
   Oct
2. Apiecionek, L., Czerniak, J.M., Zarzycki, H.: Protection tool for distributed denial of services
   attack. Beyond Databases, Architectures and Structures, BDAS, 424, pp. 405–414 (2014)
3. Czerniak, J.: Evolutionary approach to data discretization for rough sets theory. Fundamenta
   Informaticae **1–2**, 43–61 (2009)

4. Czerniak, J., Ewald, D., Macko, M., Śmigielski, G., Tyszczuk, K.: Approach to the monitoring of energy consumption in eco-grinder based on abc optimization. Beyond Databases, Architectures and Structures, pp. 516–529 (2015)
5. Czerniak, J.M., Dobrosielski, W., Zarzycki, H., Apiecionek, L.: A proposal of the new owlant method for determining the distance between terms in ontology. Intelligent Systems '2014 Vol. 2: Tools, Architectures, Systems, Applications, vol. 323, pp. 235–246 (2015)
6. Czerniak, J., Apiecionek, L., Zarzycki, H.: Application of ordered fuzzy numbers in a new ofnant algorithm based on ant colony optimization. Commun. Comput. Inf. Sci. **424**, 259–270 (2014)
7. Dygdała, R., Stefański, K., Śmigielski, G., Lewandowski, D., Kaczorowski, M.: Aerosol produced by explosive detonation. Meas. Autom. Monit. **53**(9), 357–360 (2007)
8. Ewald, D., Czerniak, J.M., Zarzycki, H.: Approach to solve a criteria problem of the abc algorithm used to the wbdp multicriteria optimization. Intelligent Systems'2014, Vol 1: Mathematical Foundations, Theory, Analyses, vol. 322, pp. 129–137 (2015)
9. Ganesan, P.K., Angryk, R., Banda, J., Wylie, T., Schuh, M.: New Trends in Databases and Information Systems. Spatiotemporal Co-occurrence Rules, pp. 27–35. Springer, Heidelberg (2014)
10. Gordon Leishman, J.: Principles of Helicopter Aerodynamics. Cambridge University Press, Cambridge (2002)
11. Karaboga, D., Basturk, B.: A powerful and efficient algorithm for numerical function optimization: Artificial bee colony (abc) algorithm. J. Glob. Optim. **39**, 459–471 (2007)
12. Karaboga, D., Gorkemli, B.: A quick artificial bee colony (QABC) algorithm and its performance on optimization problems. Appl. Soft Comput. **23**, 227–238 (2014)
13. Kosinski, W., Prokopowicz, P., Slezak, D.: Fuzzy reals with algebraic operations: Algorithmic approach. In: Proceedings of the Intelligent Information Systems 2002, pp. 311–320 (2002)
14. Kowalski, P., Łukasik, S.: Experimental study of selected parameters of the krill herd algorithm, vol. 1, pp. 473–477 (2014)
15. Liu, Z., Kim, A.K., Carpenter, D.: Extinguishment of large cooking oil pool fires by the use of water mist system. Combustion Institute/Canada Section, Spring Technical Meeting pp. 1–6 (2004)
16. Marbac-Lourdelle, M.: Model-based clustering for categorical and mixed data sets (2014)
17. Mikolajewska, E., Mikolajewski, D.: Exoskeletons in neurological diseases - current and potential future applications. Adv. Clin. Exp. Med. **20**(2), 227–233 (2011)
18. Plucński, M.: Mini-models-local regression models for the function approximation learning artificial intelligence and soft computing. In: Rutkowski, L. et al. (ed.) Lecture Notes in Artificial Intelligence, vol. 7268, pp. 160–167 (2012)
19. Reina, M.D., Trianni., V.: Towards a cognitive design pattern for collective decision-making. In: Swarm Intelligence - Proceedings of Ants 2014 - Ninth International Conference. Lecture Notes in Computer Science, vol. 8667, pp. 194–205 (2014)
20. Roeva, O., Slavov, T.: Firefly algorithm tuning of pid controller for glucose concentration control during e. coli fed-batch cultivation process. In: Proceedings of the Federated Conference on Computer Science and Information Systems (2012)
21. Sameon, D., Shamsuddin, S., Sallehuddin, R., Zainal, A.: Compact classification of optimized boolean, reasoning with particle swarm optimization. Intell. Data Anal. IOS Press **16**, 915–931 (2012)
22. Śmigielski, G., Dygdała, R.S., Lewandowski, D., Kunz, M., Stefański, K.: High precision delivery of a water capsule. theoretical model, numerical description, and control system. IMEKO XIX World Congress, Fundamental and Applied Metrology, pp. 2208–2213 (2009)
23. Stebnovskii, S.V.: Pulsed dispersion as the critical regime of destruction of a liquid volume. Combust., Explos., Shock Waves **44**(2), 228–238 (2008)

Practical Application of OFN Arithmetics in a Crisis Control Center Monitoring

Jacek M. Czerniak, Wojciech T. Dobrosielski, Łukasz Apiecionek,
Dawid Ewald and Marcin Paprzycki

Abstract In this paper there is a comparision of fuzzy arithmetic calculations in two different notations. The first is well-known L-R notation proposed by Dubois-Prade. It is well known for the researchers who deal with fuzzy logic. The second if OFN notation. This notation was introduced by Kosiński. The same data was used for comparative calculations using the benchmark "Dam and Crisis control center paradox". There was an observation of water level at the dam with two trends: when the water level increases and decreases. This could cause the different short-term forecast for mentioned situation. The result which was achieved using two different fuzzy arithmetic showed, that OFN notation is sensitive to the trend differences. The authors showed that it provides an extra value which is a relationship between the fuzzy logic and the trend of the observer phenomena.

Keywords Fuzzy logic · Fuzzy number · Ordered fuzzy numbers

J.M. Czerniak (✉) · W.T. Dobrosielski · Ł. Apiecionek · D. Ewald
Institute of Technology, Casimir the Great University in Bydgoszcz, ul. Chodkiewicza 30,
85-064 Bydgoszcz, Poland
e-mail: jczerniak@ukw.edu.pl

W.T. Dobrosielski
e-mail: wdobrosielski@ukw.edu.pl

Ł. Apiecionek
e-mail: lapiecionek@ukw.edu.pl

D. Ewald
e-mail: dawidewald@ukw.edu.pl

M. Paprzycki
Systems Research Institute of the Polish Academy of Sciences, ul. Newelska 6,
01-447 Warsaw, Poland
e-mail: marcin.paprzycki@ibspan.waw.pl

© Springer International Publishing Switzerland 2016
S. Fidanova (ed.), *Recent Advances in Computational Optimization*,
Studies in Computational Intelligence 655, DOI 10.1007/978-3-319-40132-4_4

1 Introduction

The history of artificial intelligence shows that new ideas were often inspired by natural phenomena. Many tourists come back with passion to beaches at the Oceanside and many sailors sail on tide water. The phenomenon of high and low tides, although well known, has been stimulating imagination and provoking reflexion on the perfection of the Creation for many ages. The casual observer in unable to precisely specify water level decline, but he or she can easily describe it using fuzzy concepts such as "less and less", "little" and "a bit". The same applies to increase of water in the observed basin. The observer can describe it such linguistic terms like "more", "lots of" or "very much". Such linguistic description of reality is characteristic to powerful and dynamically developing discipline of artificial intelligence like fuzzy logic. The author of Fuzzy logic is an American professor of the Columbia University in New York City and of Berkeley University in California—Lotfi A. Zadeh, who published the paper entitled "Fuzzy sets" in the journal "Information and Control" in 1965 [31]. He defined the term of a fuzzy set there, thanks to which imprecise data could be described using values from the interval (0,1). The number assigned to them represents their degree of membership in this set. It is worth mentioning that in his theory L. Zadeh used the article on 3-valued logic published 45 years before by a Pole - Jan Łukasiewicz [24, 25]. That is why many scientists in the world regard this Pole as the "father" of fuzzy logic. Next decades saw rapid development of fuzzy logic. Another milestones of the history of that discipline should necessarily mention L-R representation of fuzzy numbers proposed by D. Dubois and H. Prade [12–14], which enjoys great successes today. Coming back to the original analogy, one can see some trend, i.e. general increase during rising tide or decrease during low tide, regardless of momentary fluctuations of the water surface level. This resembles a number of macro and micro-economic mechanisms where trends and time series can be observed. The most obvious example of that seem to be bull and bear market on stock exchanges, which indicate to the general trend, while shares of individual companies may temporarily fall or rise. The aim is to capture the environmental context of changes in economy or another limited part of reality. Changes in an object described using fuzzy logic seem to be thoroughly studied in many papers. But it is not necessarily the case as regards linking those changes with trend. Perhaps this might be the opportunity to apply generalization of fuzzy logic which are, in the opinion of authors of that concept, W. Kosiński [19] and his team, Ordered Fuzzy Numbers.

As the basis for experiments, let us assume the example of the dam and the impounding basin presented in the figure below (Fig. 1). Letter A indicates the water level measured during last evening measurement. Then there was a rapid surge of water in the night. Whereas morning measurement was marked using letter C. Measurements were imprecise to some extent due to rapid changes of weather conditions. It is also known that during the last measurement the safety valve $z2$, was open and then the valve $z1$ activated. The management of the dam faces the problem of reporting rapid surges of water to the disaster recovery centre.

Fig. 1 The diagram of water flow in the impounding basin

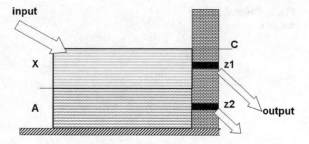

2 Theoretical Background Description of OFN

2.1 Some Definitions of OFN

Each operation on fuzzy numbers, regardless if it is addition, subtraction, division or multiplication, can increases the carrier value. Several operations performed on given L-R numbers can result in numbers that are too broad and as a result they can become less useful. Solving equations using conventional operations on fuzzy numbers [21, 22] is usually impossible either. An $A + X = C$ equation can always be solved using conventional operations on fuzzy numbers only when A is a real number. First attempts to redefine new operations on fuzzy numbers were undertaken at the beginning of the 1990-ties by Witold Kosiński and his Ph.D. student—P. Słysz [18]. Further studies of W. Kosiński published in cooperation with P. Prokopowicz and D. Ślęzak [20, 22, 23] led to introduction of the **ordered fuzzy numbers model—OFN**.

Definition 1 An **ordered fuzzy number** A was identified with an ordered pair of continuous real functions defined on the interval [0, 1], i.e., $A = (f, g)$ with $f, g : [0, 1] \longrightarrow R$ as continuous functions.

We call f and g the up and down-parts of the fuzzy number A, respectively. To be in agreement with the classical denotation of fuzzy sets (numbers), the independent variable of both functions f and g is denoted by y, and their values by x [19].

Continuity of those two parts shows that their images are limited by specific intervals. They are named respectively: UP and $DOWN$. The limits (real numbers) of those intervals were marked using the following symbols: $UP = (l_A, l_A^-)$ and $DOWN = (l_A^+, p_A)$. If both functions that are parts of the fuzzy number are strictly monotonic, then there are their inverse functions x_{up}^{-1} and x_{down}^{-1} defined in respective intervals UP and $DOWN$. Then the following assignment is valid:

$$l_A := x_{up}(0), \quad l_A^- := x_{up}(1),$$
$$l_A^+ := x_{down}(1), \quad p_A := x_{down}(0) \tag{1}$$

Fig. 2 Ordered fuzzy
number

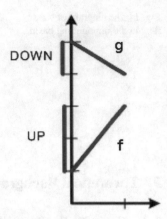

If a constant function equal to 1 is added within the interval $[1_A^-, 1_A^+]$ we get UP
and DOWN with one interval (Fig. 2), which can be treated as a carrier. Then the
membership function $\mu_A(x)$ of the fuzzy set defined on the R set is defined by the
following formulas:

$$\begin{aligned}
\mu_A(x) &= 0 && \text{for } x \notin [l_A, p_A] \\
\mu_A(x) &= x_{up}^{-1}(x) && \text{for } x \in UP \\
\mu_A(x) &= x_{down}^{-1}(x) && \text{for } x \in DOWN.
\end{aligned} \tag{2}$$

The fuzzy set defined in that way gets an additional property which is called order.
Whereas the following interval is the carrier (Fig. 3):

$$UP \cup [1_A^+, 1_A^-] \cup DOWN \tag{3}$$

The limit values for up and down parts are:

$$\begin{aligned}
\mu_A(l_A) &= 0 \\
\mu_A(1_A^-) &= 1 \\
\mu_A(1_A^+) &= 1 \\
\mu_A(p_A) &= 0
\end{aligned} \tag{4}$$

Fig. 3 OFN presented in a
way referring to fuzzy
numbers

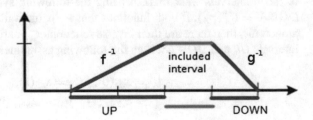

Generally, it can be assumed that ordered fuzzy numbers are of trapezoid form. Each of them can be defined using four real numbers:

$$A = (l_A, 1_A^-, 1_A^+, p_A).$$ (5)

The figures below (Fig. 4) show sample ordered fuzzy numbers including their characteristic points.

Functions f_A, g_A correspond to parts up_A, $down_A \subseteq R^2$ respectively, so that:

$$up_A = (f_A(y), y) : y \in [0, 1]$$ (6)

$$down_A = (g_A(y), y) : y \in [0, 1]$$ (7)

The orientation corresponds to the order of graphs f_A and g_A. The figure below (Fig. 5) shows the graphic interpretation of two opposite fuzzy numbers and the real number χ_0. Opposite numbers are reversely ordered [28].

Definition 2 A **membership function of an ordered fuzzy number** A is the function $\mu_A : R \to [0, 1]$ defined for $x \in R$ as follows [19, 23]:

$$\mu(x) = \begin{cases} f^{-1}(x) \text{ if } x \in [f(0), f(1)] = [l_A, 1_A^-] \\ g^{-1}(x) \text{ if } x \in [g(1), g(0)] = [l_A^+, p1_A] \\ 1 \text{ if } x \in [l_A^-, 1_A^+] \end{cases}$$ (8)

Fig. 4 Fuzzy number that is ordered, **a** positively, **b** negatively

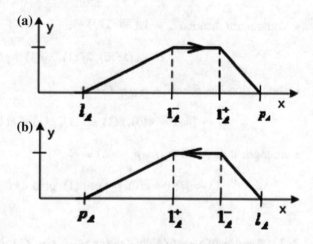

Fig. 5 Opposite numbers and the real number

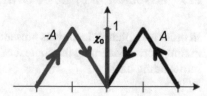

The above membership function can be used in the control rules similarly to the way membership of classic fuzzy numbers is used. All quantities that can be found in the fuzzy control describe selected part of the reality. Process of determining this value is called **fuzzy observation**.

2.2 Arithmetic Operations in OFN

The operation of adding two pairs of such functions is defined as the pair-wise addition of their elements, i.e., if $(f1, g1)$ and $(f2, g2)$ are two ordered fuzzy numbers, then $(f1 + f2, g1 + g2)$ will be just their sum. It is interesting to notice that as long as we are dealing with an ordered fuzzy number represented by pairs of affine functions of the variable $y \in [0, 1]$, its so-called classical counterpart, i.e., a membership function of the variable x is just of trapezoidal type. For any pair of affine functions (f, g) of $y \in [0, 1]$ we form a quaternion of real numbers according to the rule $[f(0), f(1), g(1), g(0)]$ which correspond to the four numbers $[l_A, 1_A^-, 1_A^+, p_A]$ as was mentioned in previous paragraph. If $(f, g) = A$ is a base pair of affine functions and $(e, h) = B$ is another pair of affine functions, then the set of typical operation will be uniquely represented by the following formulas respectively:

- addition $A + B = (f + e, g + h) = C$,

$$C \rightarrow [f(0) + e(0), f(1) + e(1), g(1) + h(1), g(0) + h(0)] \qquad (9)$$

- scalar multiplication $C = \lambda A = (\lambda f, \lambda g)$,

$$C \rightarrow [\lambda f(0), \lambda f(1), \lambda g(1), \lambda g(0)] \qquad (10)$$

- subtraction $A - B = (f - e, g - h) = C$

$$C \rightarrow [f(0) - e(0), f(1) - e(1), g(1) - h(1), g(0) - h(0)] \qquad (11)$$

- multiplication $A * B = (f * e, g * h) = C$

$$C \rightarrow [f(0) * e(0), f(1) * e(1), g(1) * h(1), g(0) * h(0)] \qquad (12)$$

2.3 Association of OFN Order with the Environmental Trend

In order to explain calculations presented in this sections, authors made the following assumptions concerning the context of changes taking place in the studied object (the impounding basin).

- **close context**—understood us the trend visible locally in the basin. It defines the trend of the object, i.e. if it is gradually filled or if the water level gradually falls. It is defined locally by the management of the dam,
- **further (environmental) context** - understood as the global trend for the observed area. It specifies trend of specific section of the river, intensity of precipitation as well as the set of other regional data which give better image of the environment. It is defined in the disaster recovery centre.

Two context types described above will be associated with the order of OFN numbers as follows. As the current water level state is reported to the disaster recovery centre, numbers that represent that state will always be oriented according to the environmental trend (reported to the centre). Whereas order of fuzzy numbers that represent changes of the water level in the impounding basin will be consistent with the local trend defined by the management of the dam. As a result, the order will be positive when the basin will be gradually filled, regardless the rate of that process. Whereas negative order can be observed when the outflow of water from the basin starts. Trend order changes themselves, as well as boundary conditions of the moment when they should occur, will make a separate process defined be the disaster recovery centre and the dam management respectively. Nothing stands in the way to ultimately use known segmentation methods for those processes and to determine the trend. Such issues are currently present in numerous publications and their detailed description is beyond the scope of this paper. Authors of that study assumed that the order equivalent to the trend changes is provided by a trusted third party and represents the expert's opinion concerning short-term weather forecast in the region important for the water level in the impounding basin as well as for the basin itself.

3   An Experimental Comparison of Fuzzy Numbers Arithmetic

3.1   Crisis Control Center in OFN Notation

In most cases the crisis control center action comes down to minimizing the effects of anomalies or sudden crises of a certain weather phenomenon. The crisis control center operation includes also preventive elements aimed to eliminate and, in most cases, to minimize potential losses. The above statements border also on the areas of economic sciences and financial risk management and [1–3, 7–10, 15, 26, 27].

Anticipating potential threats is a multistage process which is based on the assessment of premises affecting a given domain. Identification of risk factors in the context of crisis control center in this paper is limited to the evaluation of only two premises. These will include attributes defining the volume of atmospheric precipitation and the current amount of water in the reservoir, respectively. We can assume that they shall be colloquially expressed in highly simplified form as units. It is also evident

that the control of the dam by the crisis control center may be affected by other variables which were not discussed in this paper. Formulation of the forecast of the water
table level in the dam reservoir for a given period of time involves mainly the opinion
of experts. Typical opinion of an expert may include overlapping fuzzy classifying
statements. An expert may say that precipitation fits within the limits of a given fuzzy
quantity and also includes related components. The main information will be sup-
plemented by additional data on the scope or direction of changes (tendency, trend)
in precipitation. The above assessments, including two to three component items,
shall result in more and more complex analyzing systems.

The use of fuzzy logic, and in particular the ordered fuzzy numbers discussed in
previous chapters, turned out to be a good solution. The advantage of using OFN
number is the possibility to present the linguistic variable in two or three aspects of
a given problem. This approach allows to eliminate the problem with inconvenient
establishing of new quantities as well as to make use of OFN advantages. Hence
we can precisely present the fact that the estimated precipitation will be at the level
of six units with growing trend and also that the recorded growth amounts to eight
units. For better illustration, the forecast is shown in the figures (Figs. 6 and 7). For
one of many scenarios of the crisis control center operation one can assume that the
number A (see Fig. 6) corresponds to the volume of precipitation in the region. The
forecast precipitation volume is the kernel value, i.e. 6 units in this example. Another
information is that precipitation will be characterized by growing trend. This results

Fig. 6 An example of a
positive trend in the OFN
number

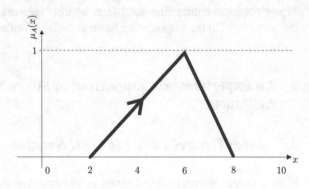

Fig. 7 An example of a
negative trend in the OFN
number

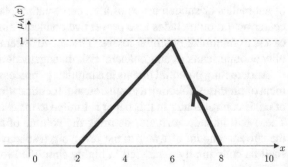

from the fact that the number A shown in figure (Fig. 6) is of positive sense. In this case we interpret this fact as the indication of growing trend of precipitation. The last information that can be read from the discussed OFN number is the amount of growth itself that is determined as the carrier width of A number. Focusing on the forecast trend itself and the value of changes, we have a reference to a classic vector. Where the trend of OFN number is the vector direction the module is the value. On the other hand, the figure (Fig. 7) refers to the number of a negative trend. In this way the expert highlights the fact that a given factor is characterized by a falling trend.

Based on the dam reservoir, let us consider the real situation that the crisis control center may have to deal with. Due to safety procedures, the decision-making center assumed that the optimal level of the water table in the reservoir is 100 units. Exceeding that value may lead to breaking the dam and flooding of the surrounding areas. The quantities analyzed in the modeled system shall include such linguistic variables as: the water table level: A, the volume of atmospheric precipitation: B. Therefore, the problem of the crisis control team may formally be represented as the equation $A + B = C$, where C is the safe value of the water table level in the reservoir. In the analyzed example, we know the current (number B) and the safe (number C) reservoir level. The ultimate aim is not to exceed the safe C value, for which the sum $A + B$ will be exactly 100. Values less than 100 suggest that the water level is too low for efficient operation of power turbines, while higher values may damage the reservoir itself. It should be noted that classic convex fuzzy numbers used in this model do not resolve the equation. Whereas in the model defined by ordered OFN numbers the equation has a solution that meets the reservoir safety condition (the sum equal to 100). This is due to the OFN number order itself as well as due to the definition of the adding operation for OFN numbers. Formal details are presented in part ... and ... .

As illustrated here, the application of ordered fuzzy numbers and the use of their mathematical methods allows more precise control of a given process. The following facts make OFN numbers the method of choice: no increase in fuzziness, i.e. arithmetic operations do not cause widening of fuzzy number carriers; the possibility of back inference, i.e. it is possible to reconstruct the sequence of the event on the basis of input data and results. The aforementioned advantages of OFN mathematical methods are clearly visible in the above examples where the very idea of OFN was referred to and in the part describing the hypothetical situations of the dam reservoir control. The application described by the authors of this article is just one of many possible approaches to the application of ordered OFN numbers. Authors of other papers presented examples of OFN applications in economics, inventory management as well as in areas concerning the rate of return on the investment.

3.2 Elementary Arithmetic Operations

To automate computational experiments, authors of this study developed dedicated programme Ordered FN, which also efficiently supports graphic interpretation of

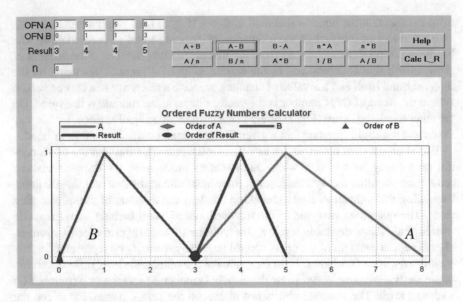

Fig. 8 Main screen of the programme

operations being performed on ordered fuzzy numbers. Additionally, it is equipped with the Calc L-R module. Ordered FN is equipped with an additional module which is started using the Calc L-R button. It includes procedures to calculate the sum, the difference and the product of L-R numbers. This allows to compare some results obtained from operation on L-R and OFN numbers. The introduced fuzzy number has a form of (m, α, β) where α and β are left and right-side dispersions. Shown here is an example of the software application. It includes the summary of simple arithmetic operation results in OFN and L-R notation. As shown in the attached figure (Fig. 8), subtraction of 3-3 is different for L-R than for OFN numbers.

3.3 Comparison of Calculations on L-R and OFN Numbers

Majority of operation performed on L-R numbers, regardless if it is addition or subtraction, increases the carrier value, i.e. the areas of non-accuracy. Hence performance of several operations on those numbers can cause such big fuzziness that the resulting quantity will be useless. It is impossible to solve an $A + X = C$ equation using L-R representation through operations because for that interpretation $X + A + (-A) \neq X$ and $X * A * A^{-1} \neq X$. Whereas every inverse operation on L-R numbers will increase the carrier. It is often impossible to solve the equations using analytical (computations) method either. However, it is possible to break the stalemate using certain empirical methods. The situation is totally different in case of operation on ordered

Table 1 Example Table 1

Solution versions	
Version I	S olution using computational method of OFN arithmetic
	$A(0, 1, 1, 3) + X = C(3, 5, 5, 8)$
	$X = C - A$
	$X = (3, 4, 4, 5)$
	Verification: $A + X = C$
	$A(0, 1, 1, 3) + X(3, 4, 4, 5) = C(3, 5, 5, 8)$
Version II	Solution using computational method of L-R arithmetic
	$A(1, 1, 2) + X = C(5, 2, 3)$
	$X = C - A$
	$X = (4, 4, 4)$
	Verification: $A + X = C$
	$A(1, 1, 2) + X(4, 4, 4) \neq C(5, 2, 3)$

fuzzy numbers. It is possible to solve the above mentioned equation using an analytical method.

Example 1 Problem: There was a rapid surge of water in the impounding basin at night. The management of the dam has to send reports to the disaster recovery centre including the value of the water level change comparing to the previous state. Unfavourable weather conditions do not allow for precise measurement.

Data
$A(1,1,2)$-Previous measurement [mln m^3]
$C(5,2,3)$-Current measurement in [mln m^3]

Mathematic interpretation: Hence the problem comes down to determining X number that satisfies the equation $A + X = C$ as $A(1, 1, 2) + X = C(5, 2, 3)$.

As a result of operations on L-R numbers we obtain the outcome $(4, 4, 4)$. However, the verification through addition $(A + X)$ gives the result different from C. Correct result $(4, 1, 1)$ can only be achieved using empirical method. It is problematic and not always feasible (Table 1).

4 Conclusion

Operations performed on ordered fuzzy numbers are often more accurate than operations performed on classic fuzzy numbers. Results of operations performed on them are the same as those obtained from operations on real numbers. Performing multiple not necessarily causes large increase of the carrier. The situation is different for L-R fuzzy numbers, where several operations often lead to numbers of high fuzziness. An infinitesimal carrier is interpreted as a real number and thus for OFN numbers one can apply commutative and associative property of multiplication over addition. The possibility to perform back inference on them allows to reproduce input data by solving an appropriate equation. This very property is an added value that makes this fuzzy logic extension worth to promulgate. Calculations performed on ordered fuzzy numbers are easy and accurate. It is worth mention the multiplication here, where the same procedure is used for all ordered fuzzy numbers regardless their sign. Whereas multiplication of L-R numbers is different for two positive numbers than for two negative ones. Another completely different procedure is used for multiplication of numbers of indefinite signs and for fuzzy zeros. It also seems to be very interesting to associate OFN numbers with trend of changes taking place for studied part of the reality. We are convinced that new applications of this property of OFN, shown here in the example of the fuzzy observation of the impounding basin in unfavourable weather conditions, will be introduced with the passing of time. Hence it seems that introduction of OFN gives new possibilities for designers of highly dynamic systems. With this approach it is possible to define trend of changes, which gives new possibilities for the development of fuzzy control and it charts new ways of research in the fuzzy logic discipline. Broadening it by the theory of ordered fuzzy numbers seems to allow for more efficient use of imprecise operations. Simple algorithmization of ordered fuzzy numbers allows to use them in a new control model. It also inspires researchers to search for new solutions. Authors did not use defuzzyfication operators [4, 11] in this paper, which in themselves are interesting subject of many researches. They will also contribute to development of the comparative calculator created here. Although authors of this study are not so enthusiastic like the creators of OFN as regards excellent prospects of this new fuzzy logic idea, but they are impressed by possibilities for arithmetic operations performed using this notation. Even sceptics who treat OFN with reserve as it is generalization of fuzzy logic, can benefit from this arithmetic. After all OFN can be treated as internal representation of fuzzy numbers (heedless of it's authors' intention). With this new kind of notation for fuzzy numbers and fuzzy control, it is possible to achieve clear and easily interpreted calculation, which can be arithmetically verified regardless of the input data type. Perhaps the OFN idea will become another paradigm of fuzzy logic, just like the object oriented programming paradigm has become dominant in software engineering after the structured programming paradigm. Whichever scenario wins, at least some aspects of OFN arithmetic seem to be hard to ignore a priori.

References

1. Angryk, R.A., Czerniak, J.: Heuristic algorithm for interpretation of multi-valued attributes in similarity-based fuzzy relational databases. Int. J. Approx. Reason. **51**(8), 895–911 (2010)
2. Apiecionek, L., Czerniak, J.M., Dobrosielski, W.T.: Quality of services method as a ddos protection tool. Intelligent Systems'2014, Vol 2: Tools, Architectures, Systems, Applications, vol. 323, pp. 225–234 (2015)

3. Apiecionek, L., Czerniak, J.M., Zarzycki, H.: Protection tool for distributed denial of services attack. Beyond Databases, Architectures And Structures, BDAS, vol. (424), pp. 405–414 (2014)
4. Bednarek, T., Kosiński, W., Węgrzyn-Wolska, K.: On orientation sensitive defuzzification functionals. In: Proceedings of the Artificial Intelligence and Soft Computing, pp. 653–664. Springer, Heidelberg (2014)
5. Bošnjak, I., Madarász, R., Vojvodić, G.: Algebras of fuzzy sets. Fuzzy Sets Syst. **160**(20), 2979–2988 (2009)
6. Couso, I., Montes, S.: An axiomatic definition of fuzzy divergence measures. Int. J. Uncertain., Fuzziness Knowl.-Based Syst. **16**(01), 1–17 (2008)
7. Czerniak, J., Zarzycki, H.: Application of rough sets in the presumptive diagnosis of urinary system diseases. Artif. Intell. Secur. Comput. Syst. **752**, 41–51 (2003)
8. Czerniak, J.: Evolutionary approach to data discretization for rough sets theory. Fundamenta Informaticae **92**(1–2), 43–61 (2009)
9. Czerniak, J.M., Apiecionek, L., Zarzycki, H.: Application of ordered fuzzy numbers in a new ofnant algorithm based on ant colony optimization. Beyond Databases, Architectures And Structures, BDAS, vol. 424, pp. 259–270 (2014)
10. Czerniak, J.M., Dobrosielski, W., Zarzycki, H., Apiecionek, L.: A proposal of the new owlant method for determining the distance between terms in ontology. Intelligent Systems'2014, Vol 2: Tools, Architectures, Systems, Applications, vol. 323, pp. 235–246 (2015)
11. Dobrosielski, W.T., Szczepański, J., Zarzycki, H.: Novel Developments in Uncertainty Representation and Processing: Advances in Intuitionistic Fuzzy Sets and Generalized Nets. In: Proceedings of 14th International Conference on Intuitionistic Fuzzy Sets and Generalized Nets, chap. A Proposal for a Method of Defuzzification Based on the Golden Ratio—GR, pp. 75–84. Springer International Publishing, Cham (2016). http://dx.doi.org/10.1007/978-3-319-26211-6_7
12. Dubois, D., Prade, H.: Operations on fuzzy numbers. Int. J. Syst. Sci. **9**(6), 613–626 (1978)
13. Dubois, D., Prade, H.: Fuzzy elements in a fuzzy set. In: Proceedings of the 10th International Fuzzy Systems Association. (IFSA) Cong. pp. 55–60 (2005)
14. Dubois, D., Prade, H.: Gradual elements in a fuzzy set. Soft Comput. **12**(2), 165–175 (2008)
15. Ewald, D., Czerniak, J.M., Zarzycki, H.: Approach to solve a criteria problem of the abc algorithm used to the wbdp multicriteria optimization. Intelligent Systems'2014, Vol 1: Mathematical Foundations, Theory, Analyses, vol. 322, pp. 129–137 (2015)
16. Gerla, G.: Fuzzy logic programming and fuzzy control. Studia Logica **79**(2), 231–254 (2005)
17. Gottwald, S.: Mathematical aspects of fuzzy sets and fuzzy logic: some reflections after 40 years. Fuzzy Sets Syst. **156**(3), 357–364 (2005)
18. Kosiński, W., Slysz, P.: Fuzzy numbers and their quotient space with algebraic operations. Bull. Polish Acad. Sci.Ser. Tech. Sci. **41**, 285–295 (1993)
19. Kosiński, W.: On fuzzy number calculus. Int. J. Appl. Math. Comput. Sci **16**(1), 51–57 (2006)
20. Kosiński, W., Prokopowicz, P., Frischmuth, K.: On algebra of ordered fuzzy numbers. Soft Computing Foundations and Theoretical Aspects, EXIT (2004)
21. Kosiński, W., Prokopowicz, P., Ślęzak, D.: Fuzzy reals with algebraic operations: Algorithmic approach. In: Proceedings of the Intelligent Information Systems 2002, pp. 311–320 (2002)
22. Kosiński, W., Prokopowicz, P., Ślęzak, D.: On algebraic operations on fuzzy numbers. Intelligent Information Processing and Web Mining, pp. 353–362. Springer, Heidelberg(2003)
23. Kosiński, W., Prokopowicz, P., Ślęzak, D.: Ordered fuzzy numbers. Bull. Polish Acad. Sci.Ser. Math. **51**(3), 327–338 (2003)
24. Łukasiewicz, J.: Elementy logiki matematycznej (1929)
25. Łukasiewicz, J.: O logice trójwartościowej (1988)
26. Mikolajewska, E., Mikolajewski, D.: Exoskeletons in neurological diseases - current and potential future applications. Adv. Clin. Exp. Med. **20**(2), 227–233 (2011)
27. Nawarycz, T., Pytel, K., Gazicki-Lipman, M., Drygas, W., Ostrowska-Nawarycz, L.: A fuzzy logic approach to the evaluation of health risks associated with obesity. In: Ganzha, M., Maciaszek, L.M.P. (eds.) Proceedings of the 2013 Federated Conference on Computer Science and Information Systems. pp. 231–234. IEEE (2013)

28. Prokopowicz, P.: Algorytmization of Operations on Fuzzy Numbers and its Applications (in Polish). Ph.D. thesis (2005)
29. Rebiasz, B., Gaweł, B., Skalna, I.: Fuzzy multi-attribute evaluation of investments. In: Ganzha, M., Maciaszek, L.M.P. (eds.) Proceedings of the 2013 Federated Conference on Computer Science and Information Systems. pp. 977–980. IEEE (2013)
30. Walker, C.L., Walker, E.A.: The algebra of fuzzy truth values. Fuzzy Sets Syst. **149**(2), 309–347 (2005)
31. Zadeh, L.: Fuzzy sets. Inf. Control **8**(3), 338–353 (1965)
32. Zadeh, L.A.: Is there a need for fuzzy logic? Inf. Sci. **178**(13), 2751–2779 (2008)

Forecasting Indoor Temperature Using Fuzzy Cognitive Maps with Structure Optimization Genetic Algorithm

Katarzyna Poczęta, Alexander Yastrebov and Elpiniki I. Papageorgiou

Abstract Fuzzy cognitive map (FCM) is a soft computing methodology that allows to describe the analyzed problem as a set of nodes (concepts) and connections (links) between them. In this paper the Structure Optimization Genetic Algorithm (SOGA) for FCMs learning is presented for prediction of indoor temperature. The proposed approach allows to automatically construct and optimize the FCM model on the basis of historical multivariate time series. The SOGA defines a new learning error function with an additional penalty for coping with the high complexity present in an FCM with a large number of concepts and connections between them. The aim of this study is the analysis of usefulness of the Structure Optimization Genetic Algorithm for fuzzy cognitive maps learning on the example of forecasting the indoor temperature of a house. A comparative analysis of the SOGA with other well-known FCM learning algorithms (Real-Coded Genetic Algorithm and Multi-Step Gradient Method) was performed with the use of ISEMK (Intelligent Expert System based on Cognitive Maps) software tool. The obtained results show that the use of SOGA allows to significantly reduce the structure of the FCM model by selecting the most important concepts, connections between them and keeping a high forecasting accuracy.

K. Poczęta (✉) · A. Yastrebov
Kielce University of Technology, al. Tysiąclecia Państwa Polskiego 7,
25-314 Kielce, Poland
e-mail: k.piotrowska@tu.kielce.pl

A. Yastrebov
e-mail: a.jastriebow@tu.kielce.pl

E.I. Papageorgiou
Technological Educational Institute (T.E.I.) of Central Greece,
3rd Km Old National Road Lamia-Athens, 35100 Lamia, Greece
e-mail: epapageorgiou@teiste.gr; elpiniki.papageorgiou@uhasselt.be

E.I. Papageorgiou
Faculty of Business Economics, Hasselt University,
Campus Diepenbeek Agoralaan Gebouw D, 3590 Diepenbeek, Belgium

© Springer International Publishing Switzerland 2016
S. Fidanova (ed.), *Recent Advances in Computational Optimization*,
Studies in Computational Intelligence 655, DOI 10.1007/978-3-319-40132-4_5

1 Introduction

Fuzzy cognitive map (FCM) [15] is a soft computing methodology combining the advantages of fuzzy logic and artificial neural networks. It allows to visualize and analyze problem as a set of nodes (concepts) and links (connections between them). One of the most important aspect connected with the use of fuzzy cognitive maps is their ability to learn on the basis of historical data [19]. Supervised [14, 23] and population-based [1, 7, 8, 27] methods can be used to determine the weights of the connections between concepts.

Fuzzy cognitive maps are an effective tool for modeling dynamic decision support systems [17]. They were applied to many different areas, such as prediction of pulmonary infection [18], scenario planning for the national wind energy sector [2] or integrated waste management [4]. Carvalho discussed possible use of FCM as tool for modeling and simulating complex social, economic and political systems [5]. The use of FCMs as pattern classifiers is presented in [21, 22]. An innovative method for forecasting artificial emotions and designing an affective decision system on the basic of fuzzy cognitive map is proposed in [25]. The application of fuzzy cognitive maps to univariate time series modeling is discussed in [12, 13, 16]. Prediction of work of complex and imprecise systems on the basis of FCM is described in [26].

In practical applications to solve certain classes of problems (e.g. data analysis, prediction or diagnosis), finding the most significant concepts and connections plays an important role. It can be based on expert knowledge at all stages of the analysis: designing the structure of the FCM model, determining the weights of the relationships and selecting input data. Supervised and population-based algorithms enable the automatic construction of fuzzy cognitive map on the basis of data selected by the experts or all available input data. However, modeling of complex systems can be difficult task due to the large amount of the information about analyzed problem. Fuzzy cognitive maps with the large number of concepts and connections between them can be difficult to interpret and impractical to use as the number of parameters to be established grows quadratically with the size of the FCM model [12]. In [13] nodes selection criteria for FCM designed to model univariate time series are proposed. Also some simplifications strategies by posteriori removing nodes and weights are presented [12]. In [20, 24], the Structure Optimization Genetic Algorithm (SOGA) [20, 24] was proposed. It enables fully automatic construction of the FCM model by selection of crucial concepts and determining the relationships between them on the basis of available historical data.

This paper is focused on the construction of the FCM model for prediction of the indoor temperature using the Structure Optimization Genetic Algorithm. Forecasting of temperature in the houses could help to reduce power consumption by determining proper control actions. Neural networks based predictive control for the thermal comfort defined by indoor temperature and energy savings was proposed in [6]. In [28], a system for forecasting of indoor temperature in the near future based on artificial neural networks was analyzed.

The research in this paper has been done with the use of data collected from a monitor system mounted in a domotic house [28]. The proposed approach is compared with well-known methods for fuzzy cognitive maps learning: the Multi-Step Gradient Method (MGM) [14] and the Real-Coded Genetic Algorithm (RCGA) [27]. Simulations have been accomplished with the use of the developed ISEMK (Intelligent Expert System based on Cognitive Maps) software tool.

The paper is organized as follows. In Sect. 2 fuzzy cognitive maps are described. Section 3 describes the method for fuzzy cognitive maps learning. Section 4 describes the developed software tool ISEMK. Section 5 presents selected results of simulation analysis of the Structure Optimization Genetic Algorithm. The last Section contains a summary of the paper.

2 Fuzzy Cognitive Maps

The structure of FCM is based on a directed graph:

$$< X, W >,$$ (1)

where $X = [X_1, \ldots, X_n]^T$ is the set of the concepts significant for the analyzed problem, W is weights matrix, $W_{j,i}$ is the weight of the connection between the jth concept and the ith concept, taking on the values from the range $[-1, 1]$, $W_{i,i} = 0$. Value of -1 means full negative influence, 1 denotes full positive influence and 0 means no causal effect [15].

Concepts obtain values in the range between $[0, 1]$ so they can be used in time series prediction. The values of concepts can be calculated according to the formulas:

$$X_i(t + 1) = F\left(X_i(t) + \sum_{j=1}^{n} W_{j,i} \times X_j(t)\right),$$ (2)

$$X_i(t + 1) = F\left(\sum_{j=1}^{n} W_{j,i} \times X_j(t)\right),$$ (3)

where t is discrete time, $t = 0, 1, 2, \ldots, T$, T is end time of simulation, $X_i(t)$ is the value of the ith concept, $i = 1, 2, \ldots, n$, n is the number of concepts, $F(x)$ is a transformation function, which can be chosen in the form:

$$F(x) = \frac{1}{1 + e^{-cx}},$$ (4)

where $c > 0$ is a parameter.

Fuzzy cognitive map can be automatic constructed with the use of supervised and population-based learning algorithms. In the next section, selected methods of FCMs learning are described.

3 Fuzzy Cognitive Maps Learning

The aim of the FCM learning process is to estimate the weights matrix W. In the paper the Structure Optimization Genetic Algorithm for fuzzy cognitive maps learning is analyzed. Performance of the developed approach is compared with the Real-Coded Genetic Algorithm and the Multi-Step Gradient Method. Description of these methods is presented below.

3.1 Real-Coded Genetic Algorithm

Real-Coded Genetic Algorithm defines each chromosome as a floating-point vector, expressed as follows [27]:

$$W' = [W_{1,2}, \ldots, W_{1,n}, W_{2,1}, W_{2,3}, \ldots, W_{2,n}, \ldots, W_{n,n-1}]^T, \tag{5}$$

where $W_{j,i}$ is the weight of the connection between the jth and the ith concept.

Additionally, two parameters of the dynamic model of every individual were introduced:

- c—parameter of the transformation function (4),
- d—type of the dynamic model, $d = 0$ means that the model type (2) is set, $d = 1$ means that the model type (3) is set.

Each chromosome in the population is decoded into a candidate FCM and its quality is evaluated on the basis of a fitness function according to the objective [9]. The aim of the analyzed learning process is to optimize the weights matrix with respect to the prediction accuracy. Fitness function can be described as follows:

$$fitness_p(J(l)) = \frac{1}{a \times J(l) + 1}, \tag{6}$$

where a is a parameter, $a > 0$, p is the number of chromosome, $p = 1, \ldots, P$, P is the population size, l is the number of population, $l = 1, \ldots, L$, L is the maximum number of populations, $J(l)$ is the learning error function, described as follows:

$$J(l) = \frac{1}{(T-1)n_o} \sum_{t=1}^{T-1} \sum_{i=1}^{n_o} (Z_i^o(t) - X_i^o(t))^2, \tag{7}$$

where t is discrete time of learning, T is the number of the learning records, $Z(t) = [Z_1(t), \ldots, Z_n(t)]^T$ is the desired FCM response for the initial vector $Z(t-1)$, $X(t) = [X_1(t), \ldots, X_n(t)]^T$ is the FCM response for the initial vector $Z(t-1)$, n is the number of the all concepts, n_o is the number of the output concepts, $X_i^o(t)$ is the value of the ith output concept, $Z_i^o(t)$ is the reference value of the ith output concept.

Each population is assigned a probability of reproduction. According to the assigned probabilities parents are selected and new population of chromosomes is generated. Chromosomes with above average fitness tend to receive more copies than those with below average fitness. In the analysis a ranking method of selection was used. Selecting a copy of the chromosome into the next population is based on ranking by fitness [3, 9].

The crossover operator is a method for sharing information between parents to form new chromosomes. It can be applied to random pairs of chromosomes and the likelihood of crossover depends on probability defined by the crossover probability P_c. The popular crossover operator is the uniform crossover [11].

The mutation operator modifies elements of a selected chromosome with a probability defined by the mutation probability P_m. The use of mutation prevents the premature convergence of genetic algorithm to suboptimal solutions [11]. In the analysis random mutation was used. To ensure the survival of the best chromosome in the population, elite strategy was applied. It retains the best chromosome in the population [9].

The learning process stops when the maximum number of populations L is reached or the condition (8) is met.

$$fitness_{best}(J(l)) > fitness_{max}, \tag{8}$$

where $fitness_{best}(J(l))$ is the fitness function value for the best chromosome, $fitness_{max}$ is a parameter.

3.2 Structure Optimization Genetic Algorithm

In this paper a new Structure Optimization Genetic Algorithm is proposed, which allows to select the most important for prediction task concepts and connections between them. SOGA defines each chromosome as a floating-point vector type (5), parameters of the dynamic model c, d and a binary vector expressed as follows [24]:

$$C' = [C_1, C_2, \ldots, C_n]^T, \tag{9}$$

where C_i is the information about including the ith concept to the candidate FCM model, whereas $C_i = 1$ means that the candidate FCM model contains the ith concept, $C_i = 0$ means that the candidate FCM model does not contain the ith concept.

The quality of each population is calculated based on an original fitness function, described as follows:

$$fitness_p(J'(l)) = \frac{1}{a \cdot J'(l) + 1},\tag{10}$$

where a is a parameter, $a > 0$, p is the number of the chromosome, l is the number of population, $l = 1, \ldots, L$, L is the maximum number of populations, $J'(l)$ is the new learning error function with an additional penalty for highly complexity of FCM understood as a large number of concepts and non-zero connections between them [10, 24]:

$$J'(l) = J(l) + b_1 \times \frac{n_r}{n^2} \times J(l) + b_2 \times \frac{n_c}{n} \times J(l),\tag{11}$$

where t is discrete time of learning, T is the number of the learning records, b_1, b_2 are the parameters, $b_1 > 0$, $b_2 > 0$, n_r is the number of the non-zero weights of connections, n_c is the number of the concepts in the candidate FCM model, n is the number of the all possible concepts, $J(l)$ is the learning error function type (11).

Figure 1 illustrates the steps of the learning and analysis of the FCM in modeling prediction systems with the use of population-based algorithms (SOGA and RCGA).

In the paper the proposed algorithm was compared with the Real-Coded Genetic Algorithm and also with supervised learning based on Multi-Step Gradient Method.

3.3   Multi-step Gradient Method

Multi-step algorithms of FCM learning are some kind of generalization of known one-step methods. Effectiveness of these methods in modeling of decision support systems was presented in [14, 23]. Multi-step supervised learning based on gradient method is described by the equation:

$$W_{j,i}(t + 1) = P_{[-1,1]} \left(\sum_{k=0}^{m_1} \alpha_k \times W_{j,i}(t - k) \right.$$
$$\left. + \sum_{l=0}^{m_2} (\beta_l \times \eta_l(t) \times (Z_i(t - l) - X_i(t - l)) \times y_{j,i}(t - l)) \right),\tag{12}$$

where $\alpha_k, \beta_l, \eta_l$ are learning parameters, $k = 1, \ldots, m_1$; $l = 1, \ldots, m_2$, m_1, m_2 are the number of the steps of the method, t is a time of learning, $t = 0, 1, \ldots, T - 1$, T is end time of learning, $X_i(t)$ is the value of the ith concept, $Z_i(t)$ is the reference

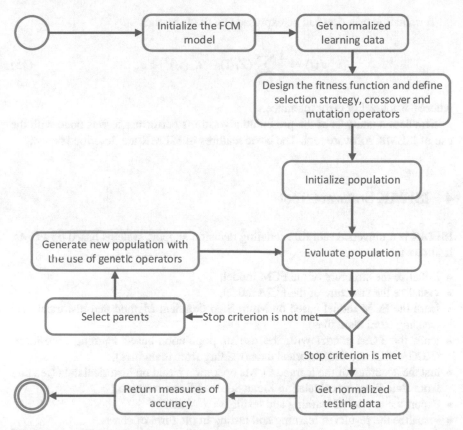

Fig. 1 Activity diagram for population-based learning algorithm

value of the ith concept, $y_{j,i}(t)$ is a sensitivity function, $P_{[-1,1]}(x)$ is an operator of design for the set $[-1,1]$, described as follows:

$$P_{[-1,1]}(x) = \begin{cases} 1 & \text{for } x \geq 1 \\ x & \text{for } -1 < x < 1 \\ -1 & \text{for } x \leq -1 \end{cases}, \tag{13}$$

Sensitivity function $y_{j,i}(t)$ is described by the equation:

$$y_{j,i}(t+1) = \left(y_{j,i}(t) + X_j(t) \right) \times F' \left(X_i(t) + \sum_{j=1}^{n} W_{j,i} \times X_j(t) \right), \tag{14}$$

where $F'(x)$ is derivative of the transformation function.

Termination criterion can be expressed by the formula:

$$J(t) = \frac{1}{n} \sum_{i=1}^{n} (Z_i(t) - X_i(t))^2 < e,$$ (15)

where e is a level of error tolerance.

Simulation analysis of the presented algorithms performance was done with the use of ISEMK software tool. The basic features of ISEMK are described below.

4 ISEMK Software Tool

ISEMK is a universal tool for modeling decision support systems based on FCMs. It allows to:

- initialize the structure of the FCM model,
- visualize the structure of the FCM model,
- learn the FCM model based on Multi-Step Gradient Method and historical data (reading from .data files),
- learn the FCM model with the use of population-based learning algorithms (RCGA, SOGA) and historical data (reading from .data files),
- test the accuracy of the learned FCMs operation based on historical data (reading from .data files) by calculating Mean Squared Error measure,
- export the results of learning and testing to .csv files,
- visualize the results of learning and testing in the form of charts.

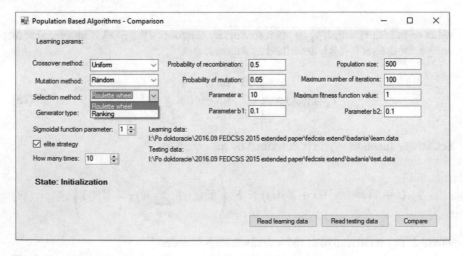

Fig. 2 Exemplary visualization of population-based learning initialization

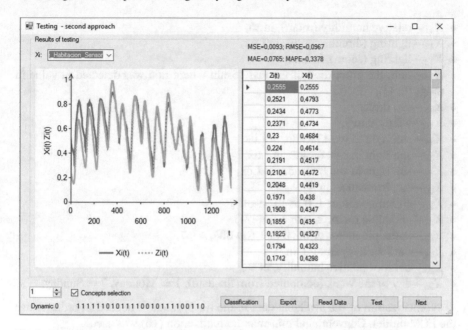

Fig. 3 Exemplary visualization of testing of learned FCM

Figure 2 shows an exemplary initialization of the population-based learning process. Figure 3 shows an exemplary visualization of testing of the learned FCM operation.

5 Simulation Results

To evaluate the proposed Structure Optimization Genetic Algorithm, historical data taken from the UCI Machine Learning Repository [28] were used. The dataset is collected from a monitor system mounted in a domotic house. It corresponds to approximately 40 days of monitoring data and has the following attributes:

- X_1—day,
- X_2—month,
- X_3—year,
- X_4—time,
- X_5—indoor temperature (dinning-room), in °C—the output concept,
- X_6—indoor temperature (room), in °C—the output concept,
- X_7—weather forecast temperature, in °C,
- X_8—carbon dioxide in ppm (dinning room),
- X_9—carbon dioxide in ppm (room),
- X_{10}—relative humidity (dinning room), in %,

- X_{11}—relative humidity (room), in %,
- X_{12}—lighting (dinning room), in Lux,
- X_{13}—lighting (room), in Lux,
- X_{14}—rain, the proportion of the last 15 min where rain was detected (a value in range [0, 1]),
- X_{15}—sun dusk,
- X_{16}—wind, in m/s,
- X_{17}—sun light in west facade, in Lux,
- X_{18}—sun light in east facade, in Lux,
- X_{19}—sun light in south facade, in Lux,
- X_{20}—sun irradiance, in $\frac{W}{m^2}$,
- X_{21}—enthalpic motor 1, 0 or 1 (on-off),
- X_{22}—enthalpic motor 2, 0 or 1 (on-off),
- X_{23}—enthalpic motor turbo, 0 or 1 (on-off),
- X_{24}—outdoor temperature, in °C,
- X_{25}—outdoor relative humidity, in %,
- X_{26}—day of the week (computed from the date), 1 = Monday, 7 = Sunday.

Normalization of the available data in the [0, 1] range is needed in order to use the FCM model. Conventional min-max normalization (16) was used.

$$f(x) = \frac{x - min}{max - min},\tag{16}$$

where x is an input numeric value, min is the minimum of the dataset, max is the maximum of the dataset.

The aim of the simulation analysis is:

- construction of the FCM model based on the historical data,
- selection of the most significant concepts form the available attributes and connections between them,
- one-step-ahead prediction of the indoor temperature in dining-room (X_5) and room (X_6),
- comparison of the results of the SOGA algorithm with the well-known methods for fuzzy cognitive maps learning.

The dataset was divided into two subsets: learning (2765 records) and testing (1373 records) data. The learning process was accomplished with the use of Multi-Step Gradient Method, Real-Coded Genetic Algorithm and Structure Optimization Genetic Algorithm.

Mean Squared Error for the output concepts MSE (17), the number of the concepts n_c and connections between them n_r were used to estimate the performance of the FCM learning algorithms.

$$MSE = \frac{1}{n_o(T-1)} \sum_{t=1}^{T-1} \sum_{i=1}^{n_o} \left(Z_i^o(t) - X_i^o(t) \right)^2,\tag{17}$$

Fig. 4 The structure of the
FCM learned with the use of
RCGA

where t is time of testing, $t = 0, 1, \ldots, T - 1$, T is the number of the test records, $Z(t) = [Z_1(t), \ldots, Z_n(t)]^T$ is the desired FCM response for the initial vector $Z(t - 1)$, $X(t) = [X_1(t), \ldots, X_n(t)]^T$ is the FCM response for the initial vector $Z(t - 1)$, $X_i^o(t)$ is the value of the ith output concept, $Z_i^o(t)$ is the reference value of the ith output concept, n_o is the number of the output concepts.

Experiments were carried out for various learning parameters determined using experimental trial and error method. For every configuration of the RCGA and SOGA 10 experiments were performed and the average values of MSE, n_c and n_r with standard deviations were calculated. Selected results of the analysis are presented below.

Standard supervised and population-based learning methods allow to obtain the FCM model based on all available attributes. As a sample result of learning by the Multi-step Gradient Method, the map with 26 concepts and 537 non-zero connections was obtained. The Real-Coded Genetic Algorithm returned the model with 26 concepts and the average of 429 links between them. Figure 4 presents the structure of the FCM learned with the use of RCGA (for the following parameters: $P = 100$, $L = 100$, ranking selection, uniform crossover, random mutation, elite strategy, $P_c = 0.8$, $P_m = 0.01$, $a = 1000$, $h_{max} = 0.999$, $c = 1$, $d = 0$). Such complex structures are unreadable and can be difficult to be further analyzed and interpreted.

Fig. 5 The structure of the
FCM learned with the use of
SOGA

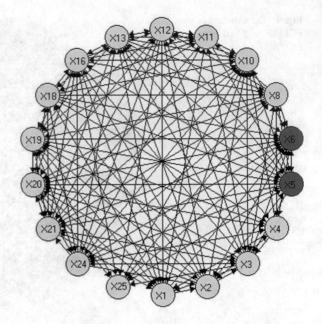

The developed Structure Optimization Genetic Algorithm allows to reduce the size
of the initial FCM model by selecting only the most significant for the prediction task
concepts and connection between them. Figure 5 presents the structure of the FCM
learned with the use of SOGA (for the following parameters: $P = 100$, $L = 100$,
ranking selection, uniform crossover, random mutation, $P_c = 0.8$, $P_m = 0.01$, $a = 1000$, $h_{max} = 0.999$, $c = 1$, $d = 0$, $b_1 = 0.1$, $b_2 = 0.05$). Table 1 shows the weights
matrix for this map. As a sample result of learning using the SOGA algorithm the
FCM with 18 concepts and 174 non-zero connections was obtained.

Figures 6 and 7 show the exemplary results of testing of the learned FCMs operation. The SOGA method is the best fitted with the analyzed testing data. Table 2 shows
selected results of the comparative analysis of the Multi-Step Gradient Method, the
Real-Coded Genetic Algorithm and the Structure Optimization Genetic Algorithm.

The lowest average value of the $MSE = 0.013$ was obtained for the Structure
Optimization Genetic Algorithm, which confirms the usefulness of the proposed
approach in forecasting indoor temperature. The SOGA algorithm returned the model
with the average of 16 concepts and the average of 150 links between them. The
results show the superiority of the SOGA. It allows to simplify the structure of the
initialize FCM model by selecting the most significant concepts and connections
between them and gives a high forecasting accuracy in the selected case study.

Table 1 Exemplary weights matrix for the map learned with the use of SOGA

	X_1	X_2	X_3	X_4	X_5	X_6	X_8	X_{10}	X_{11}	X_{12}	X_{13}	X_{16}	X_{18}	X_{19}	X_{20}	X_{21}	X_{24}	X_{25}
X_1	0	−0.1	0	0	0	−0.5	0	0.7	0.3	0	−0.8	0	−0.9	0	0	0	0	0
X_2	0	0	−0.6	0.3	0.1	−0.2	0	0.3	0	0	−0.8	−0.9	−0.7	−0.8	0.5	−0.3	0	0
X_3	−0.2	0	0	0	0	−0.1	0.2	1	−0.3	0.7	0.5	0	0	−1	−0.9	0.9	−0.9	0.1
X_4	0.9	0.2	−0.8	0	0	0.2	0	−0.4	0	0	−0.9	0	0	0.1	0.2	0	0.3	0
X_5	0	−0.5	0.4	0	0	1	0.7	0.4	0	−0.3	−0.5	−0.4	0	0	−0.7	−0.5	0.1	0.4
X_6	0	0.3	0	−0.4	0.9	0	0	−0.2	−0.7	0.9	0	0	0	0	0.3	1	−0.1	0.4
X_8	0	0	0.4	0.8	−0.1	−0.5	0	0	0	0	0	0	0	0.2	−0.9	0.9	0	0.2
X_{10}	−0.5	0	0	0	0	−0.9	0	0	0	−0.3	0	−1	0	0	−0.7	0	0.3	0
X_{11}	0	0	0.7	0	−0.6	0	0	0.2	0.3	0	−0.5	0	0.5	0	0	0.8	0	0
X_{12}	−1	−0.6	0	0	−0.2	0	0	0.8	−0.8	0	0.8	0.9	0	0.8	−0.6	0	0.1	−0.3
X_{13}	−0.9	−1	0.8	0.4	0	−0.5	−0.5	0.5	−0.8	−0.2	0	0	−0.7	−0.4	0	0	0	0
X_{16}	0.6	0	0.1	0	−0.7	0	0.4	0	−0.9	−0.9	−0.9	0.4	0	−0.9	−0.1	−0.1	0	−1
X_{18}	0.6	0	0	0.2	−1	−0.5	0.4	−0.6	−0.2	0	0.6	0	0	0	0	0	0	0
X_{19}	−0.3	0.3	0	−0.4	0.5	0.8	0	0	0	0	0	0.8	0	0	−0.4	0.3	−0.5	−0.7
X_{20}	0.1	0	0.8	0.4	0	−0.4	0	0.5	0	0.2	−0.4	0.8	0.7	0	0	0.9	−0.3	0.4
X_{21}	−0.6	0	−0.8	0	0	−0.4	0.5	−0.1	0	0.5	−0.8	0.8	0.2	−0.4	0.7	0	0.7	0
X_{24}	0	0.8	0.1	0.8	0.6	0.9	−0.7	0.3	−0.3	0	0	−0.5	0	0.3	0	0.9	0	0
X_{25}	0	0	0	0	−1	−0.5	0.8	−0.9	0	0	0	0	0	0.3	0.1	0	0	0

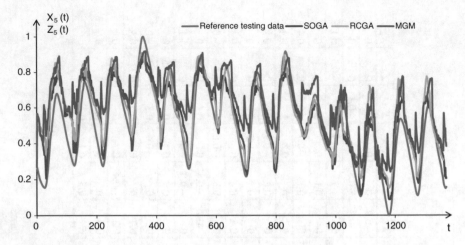

Fig. 6 Obtained values $X_5(t)$ and the desired values $Z_5(t)$ during testing

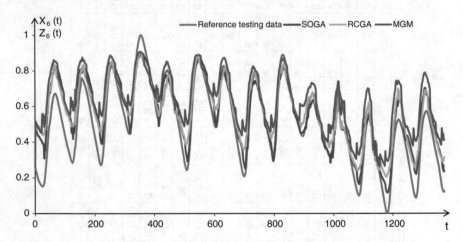

Fig. 7 Obtained values $X_6(t)$ and the desired values $Z_6(t)$ during testing

6 Conclusion

In this paper we present the approach for fuzzy cognitive maps learning allowing selection of the most important for the analyzed tasks concepts (input data) and connections between them. The proposed SOGA is described together with well-known methods for FCM learning. Comparative analysis of SOGA, RCGA and MGM was performed on the example of prediction of the indoor temperature. Simulation research was done in ISEMK. The experimental results show that the proposed approach can significantly reduce the size of FCM by selecting the most important concepts and connections between them keeping the low values of MSE.

Table 2 Chosen results of analysis of the MGM, RCGA, SOGA

Method	Learning parameters	MSE^1	n_c	n_r
MGM	$m_1 = 1, m_2 = 0, \alpha_0 = 1.5, \alpha_1 = -0.5,$ $\beta_0 = 10, \lambda_0 = 100, c = 2, d = 0,$ $e = 0.00001$	0.0256	26	537
RCGA	$P = 100, L = 100, a = 1000, c = 5,$ $d = 0, fitness_{max} = 0.999,$ uniform crossover, $P_c = 0.8$, random mutation, $P_m = 0.01$, ranking selection, elite strategy	0.015 ± 0.003	26	429 ± 12
SOGA	$P = 100, L = 100, a = 1000, c = 5,$ $d = 0, fitness_{max} = 0.999, b_1 = 0.1,$ $b_2 = 0.05,$ uniform crossover, $P_c = 0.8$, random mutation, $P_m = 0.01$, ranking selection, elite strategy	0.013 ± 0.002	16 ± 2	150 ± 37

^{1}For the RCGA and SOGA algorithms, the average values and standard deviations of MSE, n_r and n_c were given

References

1. Ahmadi, S., Forouzideh, N., Alizadeh, S., Papageorgiou, E.: Learning fuzzy cognitive maps using imperialist competitive algorithm. Neural Comput. Appl. **26**(6), 1333–1354 (2015)
2. Amer, M., Jetter, A.J., Daim, T.U.: Scenario planning for the national wind energy sector through fuzzy cognitive maps. In: Proceedings of PICMET'13, pp. 2153–2162 (2013)
3. Arabas, J.: Lectures on Genetic Algorithms. WNT, Warsaw (2001)
4. Buruzs, A., Hatwágner, M.F., Torma, A., Kóczy, L.T.: Expert based system design for integrated waste management. Int. Sch. Sci. Res. Innov. **8**(12), 685–693 (2014)
5. Carvalho, J.P.: On the semantics and the use of fuzzy cognitive maps and dynamic cognitive maps in social sciences. Fuzzy Sets Syst. **214**, 6–19 (2013)
6. Ferreira, P., Ruano, A., Silva, S., Conceicao, E.: Neural networks based predictive control for thermal comfort and energy savings in public buildings. Energy Build. **55**, 238–251 (2012)
7. Froelich, W., Salmeron, J.: Evolutionary learning of fuzzy grey cognitive maps for the forecasting of multivariate, interval-valued time series. Int. J. Approx. Reason. **55**, 1319–1335 (2014)
8. Froelich, W., Papageorgiou, E.I., Samarinasc, M., Skriapasc, K.: Application of evolutionary fuzzy cognitive maps to the long-term prediction of prostate cancer. Appl. Soft Comput. **12**, 3810–3817 (2012)
9. Fogel, D.B.: Evolutionary Computation. Toward a New Philosophy of Machine Inteligence, 3rd edn. Wiley, Hoboken (2006)
10. Grad, L.: An example of feed forward neural network structure optimisation with genetic algorithm. BIULETYN INSTYTUTU AUTOMATYKI I ROBOTYKI **23**, 31–41 (2006)
11. Herrera, F., Lozano, M., Verdegay, J.L.: Tackling real-coded genetic algorithms: operators and tools for behavioural analysis. Artif. Intell. Rev. **12**, 265–319 (1998)
12. Homenda, W., Jastrzebska, A., Pedrycz, W.: Time series modeling with fuzzy cognitive maps: simplification strategies. The case of a posteriori removal of nodes and weights. Lect. Notes Comput. Sci. LNCS **8838**, 409–420 (2014)
13. Homenda, W., Jastrzebska, A., Pedrycz, W.: Nodes selection criteria for fuzzy cognitive maps designed to model time series. Adv. Intell. Syst. Comput. **323**, 859–870 (2015)
14. Jastriebow, A., Poczeta, K.: Analysis of multi-step algorithms for cognitive maps learning. Bull. Pol. Acad. Sci. Tech. Sci. **62**(4), 735–741 (2014)

15. Kosko, B.: Fuzzy cognitive maps. Int. J. Man-Mach. Stud. **24**(1), 65–75 (1986)
16. Lu, W., Pedrycz, W., Liu, X., Yang, J., Li, P.: The modeling of time series based on fuzzy information granules. Expert Syst. Appl. **41**, 3799–3808 (2014)
17. Papageorgiou, E.I.: Fuzzy Cognitive Maps for Applied Sciences and Engineering From Fundamentals to Extensions and Learning Algorithms. Intelligent Systems Reference Library, vol. 54. Springer, Berlin (2014)
18. Papageorgiou, E.I., Froelich, W.: Multi-step prediction of pulmonary infection with the use of evolutionary fuzzy cognitive maps. Neurocomputing **92**, 28–35 (2012)
19. Papageorgiou, E.I., Salmeron, J.L.: A review of fuzzy cognitive maps research during the last decade. IEEE Trans. Fuzzy Syst. **21**(1), 66–79 (2013)
20. Papageorgiou, E.I., Poczeta, K., Laspidou, C.: Application of fuzzy cognitive maps to water demand prediction. In: IEEE International Conference on Fuzzy Systems (FUZZ-IEEE), Istanbul, pp. 1–8 (2015)
21. Papakostas, G.A., Koulouriotis, D.E.: Classifying patterns using fuzzy cognitive maps. In: Glykas, M. (ed.) Fuzzy Cognitive Maps, STUDFUZZ, vol. 247, pp. 291–306. Springer, Berlin (2010)
22. Papakostas, G.A., Koulouriotis, D.E., Polydoros, A.S., Tourassis, V.D.: Towards Hebbian learning of fuzzy cognitive maps in pattern classification problems. Expert Syst. Appl. **39**, 10620–10629 (2012)
23. Poczeta, K., Yastrebov, A.: Analysis of fuzzy cognitive maps with multi-step learning algorithms in valuation of owner-occupied homes. In: IEEE International Conference on Fuzzy Systems (FUZZ-IEEE), Beijing, China, pp. 1029–1035 (2014)
24. Poczeta, K., Yastrebov, A., Papageorgiou, E.I.: Learning fuzzy cognitive maps using structure optimization genetic algorithm. In: Ganzha, M., Maciaszek, L., Paprzycki, M. (eds.) Annals of Computer Science and Information Systems. Proceedings of the Federated Conference on Computer Science and Information Systems, vol. 5, pp. 547–554. IEEE (2015)
25. Salmeron, J.L.: Fuzzy cognitive maps for artificial emotions forecasting. Appl. Soft Comput. **12**, 3704–3710 (2012)
26. Słoń, G.: Application of models of relational fuzzy cognitive maps for prediction of work of complex systems. Lecture Notes in Artificial Intelligence LNAI, vol. 8467, pp. 307–318. Springer, Berlin (2014)
27. Stach, W., Kurgan, L., Pedrycz, W., Reformat, M.: Genetic learning of fuzzy cognitive maps. Fuzzy Sets Syst. **153**(3), 371–401 (2005)
28. Zamora-Martínez, F., Romeu, P., Botella-Rocamora, P., Pardo, J.: On-line learning of indoor temperature forecasting models towards energy efficiency. Energy Build. **83**, 162–172 (2014)

Correlation Clustering by Contraction, a More Effective Method

László Aszalós and Tamás Mihálydeák

Abstract In this article we propose two effective methods to produce a near optimal solution for the problem of correlation clustering. We study their properties at different circumstances, and show that the inner structure generated by a tolerance relation has effect on the accuracy of the methods. Finally, we show that there is no royal road to the sequence of clusterings.

1 Introduction

It is common, that by applying simple rules we can generate rather complicated structures (e.g. the crystallization). The same is true for animals, some simple reactions produce rather complicated and effective behaviour patterns. The latter is the foundation of several optimization methods, the swarm intelligence is an emerging field for solving complex optimization problems, e.g. ant colony optimization for the travelling salesmen problem [10] or harmony search for water distribution networks [12].

In this paper we propose a solving method for correlation clustering by using two simple steps.

Clustering is an important tool of unsupervised learning. Its task is to group the objects in such a way, that the objects in one group (cluster) are similar, and the objects from different groups are dissimilar, so it generates a partition and hence an equivalence relation. We can say that the objects aim to achieve a relatively stable situation in which they are in a cluster containing minimal number of dissimilar, and maximal number of similar objects. In the last fifty years, many different clustering methods were invented based on different demands.

Correlation clustering is a relatively new method in data mining, Bansal et al. published a paper in 2004, proving several of its properties, and gave a fast, but not

L. Aszalós (✉) · T. Mihálydeák
Faculty of Informatics, University of Debrecen, Debrecen, Hungary
e-mail: aszalos.laszlo@inf.unideb.hu

T. Mihálydeák
e-mail: mihalydeak.tamas@inf.unideb.hu

© Springer International Publishing Switzerland 2016
S. Fidanova (ed.), *Recent Advances in Computational Optimization*,
Studies in Computational Intelligence 655, DOI 10.1007/978-3-319-40132-4_6

quite optimal algorithm to solve the problem [7]. Naturally, correlation clustering
has a predecessor. Zahn proposed this problem in 1965, but using a very different
approach [17]. The main question is the following: *which equivalence relation is the
closest to a given tolerance (reflexive and symmetric) relation?* Bansal et al. have
shown, that this is an NP-hard problem [7]. The number of equivalence relations of
n objects, i.e. the number of partitions of a set containing n elements is given by
Bell numbers B_n, where $B_1 = 1$, $B_n = \sum_{i=1}^{n-1} \binom{n-1}{k} B_k$. It can be easily checked that
the Bell numbers grow exponentially. Therefore if $n > 15$—in a general case—we
cannot achieve the optimal partition by exhaustive search, thus we need to use some
optimization methods, which do not give optimal solutions necesserily, but help us
achieve a near-optimal one.

This kind of clustering has many applications: image segmentation [13],
identification of biologically relevant groups of genes [8], examination of social
coalitions [16], improvement of recommendation systems [11] reduction of energy
consumption [9], modelling physical processes [14], (soft) classification [2, 3], etc.

At correlation clustering, those cases where two dissimilar objects are in the
same cluster, or two similar objects are in different clusters, are treated as a conflict.
Thus the main question could be rewritten as: which partition generates the minimal
number of conflicts? It can be shown, that in the general case—where the transitivity
does not hold for the tolerance relation—the number of conflicts is a positive number
for all partitions of a given set of objects. Let us take a graph on Fig. 1, where the
similarity is denoted by solid, and dissimilarity by dashed lines. In case of people, the
similarity and dissimilarity are treated as liking or disliking each other, respectively.
Mathematically, the similarity is described as a weighted graph, where 1 denotes
similarity, and −1 denotes dissimilarity. As the absolute value of these weights are
the same, thus it is enough to only use the signs of the weights. Hence the graph
is called a *signed graph* in the literature. The lower part of Fig. 1 shows the three
optimal partitions from the five, each of them containing only one conflict. As all
partitions of the graph have at least one conflict, it is called a *frustrated graph*, and
this one is the smallest such graph.

Fig. 1 Minimal frustrated
graph and its optimal
partitions. Here the similarity
is denoted by *solid*, and
dissimilarity by *dashed lines*.
From 5 different partitions 3
is optimal (presented below),
each have one conflict: *solid
line* between regions or
dashed line inside region

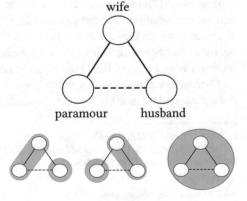

If the correlation clustering is expressed as an optimization problem, the traditional optimization methods (hill-climbing, genetic algorithm, simulated annealing, etc.) could be used in order to solve it. We have implemented and compared the results in [1]. With these methods the authors were able to determine a near optimal partition of signed graphs with 500 nodes.

In this paper we introduce two methods which were invented directly to solve the problem of correlation clustering, and they take the speciality of the problem into account. The main idea is extremely simple, but it needs several witty concepts to get a fast and effective algorithm. The first method (contraction) joins two clusters, and the other method (correction) puts one object into a different cluster. In both case these steps are only executed if they reduce the cost of the partition.

The structure of the paper is the following: Sect. 2 explains the correlation clustering and shows the result of our steps. In Sect. 3 we present a naive contraction method. Next we show how it can be improved by the correction method. After this we apply these methods for sparse graphs, and get some interesting results. In Sect. 5 we discuss the technical details. Finally we overview our plans and conclude the results.

2   Correlation Clustering

In the paper we use the following notations: V denotes the set of the objects, and $T \subset V \times V$ the tolerance relation defined on V. We handle a partition as a function $p : V \to \{1, \ldots, n\}$, there n is the number of nodes in V. (Function p is not necessary surjective.) The objects x and y are in a common cluster, if $p(x) = p(y)$. We say that objects x and y are in conflict at given tolerance relation T and partition p iff value of $c_T^p(x, y) = 1$ in (1), i.e. if they are similar and are in different clusters, or if they are dissimilar and in the same cluster.

$$c_T^p(x, y) = \begin{cases} 1 \text{ if } (x, y) \in T \text{ and } p(x) \neq p(y) \\ 1 \text{ if } (x, y) \notin T \text{ and } p(x) = p(y) \\ 0 \text{ otherwise} \end{cases} \tag{1}$$

We are ready to define the cost function of relation T according to partition p:

$$c_T(p) = \frac{1}{2} \sum c_T^p(x, y) \tag{2}$$

As relation T is symmetric, we sum c_T^p twice for each pair. Our task is to determine the value of $\min_p c_T(p)$, and one partition p' for which $c_T(p')$ is minimal. Unfortunately this exact value cannot be determined in practical cases, except for some very special tolerance relations. Hence we can get only approximative, near optimal solutions.

Our method is vaguely reminiscent to the hierarchical clustering. Hierarchical clustering has two different strategies: agglomerative and divisive. The agglomerative strategy is the "bottom up" approach: it starts with singleton clusters and merges a pair of them in one step. The divisive strategy is the "top down" approach: all the objects are put in common cluster at first, and it split one cluster in each steps. The complexity of the strategies are $O(n^3)$ and $O(2^n)$—where n is the number of the objects—hence the divisive strategy is seldom used. Otherwise for big n even the cubic complexity is awkward, so the hierarchical clustering is not applied for really big data sets.

Now we use the agglomerative approach, but instead of distance we will use attraction to select the clusters to merge.

Let assume that p and T are fixed and treat objects as electrical particles. Let us say that these object attract each other if the tolerance relation holds between them. We say that they repulse each other if the tolerance relation does not holds between them. If the two objects are the same, there is no attraction and there is no repulsion:

$$f(x, y) = \begin{cases} 0 \text{ if } x = y \\ 1 \text{ if } (x, y) \in T \text{ and } x \neq y \\ -1 \text{ if } (x, y) \notin T \text{ and } x \neq y \end{cases} \tag{3}$$

We can extend this attraction from objects to an object and a set:

$$f(x, S) = \sum_{y \in S} f(x, y) \tag{4}$$

and even into any pair of sets:

$$f(S_1, S_2) = \sum_{x \in S_1} \sum_{y \in S_2} f(x, y). \tag{5}$$

What does it mean if the attraction between two clusters S_1 and S_2 is positive? It can occur only if the sums contain more positive values than negatives ones in (5). If we join these clusters, the total cost $c_T(p)$ decreases with the value of attraction. (The positive values imply conflicts before joining, but not generate conflicts after joining. For negative values the opposite holds.) Hence our aim is to join all such pairs of clusters.

3   Contraction Method

Now we are ready to show the contraction method. The authors presented several implementations in detail [4], so here we just give the main steps:

1. Create an attraction matrix A based on the tolerance relation: $a_{x,y} = f(x, y)$.
2. Now each object (as a singleton set) belongs to one row (and column) of the attraction matrix. Select one maximal element of this matrix, and its row and

column coordinates determine the clusters to join. Denote these coordinates with x and y!

3. If the maximal element $a_{x,y}$ is not positive, the method ends here.
4. Otherwise let us add to the xth row the yth row, and delete the yth row (or fill with zeros). Repeat this with the xth and yth columns.
5. Continue at Step 2.

If we really delete the rows and columns from matrix A, then this matrix contains the values $f(S_1, S_2)$, where S_1 and S_2 are clusters. It is easy to prove that $f(S_1 \cup S_2, S_3) = f(S_1, S_3) + f(S_2, S_3)$, so our algorithm is sound.

In two cases—when the relations are degenerated—we can easily give the optimal partition. Although there is no chance to work with these relations in the practice, these relations are helpful at testing the algorithms. If the tolerance relation does not hold anywhere, then there is no reason to put two objects into one cluster, because it generates a conflict. Hence the optimal clustering contains singletons, so the total cost (number of conflicts) is zero. Otherwise if the tolerance relation holds everywhere, there is no reason to put two objects into different clusters. Therefore all the objects are in a common set, so the partition contains one set only. The total cost is zero again.

Although the cost is based on the tolerance relation, we need a simpler characterisation of the tolerance relation to compare the results. We have chosen the probability of fulfilment of the tolerance relation. More precisely

$$q = \frac{\#\{(x, y)|(x, y) \in T\}}{\#\{(x, y)|(x, y) \subset T\} + \#\{(x, y)|(x, y) \notin T\}} \tag{6}$$

Of course this value is not a perfect characterisation. Figure 2 shows, that in both cases $q = 0.5$, but the cost is different. If the tolerance relation is an equivalence relation, then the value of the cost is 0. But an equivalence relation is very rarely generated. Figure 3 shows the costs we obtained at randomly generated tolerance relations. Here q runs on horizontal axes from 0 to 1. The detailed analysis of the cost function and the size of maximal clusters can be found in [14].

At the contraction method in each step two clusters have been joined, hence the number of clusters is decreasing by one. This means, that if we have n objects, we can have at most $n - 1$ contractions, and at each step we need to choose the maximal

Fig. 2 In both cases we have 3 similarity and 3 dissimilarity, but the optimal partition gives 0 conflicts on the *left* and 1 on the *right*

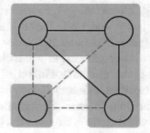
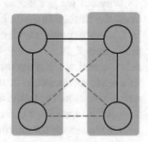

Fig. 3 Dispersion of the cost at correlation clustering

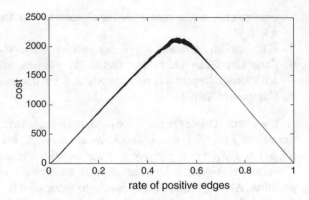

Fig. 4 Relative cost of parallel contraction according to contraction. The parallel contraction never gave better result than the original contraction

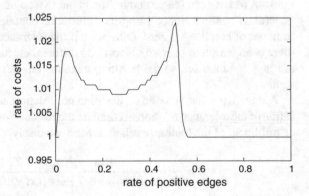

value $a_{x,y}$ among $O(n^2)$ ones. During the contraction from elements $a_{x,y}$ over all $O(n)$ change their value. Hence the traditional storing-retrieving methods cannot speed up the running of the algorithm. Therefore we used parallelism, i.e. if there are several maximal values in $a_{x,y}$, then we execute several contractions instead of only one. Let us be careful with these contractions! If $f(S_1, S_2) = f(S_2, S_3) = 1$ but $f(S_1, S_3) = -5$, then it is not worth joining all the three sets, only S_1 and S_2 or S_2 and S_3. Figure 4 shows the results. This parallel execution of the contraction never gave better result than the original. The difference is within 3 % of the cost, but if we are interested in optimization, this is a high price for the speed.

Why do we get worse results by parallel execution? We did not inadvertently draw the dashed lines on Fig. 2: *everything influences everything else*. If object x is in tolerance relation with object y, and x is in the cluster S, then there is a request to put y into S, if it is not yet there. Otherwise if x and y are not related by tolerance relation and $x \in S$, then there is a request to put y outside of S, if it is not yet there. In this sense all decisions have global effects that affect all elements. If a contraction is carried out unnecessarily, we cannot withdraw it. Do not forget, this method is a one-way method, thus we can only join clusters! The contraction is a greedy method (at Step 2 we search for the maximal element). At parallel contraction there is a contest

Fig. 5 Size of the maximal
and second maximal clusters

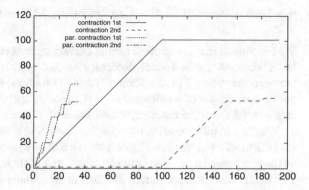

between the growing clusters, such that which can annex more objects to itself. Hence
the parallel contraction makes decisions too early, so *haste makes waste*. If there is
only one decision at once, we can postpone these decisions. Figure 5 provides insight
into the process.

Here we have 200 nodes, and q is around 0.5. The contraction method in the first
round chooses two elements (two singleton sets) and joins them. This new set throttles
the other sets. From now on, this is the biggest set, it has the biggest attraction, so
in each round it annexes a new singleton. It became bigger and more attractive. The
figure shows that this continues in the first 100 steps. If the possibilities of this giant
set are exhausted, then it permits an other set to do the same. This takes about 50
steps, until the growth of a third set can begin, and so on. Finally we have a huge
set, some middle-sized ones and many small. (In our case 4 sets have less than 10
element.)

The figure does not show that at parallel case at the fist round each element found
a pair, joined to it and no singletons were left. Next many pair joined to 4-element
sets, but according to q this process slowed down, and the growth of the maximal
set was not continuous, it left option for the smaller sets. Finally the maximal set
contained 66 element, and one set contained less than 10 element. Mostly, we had
middle-size sets.

Although the second approach is more sympathetic for us, the former is more
effective. If the tolerance relation has a clique (where its members are related by the
tolerance relation, but not related to other elements), and the traditional contraction
finds some of its element, then it feeds up these element to the whole clique. The
parallel contraction can find several elements of the clique, and parcel it among these
elements. If the clique is perfect (not related to any outside element), the contraction
will join these sub-cliques. But if the clique is not perfect, the sets originated form
the clique could grow in such a way, that their joining produces many conflicts.

4 Improvement of the Contraction Method

We mentioned that contraction cannot correct the errors of the former decisions. So let us take a new method which decreases the cost by rearranging one element. For this we need the values $f(x, S)$. From (3) and (4) follows, that $f(x, S\backslash\{x\}) = f(x, S)$, so the own cluster of an element does not require specific treatment. (We note, that this was the reason of the specific definition of $f(x, x)$.)

We need to take the attractions of the clusters to element x, and put this element into the most attractive set. Figure 1 shows a case where the husband is as attractive as the paramour, so the wife could choose either of them, or even is toing and froing. The latter does not work in real life, like at algorithms we require finiteness. Hence we restrict the case as follows: if the other cluster is *more* attractive than the current cluster, then we move the element into one of the most attractive clusters.

Let us assume that the element y from cluster S_s wants to move into cluster S_t. Denote $S_s\backslash\{y\}$ and $S_t \cap \{y\}$ with S_s' and S_t', respectively. S_s' and S_t' are the clusters after moving y. It is easy to show, that

$$f(x, S_s') = f(x, S_s) - f(x, y) \qquad f(x, S_t') = f(x, S_t) + f(x, y) \qquad (7)$$

and

$$f(\hat{S}, S_s') = f(\hat{S}, S_s) - f(y, \hat{S}) \qquad f(\hat{S}, S_t') = f(\hat{S}, S_t) + f(y, \hat{S}) \qquad (8)$$

for each element x and cluster $\hat{S}$.

Some tests have shown, that this is not enough to move elements to the most attractive clusters. Very often two clusters have been formed, which had the same size, and each element attracted a particular element from the other cluster. These two particular elements have changed places, and as the situation almost remained the same, changed places again. To stop this infinite process, we need to damp these moves. We have several options:

- Enable each move with a fixed probability, which is enough to get a finite method.
- We can examine the independence: if we had chosen an element and a cluster, we cannot use this cluster and the cluster of the element to get from or put there other elements.

Figure 6 shows that with correction after contraction we get a smaller cost for small q than if we were to use contraction alone, but for a big q the correction does not help. We examined correction and parallel correction, but there is no real difference in costs when we use both of them to correct the contraction. We can use the methods of correction and parallel correction alone, i.e. without contraction. In this case the correction gives better results than the parallel contraction. The biggest difference in cost was 5 %.

Figure 7 was generated with the same parameters as Fig. 5, only that here the corrections methods were used. The correction method at the beginning constructs pairs, and few rounds later annexes elements to these pairs, and construct bigger and

Fig. 6 Effect of the correction

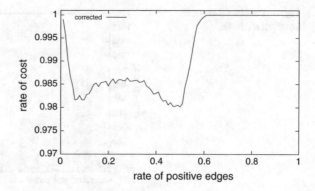

Fig. 7 Size of maximal and second maximal clusters

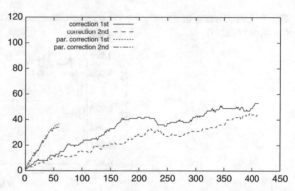

bigger clusters. Here the biggest cluster does not overwhelm the others like at Fig. 5. Sometimes the correction method realises that an annexation many rounds ago may have been a mistake, and places the element into a better cluster. The figure shows that this happens many times, so the number of rounds was 412 for 200 elements. (We note that 100, 500 and 1000 element demanded about 170, 1330 and 3100 rounds, respectively.) The parallel case has less rounds, of course; but has more than the parallel contraction has. By comparing the figures it is obvious, that we have smaller clusters than at contraction.

The contest between clusters (the parallel case) gives many middle-size clusters, and usually no small clusters. At the non-parallel case some small clusters were left, and the biggest clusters are bigger than in parallel case; briefly: the variance of cluster sizes is bigger.

Until now we examined the accuracy of the methods. But they are intended to solve real problems as fast as possible. Figure 8 shows the time needed to cluster 1000 and 2000 elements with different methods. Do not worry about these figures, they denote the total time of 101 clustering from $q = 0$ to $q = 1$ to take into account all interesting cases. From above, we know that the correction method alone gave the best results. But as the figure shows, this is the slowest method. We could combine contraction with correction, but it took 209 s at 1000 element, so it is even slower.

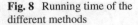

Fig. 8 Running time of the different methods

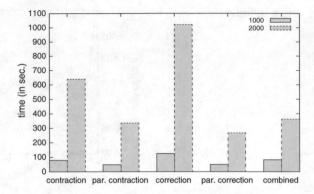

Fig. 9 Effectiveness of the combined method

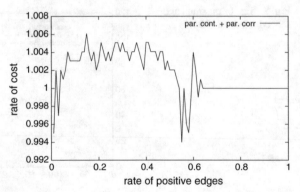

We can try the parallel methods. As we have seen before the parallel contraction is about 2% worse than the contraction, and contraction is about 2% worse than the correlation alone. If we improve its result with parallel correction, we get close to our best results (mostly within half percent) as Fig. 9 shows. The overhead of the parallel correction after parallel contraction is slight, the parallel contraction constructed a rough approximation, which is not too far from the optimum, so the parallel correction gets close to it almost immediately.

5 Partial Relations

As [14] analysed the asymptotic properties of correlation clustering of tolerance relations and our experiments verify their results, these tolerance relation are not challenging. But at partial tolerance relation there are only conjectures. The Barabási–Albert type random graphs successfully describe many natural and social networks. The correlation clustering of partial tolerance relation based on these graphs was analysed in [6, 15], but they used graphs with 140 and 500 nodes, respectively. These graphs are too small to discover tendencies, and it may be necessary to test graphs that are several magnitudes bigger. In this article we continue the path laid out by these articles, and in the following we shall use 3/2 type Barabási–Albert graphs.

The signed graph of a tolerance relation (if the relation holds between nodes then the edge has type +, and when the relation does not hold between the nodes then edge has type −) is a complete graph, and in this sense is symmetric. As Fig. 2 shows, only the signs differ, hence the variance of the costs is small by Fig. 3.

But the Barabási–Albert graphs are not symmetric, so the signs have greater effect as Fig. 10 shows. This means, that we need to repeat our experiments many times (according to the law of large numbers), to get figures that are close to reality. This figure tell us some new information: the cost functions until now were mostly symmetric, and reached their maximum around 0.5. Now this extrema is around 0.65 for 500 objects.

Most of the definitions from (1) to (6) is suitable for a partial case, too. At (3) we need to add one remark: if the relation of two objects is not defined, then there is no attraction and no repulsion. Hence we can apply all of our methods for partial tolerance relations, too. But we get some surprising results. At first, let us see our original methods on Fig. 11. The contraction method works well, at $q = 0$ and $q = 1$

Fig. 10 Dispersion of the cost at correlation clustering

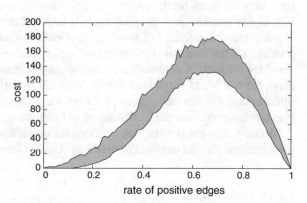

Fig. 11 Contraction and correction methods on sparse graphs

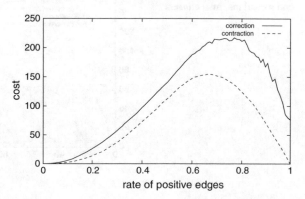

if it finds the optimal partitions with zero conflicts. But the correction alone cannot reach the same result. By examining the cases at $q = 1$, several loosely coupled clusters were generated. These clusters intern their objects, the inner attractions are bigger then the outer ones. In physics, in a similar situation the particle tunnelling through a potential barrier is the solution. For us now a contraction could be the solution.

If we calculate the cost functions for the contractions, then we can realize, that the parallel contraction gives better results, than the original contraction. What is the difference between the tolerance relation and partial tolerance relation, to get such results? At generating 3/2 type Barabási–Albert random graphs each new node connects 2 old nodes. Therefore the average degree of nodes is around 4. At signed graphs of tolerance relations the degree of nodes is $n - 1$, where n could be 1000 or more. The degree distribution in a Barabási–Albert graph follows a power law, so only a few nodes have many neighbours, and many nodes have a few neighbours (below the average). Hence to move a node into a new cluster implies only a few demands, so these decisions are mostly local. Before, the contraction method steadily annexed new and new singletons. As *everything is connected to everything else*, the small and large sets both had many neighbours. At sparse graphs the number of neighbours is proportional to the size: small sets have a few neighbours but bigs sets have many of them. So the contraction starts slowly, at the experiment presented on Fig. 12, five small set grow parallel and only later will one cluster dominate the others. Among many singletons 2–3 small set (2–4 elements) survive or emerge during the process; and the final state is similar: one huge set, more than 30 singletons and three small set. At parallel contraction on sparse graphs, many singletons join into pairs. Next, the number of singletons slowly decreases. Mostly there are many small and middle-sized clusters and one huge during the process; and in the last rounds the biggest cluster enables the increasing of an other cluster.

We have seen previously, that although the correction alone gave the best result, the combined parallel methods gave nearby result in fraction time. At partial case the

Fig. 12 Size of maximal and second maximal clusters

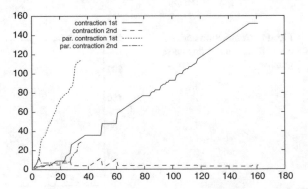

correction alone performs poorly, but by our experiments it can improve the results of the parallel contraction. We compared the original and parallel versions, and—as could be expected—the combined parallel version is the winner.

6 Technical Details

We said before, the definition of the force between objects, between object and set or between set is the same in partial and non-partial case. If the number edges of the graph is proportional to the nodes of the graph (i.e. sparse graph), it is reasonable to use a space-saving storage for the signed graph. But the values of $f(S, T)$ and $f(x, S)$ change so often that there are other aspects to consider, not only the size of the storage. We described our storage types in detail in [4], an associative array and a vector (or associative array) of associative arrays. The contraction is almost the same in both cases. Using a vector of associative arrays needs extra bookkeeping, however at many nodes the algorithm runs faster, because we only need to process two short associative arrays; while in the other case the only one—and hence huge—associative array.

At correction we need to find the neighbours of a node. If the edge is the key in the associative array, then it needs a full search. At a vector of associative arrays we got these nodes almost immediately. Figure 13 shows the running time of parallel correction alone and its improvement with parallel correction. It is obvious from the figure, that the correction needs more time when only one associative array is used. Moreover, the extra bookkeeping takes extra time for small graphs, but speeds up the process for big graphs. The lines on the figure (on a logarithmic scale) are near linear, so these methods cannot be effective for big graphs.

We are interested in asymptotic properties of correlation clustering of sparse graphs, so it raised the question whether we can use the partition we got as a solution for a nearby problem? The contraction starts with singletons. We constructed a variant which uses the previous solution for q to a problem with $q' = q + 0.01$ instead of

Fig. 13 Running time of parallel contraction and its correction

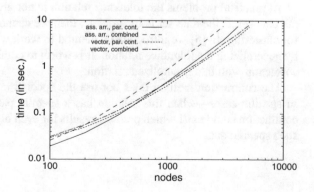

Fig. 14 Comparison of the
solving methods "start from
the scrach" (100 steps) and
"correct the previous
solution" (continuous). In
the latter case the errors sum
up giving a worse result

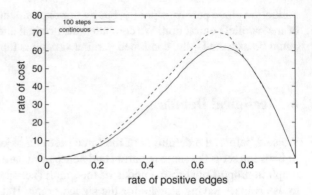

singletons, and applies the correction afterwards. In such a construction the errors
add up from step to step, and the biggest difference was 20% between the totally and
partly recalculated results! We decreased the steps without success, so constructed
a second variant which recalculated the forces, reorganized clusters as one edge
changed its sign. This reduced the biggest difference to 16%, so this direction is a
dead end: we cannot save calculations in such a way (Fig. 14).

7 Conclusion and Future Work

In this paper we introduced a method and an improvement for it. With this method we
pushed the boundaries: we are able to cluster bigger data sets than before. Remember
that the original problem is NP-hard! The parallel execution speeds up the process,
but needs correction. This combination of methods gives a tool that can be used for
practical problems.

By the experiments, we did not cancel the bounds, with this methods we cannot
solve the clustering of hundred thousands objects in reasonable time. To do this, a
new direction or several tricks may be needed.

At practical problems the tolerance relation is not given, so it is generated from
the data set. If the data set is incomplete, then we cannot use the distances directly
for clustering. In [5] we introduced a method to work with incomplete information
by generalizing the tolerance relation. It is worth to combine the contraction and its
correction with this generalized relation.

The contraction method does not use the specialities of the tolerance relation.
In specific cases—when this relation has a specific pattern—a fine tuning of the
contraction could exist which produce results for even bigger data sets. We quest for
such specific data.

References

1. Aszalós, L., Bakó, M.: Advanced search methods (in Hungarian) (2012). http://morse.inf. unideb.hu/~aszalos/diak/fka
2. Aszalós, L., Mihálydeák, T.: Rough clustering generated by correlation clustering. Rough Sets, Fuzzy Sets, Data Mining, and Granular Computing, pp. 315–324. Springer, Berlin (2013). doi:10.1109/TKDE.2007.1061
3. Aszalós, L., Mihálydeák, T.: Rough classification based on correlation clustering. Rough Sets and Knowledge Technology, pp. 399–410. Springer, Berlin (2014). doi:10.1007/978-3-319-11740-9_37
4. Aszalós, L., Mihálydeák, T.: Correlation clustering by contraction. In: Ganzha, M., Maciaszek, L.A., Paprzycki, M. (eds.) Federated Conference on Computer Science and Information Systems, FedCSIS 2015, Lódz, Poland, 13–16 September 2015, pp. 425–434. IEEE (2015). doi:10. 15439/2015F137
5. Aszalós, L., Mihálydeák, T.: Rough classification in incomplete databases by correlation clustering. In: Alonso, J.M., Bustince, H., Reformat, M. (eds.) Conference of the International Fuzzy Systems Association and the European Society for Fuzzy Logic and Technology (IFSA-EUSFLAT-15), Gijón, Spain, 30 June 2015. Atlantis Press (2015)
6. Aszalós, L., Kormos, J., Nagy, D.: Conjectures on phase transition at correlation clustering of random graphs. Ann. Univ. Sci. Bp. Sect. Comput. **42**, 37–54 (2014)
7. Bansal, N., Blum, A., Chawla, S.: Correlation clustering. Mach. Learn. **56**(1–3), 89–113 (2004). doi:10.1023/B:MACH.0000033116.57574.95
8. Bhattacharya, A., De, R.K.: Divisive correlation clustering algorithm (DCCA) for grouping of genes: detecting varying patterns in expression profiles. Bioinformatics **24**(11), 1359–1366 (2008). doi:10.1093/bioinformatics/btn133
9. Chen, Z., Yang, S., Li, L., Xie, Z.: A clustering approximation mechanism based on data spatial correlation in wireless sensor networks. Wireless Telecommunications Symposium (WTS), pp. 1–7. IEEE (2010). doi:10.1109/WTS.2010.5479626
10. Dorigo, M., Gambardella, L.M.: Ant colonies for the travelling salesman problem. BioSystems **43**(2), 73–81 (1997)
11. DuBois, T., Golbeck, J., Kleint, J., Srinivasan, A.: Improving recommendation accuracy by clustering social networks with trust. Recomm. Syst. Soc. Web **532**, 1–8 (2009). doi:10.1145/2661829.2662085
12. Geem, Z.W.: Optimal cost design of water distribution networks using harmony search. Eng. Optim. **38**(03), 259–277 (2006)
13. Kim, S., Nowozin, S., Kohli, P., Yoo, C.D.: Higher-order correlation clustering for image segmentation. In: Advances in Neural Information Processing Systems, pp. 1530–1538 (2011)
14. Néda, Z., Florian, R., Ravasz, M., Libál, A., Györgyi, G.: Phase transition in an optimal clusterization model. Phys. A: Stat. Mech. Appl. **362**(2), 357–368 (2006). doi:10.1016/j.physa.2005. 08.008
15. Néda, Z., Sumi, R., Ercsey-Ravasz, M., Varga, M., Molnár, B., Cseh, G.: Correlation clustering on networks. J. Phys. A: Math. Theor. **42**(34), 345003 (2009)
16. Yang, B., Cheung, W.K., Liu, J.: Community mining from signed social networks. IEEE Trans. Knowl. Data Eng. **19**(10), 1333–1348 (2007)
17. Zahn Jr., C.: Approximating symmetric relations by equivalence relations. J. Soc. Ind. Appl. Math. **12**(4), 840–847 (1964). doi:10.1137/0112071

Synthesis of Power Aware Adaptive Embedded Software Using Developmental Genetic Programming

Stanisław Deniziak and Leszek Ciopiński

Abstract In this paper we present a method of synthesis of adaptive schedulers for power aware real-time embedded software. We assume that the system is specified as a task graph, which should be scheduled on multi-core embedded processor with low-power processing capabilities. First, the developmental genetic programming is used to generate the scheduler and the initial schedule. The scheduler and the initial schedule are optimized taking into consideration power consumption as well as self-adaptivity capabilities. During the system execution the scheduler modifies the schedule whenever execution time of the recently finished task occurred shorter or longer than expected. The goal of rescheduling is to minimize the power consumption while all time constraints will be satisfied. We present real-life example as well as some experimental results showing advantages of our method.

1 Introduction

Besides the cost and performance, power consumption is one of the most important issues considered in the optimization of embedded systems. Design of energy efficient embedded systems is important especially for battery-operated devices. But the minimization of power consumption is also important in other systems, since it reduces the cost of running and cooling the system. Therefore, efficient synthesis methods that minimize the power consumption in embedded systems are needful.

Embedded systems are usually real-time systems, i.e. for some tasks time constraints are defined. Therefore, power optimization should take into consideration that all time requirements should be met. Performance and power consumption are orthogonal features, i.e. in general, higher performance requires more power. Hence, the optimization of embedded system should consider the trade-off between power, performance, cost and perhaps other attributes.

S. Deniziak · L. Ciopiński (✉)
Division of Computer Science, Kielce University of Technology, Kielce, Poland
e-mail: l.ciopinski@tu.kielce.pl

S. Deniziak
e-mail: s.deniziak@tu.kielce.pl

© Springer International Publishing Switzerland 2016
S. Fidanova (ed.), *Recent Advances in Computational Optimization*,
Studies in Computational Intelligence 655, DOI 10.1007/978-3-319-40132-4_7

Performance of the system may be increased by applying a distributed architecture. The function of the system is specified as a set of tasks then during the co-design process, the optimal architecture is searched. Distributed architecture may consist of different processors, dedicated hardware modules, memories, buses and other components. Recently, the advent of embedded multicore processors has created interesting alternatives to dedicated architectures. First, the co-design process may be reduced to task scheduling. Second, advanced technologies for power management, like DVFS (Digital Voltage and Frequency Scaling) or big.LITTLE [1], create new possibilities for designing low-power embedded systems.

Optimization of embedded systems is based on assumptions that certain system properties are known. For example, to estimate the performance of the system, execution times for all tasks should be known. Sometimes it is difficult to precisely predict all required information. Therefore, to guarantee the proper design, the worst case estimation is used. During the operation of the system it may occur that certain system properties may significantly differ from estimations or may dynamically change. It may be caused by too pessimistic estimation, by data-dependence or by some unpredictable events. In such cases the idea of self-adaptivity may be used to optimize some system properties.

In this paper we present the novel method for synthesis of the power-aware software for real-time embedded systems. We assume that the function of the system is specified using the task graph that should be executed by the multicore processor supporting the big.LITTLE technology. Then the power-aware scheduler is generated automatically using the developmental genetic programming (DGP). The scheduler is self-adaptive, i.e. it dynamically reschedules tasks whenever any task finished its execution earlier or later than expected. In the first case the goal of the rescheduling is the reduction of power consumption by moving some tasks to low-power cores. In the second case, the system is rescheduled to satisfy all time constraints by moving some tasks to high-performance cores. The paper continues our research presented in [2]. The main new feature of our methodology is that self-adaptivity capabilities are taken into consideration during optimization. Example and experimental results show the benefits of using our methodology.

The rest of the paper is organized as follows. Next section presents the related work. Section 3 presents the concept of the developmental genetic programming with respect to other genetic approaches. In Sect. 4 we present our method. Section 5 describes an example and experimental results. The paper ends with conclusions.

2 Related Work

Although there are a lot of synthesis methods for low-power embedded systems [3], the problem of optimal mapping of a task graph onto the multicore processor is rather a variant of the resource constrained project scheduling (RCPSP) [4] one, than the co-synthesis. Since the RCPSP is NP-complete, only heuristic approach may be applied to real-life systems. Among the proposed heuristics for solving RCPSP, ones of the most efficient are methods based on genetic algorithms [5–7].

For systems that may dynamically change during operation, some methods of rescheduling was proposed. In [8] the number of tasks that receives a new start time, after rescheduling, is minimized. Another approach [9] proposes to reschedule the remaining tasks, such that the sum of deviations of the new finishing times from the original ones is minimized. In [10] the sum of deviations of starting and finishing times of all tasks is minimized. Proactive scheduling [11] does not perform rescheduling, but it minimizes the perturbations, caused by delays, by maximization of the minimum or total free slacks of task executions.

Three different methods of applying big.LITTLE technology for minimizing the power consumption were proposed [12]. In the cluster switching, low-power cores are grouped into "little cluster", while high performance cores are arranged into "big cluster". The system uses only one cluster at a time. If at least one high performance core is required then the system switches to the "big cluster", otherwise the "little cluster" is used. Unused cluster is powered off. In CPU migration approach, low-power and high-performance cores are paired. At a time only one core is used while the other is switched off. At any time it is possible to switch paired cores. The most powerful model is a global task scheduling. In this model all cores are available at the same time.

The big.LITTLE technology is quite new and is mainly used in mobile devices. According to our best knowledge there are no applications of this technology to design low power real-time embedded systems, as well as an adaptive scheduling method for such systems. In [2] we presented our preliminary results obtained using our approach. We developed the methodology of synthesis the self-adaptive scheduler, which minimizes power consumption in real-time embedded systems using the big.LITTLE technology. The scheduler is optimized using the developmental genetic programming. It is able to reschedule tasks during runtime when it occurred that execution time of some tasks was longer or shorter then it was expected. The goal of the rescheduling is to satisfy time constraints and to minimize the power consumption. But we observed that in some cases the rescheduling was not able to improve the system. Thus, to increase the self-adaptivity capabilities of the target system, some design for self-adaptivity approach should be used.

3 Developmental Genetic Programming

Genetic algorithms (GA) [13] are very commonly used in wide spectrum of optimisation problems. Main advantage of GA approach is the possibility of getting out from the local minima of optimization criterion. Thus, GA is efficient for global optimization of complex problems, like RCPSP or multi-objective optimization of distributed real-time systems [14].

Although GA approach usually give satisfactory results it may be inefficient for hard constrained problems. In these cases a lot of individuals obtained using genetic

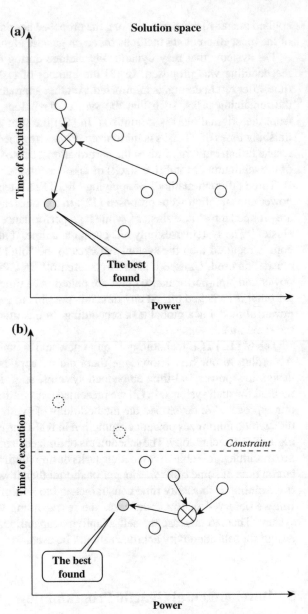

Fig. 1 Search space in GA.
a Unconstrained search
space. **b** Constrained search
space

operators correspond to not feasible solutions (e.g. schedule that exceeds required
deadline or incorrect schedule, in the RCPSP). Such individuals should not be considered during the evolution. It is provided by defining the constrained genetic operators
that produce only correct solutions. But such constrained operators may create infeasible regions in the search space. Such regions may contain optimal or close to optimal
solutions. This problem is illustrated on Fig. 1. Assume that we optimize the power

consumption of the real-time system. Due to constraint violation, the solutions above the dotted line are not valid, hence they are never produced during the evolution. But it is possible that such "forbidden" individuals may be used during the crossover or mutation to produce highly optimized solutions (Fig. 1a). Unfortunately, the search

Fig. 2 Genotype to phenotype mapping

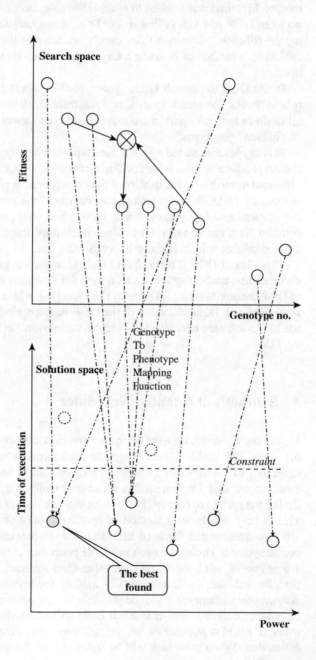

space should be limited to consider only valid solutions (Fig. 1b) and the optimal solution may never be obtained.

Above problem may be eliminated by using the Developmental Genetic Programming (DGP). DGP is an extension of the GA by adding the developmental stage. This method first time was applied to optimize analog circuits [15]. The main difference between DGP and GA is that in the DGP genotypes represent the method building the solution, while in the GA genotypes describe the solution. Thus, during the evolution, a method of building a target solution is optimized, instead of a solution itself.

In the DGP the search space (genotypes) is separated from the solution space (phenotypes). The search space is not constrained, all individuals are evolved. Thus, all of them may take part in the reproduction, crossover or mutation. There are no "forbidden" genotypes.

Phenotypes are created by using genotype-to-phenotype mapping function, which always produces a valid solution. During the evolution, the fitness of the genotype is evaluated according to the quality of the corresponding phenotype. Figure 2 presents the idea of the DGP. The search space consists of the genotypes that evolve without any restrictions. Any genotype may be mapped onto phenotype representing valid solution. This can be assured by using constrained mapping function. This function never produces solutions above the constraint line.

This idea of DGP is taken from biology, where the genotype corresponds to the chromosome containing information used for synthesis of proteins. The application of DGP occurred successful in many domains [16], where human-competitive results were obtained. High efficiency of the DGP-based optimization was also proved for hardware-software codesign [17] and cost minimization in real-time cloud computing [18].

4   Synthesis of Adaptive Scheduler

Idea of our approach is based on the observation that when the DGP will be applied for the RCPSP problem, then except the final schedule we also obtain the scheduler dedicated to the optimized system. Thus, instead of the implementation of static schedule we may implement this scheduler, which may adapt to any perturbation during the system operation. We assume that the system is specified as a task graph. This is very widely used method of specification of real-time embedded systems. We also assume that for each task, the time of execution and the average power consumption is known for each available processor core. Usually these parameters are estimated using the worst case estimation methods. During the system operation, the scheduler will dynamically modify the schedule to minimize the power consumption whenever it will be possible to move some tasks to low-power cores, i.e. when execution time of finished tasks will occur shorter than estimated. In our method it is also possible to use average execution time, instead of the worst case estimation. When same task will be delayed then the scheduler will try to find the new schedule that satisfies the time requirements. We use ARM multicore processors with big.LITTLE technology for implementation of the target systems. Such system consists of two processors, usually quad-core. The first of them has higher performance (about 40%), but consumes more power. The second one is slower, but it is optimized to use much less energy (about 75%), to execute the same task. The goal of optimization is to find the makespan for which the power consumption is as small as possible, during the whole time period. It is assumed that the system is multirate, i.e. after finishing execution of all tasks, it starts again. Thus, finding the most energy efficient schedule does not guarantee, that in case of disruption of task execution time, the power consumption will not increase. To prevent this case, faster solutions are preferred during optimization.

4.1 Task Graph

The function of an embedded system is specified as a set of tasks. Between certain tasks may be the relationship that specifies the order of their execution. This may be specified as a task graph, which is the acyclic directed graph where nodes correspond to tasks and edges describe required order of execution. A sample task graph is given on Fig. 3.

4.2 Resources

Estimated execution parameters are given in a library of available resources. A resource is a core of a processor, which is able to execute a task. For each core the execution time and the power consumption are given. Part of a sample ARM Cortex-A15/Cortex-A7 database, for task graph from Fig. 3, is presented in Table 1.

Fig. 3 A sample task graph

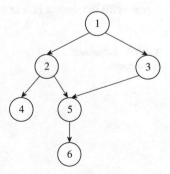

Table 1 A sample library of resources

Task #	Core #	Execution time (ns)	Power consumption (mJ)
1	0	537	5
1	1	537	5
1	2	537	5
1	3	537	5
1	4	671	3
1	5	671	3
1	6	671	3
1	7	671	3
2	0	1072	11
...	...	...	...
6	7	176	1

4.3 Strategies of Scheduling

The scheduler creates a makespan in two steps:

1. task assignment: tasks are assigned to cores according to preferences specified for each group of tasks (Table 2),
2. task scheduling: this step is executed only when more than one task is assigned to the same core. During this step a selected group of tasks is scheduled using the scheduling strategy specified for this group (Table 2).

Initial population consists of randomly generated genotypes. During initialization, preferences defining the decision table for the scheduler are assigned to each gene. Table 2 contains the set of possible preferences that the scheduler may choose. The last column in Table 2 shows a probability of the selection.

The first option prefers the core with the highest performance. Second one prefers a core with the lowest power consumption. Third option prefers a core with the best

Table 2 Scheduler's preferences

Step	Option	P
1	a. The highest performance	0.143
	b. The lowest power	0.143
	c. The lowest time * power	0.143
	d. Determination by second gene	0.143
	e. The fastest starting core	0.143
	f. The fastest finishing core	0.143
	g. The most energy efficient core from the fastest starting cores	0.143
2	List scheduling	1

ratio of the power consumption to the time of execution. Fourth option allows using a core, that cannot be obtained as a result of the remaining options. The next option prefers a core, which could start an execution of the task as soon as possible (other cores might be busy). The 'f' option prefers a core which could be the first to finish a task (be freed). The last option chooses the most energy efficient resource from a group of ones, which meet the criteria of 'e' strategy. For the second step only one option is available, the list scheduling method.

4.4 Genotype

The genotype has a form of binary tree corresponding to the certain procedure of task scheduling [19]. Every node in the genotype has the structure presented on Fig. 4.

The first field *isLeaf* determines a type of the node in a tree. When the node is a leaf this field equals true. Then, the field named *"strategy"* defines the strategy of scheduling for group of tasks assigned to this node. All possible strategies are given in Table 2. In this case, information from the other fields is omitted. When the node is not a leaf, a content of the field *"strategy"* is neglected. In this case, *cutPos* contains a number describing which group of tasks should be assigned and scheduled by the left node and which one by the right one. Thus, *nextLeft* and *nextRight* must not be null pointers.

The simplest genotype consists of only one node, which is also a leaf and a root. A sample genotype and the corresponding phenotype are presented on Fig. 5.

During the evolution a genotype may grow up but the size of the tree is limited. If a tree will be too large, the performance of genotype to phenotype mapping, required for the fitness evaluation, would be slightly decreased. Size limit of the genotype also avoids constructing too many unused branches.

An initial genotype tree may grow up as an effect of genetic operators: mutation and crossover. An action associated with the mutation depends on the type of the node and is presented on the Fig. 6 [19].

Fig. 4 A node of the genotype

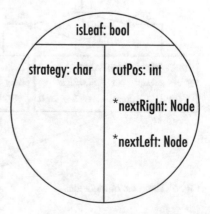

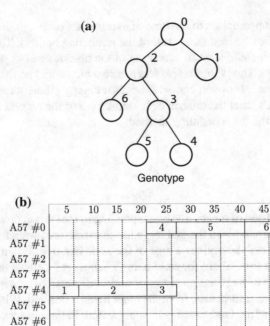

Genotype

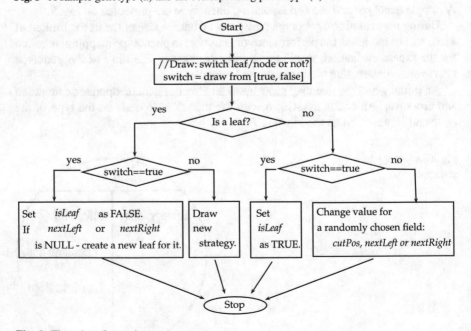

Fig. 5 A sample genotype (**a**) and the corresponding phenotype (**b**)

Phenotype

Fig. 6 The rules of mutations

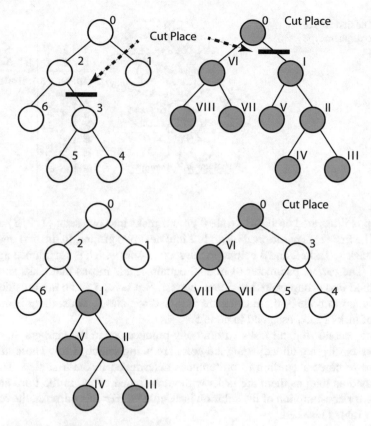

Fig. 7 An example of the crossover

The crossover is used to create new individuals that are a combination of genes of parent genotypes. First, points of cut for both trees are drawn, then the cut branches are exchanged. In this way two new genotypes are created. A sample crossover is presented on Fig. 7.

With every genotype an array is associated. Its size is equal to the number of tasks and contains indexes of cores. If for given task, strategy "d" is chosen, the core with an index taken from the array is used. During the mutation, a position in the array is randomly chosen. Then, a new index is randomly generated. During the crossover, parts of the arrays from both genotypes are swapped. The array defines the alternative scheduling strategy that is not driven by the performance or power consumption.

4.5 Genotype to Phenotype Mapping

The first step, during the genotype-to-phenotype mapping, is to assign strategies to tasks (i.e. preferences for assigning tasks to resources). For the example from Fig. 1,

Fig. 8 The first step in genotype-to-phenotype mapping

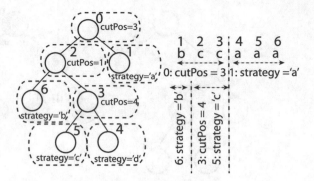

this step is illustrated on Fig. 8. Node 0 groups tasks into two sets: {1, 2, 3} and {4, 5, 6}. The first set is partitioned by node 2 into next two groups. In the first one, there is only task 1. Tasks 2 and 3 belong to the second one, which is partitioned again by node 3. The *cutPos* parameter of node 3 equals 4, this means that tasks should be partitioned into group from 2 to 5 and the rest. But tasks 5 and 6 are outside of the set assigned to node 3, these tasks are assigned to node 1. Thus, there is only one group of tasks {2, 3} assigned to node 5.

In the second step, all tasks without any predecessor in the task graph, or with predecessors having already assigned core, are being searched for. These tasks are assigned to cores according to preferences determined in the first step. This step is repeated as long as there are tasks without assigned cores. In the third step, the total power consumption of the solution is calculated. For this purpose, the resource library (Table 1) is used.

4.6 Parameters of DGP

During the evolution, new populations of schedulers are created using genetic operations: reproduction, crossover (recombination) and mutation. After the genetic operations are performed on the current population, a new population replaces the current one. The evolution is controlled by the following parameters:

- *population size*: the number of individuals in each population is always the same. The value of this parameter is determined according to the value of "number of tasks" * "number of cores",
- *reproduction size*: number of individuals created using the reproduction,
- *crossover size*: the number of individuals created using the crossover,
- *mutation size*: the number of individuals created using the mutation.

Finally, the selection of the best individuals by a tournament is chosen [13]. In this method, chromosomes (genotypes) are drawn with the same probability in quantity defined as a size of the tournament. The best one is taken to the next generation.

Hence, the tournament is repeated as many times as the number of chromosomes for a reproduction, crossover and mutation is required. A size of the tournament should be defined carefully. It should not be too high, because the selection pressure is too strong and the evolution will be too greedy. It also should not be too low, because the time of finding any better result would be too long.

4.7 Self-adaptability of the Scheduler

Finding the best makespan for low-power real-time embedded system is not the only goal of our approach. DGP methods are very effective in solving optimization problems and very often give the optimal solution. From the other side, they give results in relatively long time, thus genetic approach can not be used for rescheduling in real-time systems.

In the DGP the scheduling is performed during the genotype to phenotype mapping. This process is very fast, therefore it can be executed during the system operation. So, instead of implementing the final schedule we implement the method which creates this schedule. We observed that such approach has great self-adaptability capabilities.

Since the DGP has to consider only valid makespans. The genotype to phenotype mapping is a constrained process. If for a set of preferences defined by the genotype, it is not possible to obtain the valid phenotype then the mapping selects the next matching resource. Thus, the preferences specified by the genotype need not be strictly adhered. For example, if for given task preference suggest assigning this task to the low-power core, then if this decision will lead to an infeasible makespan, the scheduler will choose another, faster core that best matches to this preference.

Therefore, scheduler is able not only to build a correct solution, but also modify it if any unpredictable events will occur. E.g. if a task execution will be longer than expected, then the scheduler could move some tasks from a slower to a faster core, to fulfill time requirements. Similarly, if a task will be finished before its predicted end time, scheduler can move other tasks from a faster core to slower one, to save some energy.

4.8 Fitness Function

A fitness function determines the optimization goal of the DGP. In opposite to our previous approach [2], the fitness function takes into consideration the self-adaptability capabilities. We assume that the system is more self-adaptive if the laxity (difference between deadline and time of execution) is longer. Thus the fitness function is multiobjective (Eq. 1).

$$Q = \frac{SC}{WSC} - SAF \frac{SD - STT}{SD} \tag{1}$$

The symbols in Eq. 1 have the following meanings:

- Q—quality (fitness), the goal of this method is to minimize this value,
- SC—schedule cost (power consumption)—cost of executing tasks according to order and assignment described by the chromosome,
- WSC—the worst schedule cost—the cost of executing all tasks using the most energy consuming resources,
- SAF—Self-Adaptability Factor—this value controls the self-adaptability capability of the target system. If SAF is equal 0, then the cost is the only criterion applied during optimization. By increasing the value of this parameter we may increase the self-adaptivity of the system. But too high value of SAF may drive the evolution to produce too energy intensive solutions,
- SD—Schedule Deadline,
- STT—Schedule Total Time—a time necessary to execute all tasks, according to scheduling method specified by the genotype.

5 Example and Experimental Results

In the previous work [2] we showed that our approach gives better results than existing methods (e.g. based on Least-Laxity-First Scheduling Algorithm [21]). Initial makespans were usually more energy efficient, moreover target systems were able to reschedule tasks in case of some disruptions. In this work we will mainly evaluate the self-adaptivity capabilities of systems synthesized by the modified method. We present experimental results showing the influence of Self-Adaptability Factor on the target system. The results will be compared with systems optimized using our previous approach.

5.1 Task Graph and Run-Time Parameters

The sample system is a multimedia player implemented as a real-time embedded system. The specification of the system consists of 40 tasks. Figure 9 presents the task graph describing the system, details are given in [20]. We consider shared memory architecture, thus the communication between tasks may be neglected. The execution time is critical—it is typical soft real-time system. If the deadline is only slightly exceeded, the quality of the system is decreased, but the solution may be accepted. If the system exceeds hard deadline, then result is unacceptable and system should be redesigned.

We assumed, that the application will be implemented in software running on system consisting of two 4-core processors. One of them is ARM Cortex A57 and

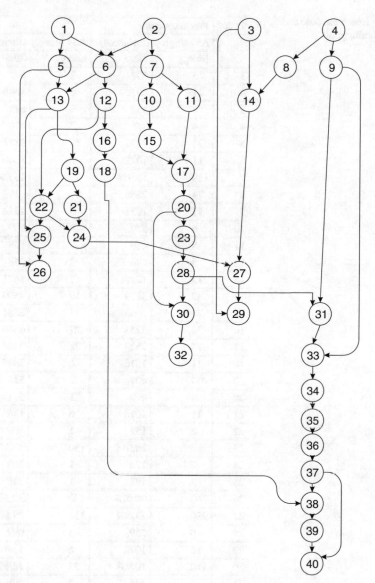

Fig. 9 Task graph of the multimedia system

the second is ARM Cortex A53. The first processor is faster for about 25%, but it consumes about 50% more power. Both processors support ARM big.LITTLE technology. Run-time parameters for both processors are given in Table 3.

The goal of our methodology is to create a self-adapting scheduler, which runs the program tasks, balancing them between the cores, to minimize the power usage. The scheduler should be able to reschedule remaining tasks, whenever any task will finish its execution before or after expected time frame.

Table 3 Execution time and power consumption

Task	Processor cores			
	A57 (high performance)		A53 (energy efficient)	
	Energy	Time	Energy	Time
1	5	537	3	671
2	11	1072	6	1340
3	5	537	3	671
4	4	376	2	470
5	73	7337	37	9171
6	11	1072	6	1340
7	110	10,958	55	13,698
8	74	7358	37	9198
9	11	1051	6	1314
10	6	559	3	699
11	5	486	3	608
12	3	286	2	358
13	13	1298	7	1623
14	37	3679	19	4599
15	21	2065	11	2581
16	53	5253	27	6566
17	75	7523	38	9404
18	11	1076	6	1345
19	4	409	2	511
20	4	409	2	511
21	11	1076	6	1345
22	2	157	1	196
23	260	26,018	130	32,523
24	2	176	1	220
25	2	197	1	246
26	260	26,018	130	32,523
27	236	23,607	118	29,509
28	6	559	3	699
29	11	1072	6	1340
30	110	10,958	55	13,698
31	5	486	3	608
32	3	286	2	358
33	11	1072	6	1340
34	4	409	2	511
35	4	409	2	511
36	236	23,607	118	29,509
37	74	7414	37	9268
38	3	253	2	316
39	2	179	1	224
40	2	176	1	220

5.2 Genetic Parameters

We assumed that the deadline for the system from Fig. 4 is equal to 100,000 ns. The Power Aware Scheduler and the optimized makespan were generated using DGP. During the experiments, the following values of genetic parameters were used:

- the evolution was stopped after 100 generations,
- each experiment was repeated 7 times,
- the population size was equal to 128,
- tournament size was equal to 10,
- the number of mutants in each generation, was equal to 20%,
- the crossover was applied for creation of 40% genotypes,
- 20% of individuals were created using reproduction.

The values of parameters described above were tuned according to method described in our previous work [19], thus we will describe it here very shortly. In the first step, we estimated an influence of the tournament size. When this parameter was too small, the evolution got stuck. When the tournament size was too big, the DGP found semi optimal solution very fast, but a further optimization was not possible. Next, the influence of crossover and mutation for obtaining the best solution has been tested. It has been done by searching for the best solution using different combination of these parameters. Thus the best values of these parameters have been chosen. Finally, the best combination of other evolution parameters has been evaluated.

5.3 Self-adaptivity Capabilities

Static scheduling is based on estimation of execution times for all tasks. During the system operation, time of execution may significantly be shorter (e.g. if the worst case estimation was applied) or longer (e.g. in case of the most likely estimation). Therefore, the system may be additionally optimized during run-time, by using self-adaptive scheduling. In this way certain system parameters (power consumption, performance) may be improved. Scheduler generated using our method consists of series of system construction functions, corresponding to each gene. These functions are flexible, i.e. design decisions are driven by preferences, they are not strictly defined. This is necessary to assure that only feasible makespan will be created. Flexibility of the system construction functions provides to self-adaptivity capabilities of the scheduler.

An example of the self-adaptivity is presented on Fig. 10. If the execution time of task 23 will be too long, then all succeeding tasks should be postponed and the system would exceed the deadline. But our scheduler adapts to the delayed end time of task 23, and despite the fact that it uses the same construction functions, some tasks (tasks 30 and 36 in our example) will be assigned to more efficient cores (Fig. 10b).

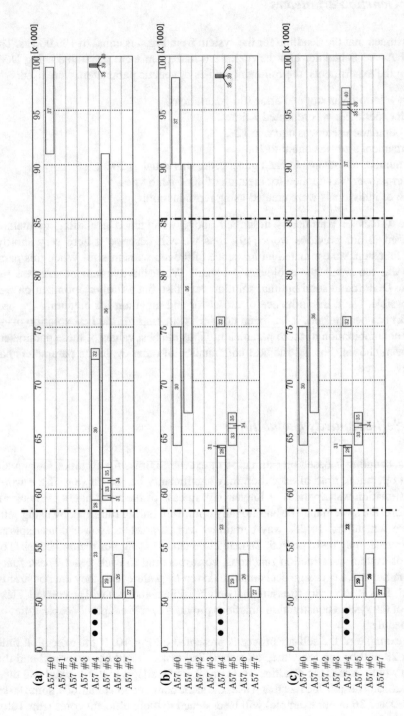

Fig. 10 Self-adaptation capabilities of power-aware scheduler. **a** Without changes. **b** Longer execution time of task 23. **c** Shorter execution time of task 36

In other case, if an execution time of a task 36 will be shorter than it was expected, the scheduler will assign some tasks (task 37 in our example) to low power core (Fig. 10c). In this way power consumption will be reduced.

To verify the capabilities of self-adaptivity of the scheduler we performed some simulations of different changes in execution times for some task.

5.4 Power-Aware Scheduling

To compare the self-adaptivity features of solutions generated by our methodology, we performed some experiments simulation delays of some tasks. The obtained results are shown in Tables 4, 5 and 6. The following columns correspond to system with predicted execution times (column 0) and systems where execution times

Table 4 Delays with deadline 90,000 ns

Case	0	1	2	3	4	5
Delay	(None)	T2 + 53 %	T8 + 20 %	T17 + 3 %	T28 + 36 %	T36 + 1,3 %
Deadline (ns)	90,000	90,000	90,000	90,000	90,000	90,000
Self-adaptability factor	Time without rescheduling					
Previous work	89,756	90,466	91,595	90,038	90,007	90,056
0	88,675	89,385	90,150	88,957	88,927	88,982
1.5	88,407	89,479	89,879	88,689	88,659	88,713
3	83,168	83,667	83,168	83,394	83,349	83,474
	Time after rescheduling					
Previous work	–	88,612	89,741	89,994	89,964	89,967
0	–	89,385	88,675	88,957	88,927	88,982
1.5	–	89,479	89,879	88,689	88,659	88,713
3	–	83,667	83,168	83,394	83,349	83,474
	Time-out (%)					
Previous work	0	0.52	1.8	0.04	0.01	0.06
0	–	0.00 %	0.17 %	0.00 %	0.00 %	0.00 %
1.5	–	0.00 %	0.00 %	0.00 %	0.00 %	0.00 %
3	–	0.00 %	0.00 %	0.00 %	0.00 %	0.00 %
	Energy (mJ)					
Previous work	1494	1586	1586	1550	1550	1496
0	1291	1291	1409	1291	1291	1291
1.5	1295	1295	1295	1295	1295	1295
3	1563	1563	1563	1563	1563	1563

Table 5 Delays with deadline 95,000 ns

Case	0	1	2	3	4
Delay	(None)	T8 + 32%	T15 + 20%	T29 + 51%	T33 + 43%
Deadline (ns)	95,000	95,000	95,000	95,000	95,000
Self-adaptability factor	Time without rescheduling				
Previous work	94,463	96,817	94,903	94,903	94,995
0	94,717	97,660	95,233	95,400	95,293
1.5	91,810	94,753	92,223	92,493	92,386
3	83,168	86,111	83,581	83,168	83,629
	Time after rescheduling				
Previous work	–	93,404	92,766	94,903	95,039
0	–	94,717	93,379	94,717	93,439
1.5	–	91,810	92,223	91,810	92,386
3	–	83,168	83,581	83,168	83,629
	Time-out (%)				
Previous work	–	1.90%	0.00%	0.00%	0.04%
0	–	2.80%	0.25%	0.42%	0.31%
1.5	–	0.00%	0.00%	0.00%	0.00%
3	–	0.00%	0.00%	0.00%	0.00%
	Energy (mJ)				
Previous work	1275	1312	1265	1275	1276
0	1086	1091	1241	1091	1241
1.5	1150	1150	1150	1150	1150
3	1517	1535	1517	1517	1517

of some tasks were longer than expected (columns 1–5). For each system, results obtained using our previous method as well as our new method with 3 different values of SAF (0, 1.5 and 3), are given. "Time without rescheduling" corresponds to the influence of the task delay to the initial schedule length. "Time after rescheduling" means the result of the self-adaptivity.

We may observe that our new fitness method allows obtaining better results. In most cases, additional rescheduling was unnecessary. But with increasing the Self-Adaptability Factor, the total amount of consumed energy was increased too. Thus the SAF $= 0 - 1.5$ seems optimal, i.e. gives energy efficient solutions that have a laxity which is usually enough to tolerate small delays.

Analyzing data from Tables 4, 5 and 6, it is visible, that Self-Adaptability Factor has significant impact on quality of obtained solutions. If the value is small or equal 0, the solutions are the most power-efficient ones. But almost in every time, when disruption occurs, a rescheduling of makespan is necessary. This problem is disappears, when the value of Self-Adaptability Factor is increasing. But the negative effect of this, it is obtaining makespans, which are usually less energy-efficient.

Table 6 Delays with deadline 100,000 ns

Case	0	1	2	3	4	5
Delay	(None)	T7: +30%	T14: +50%	T15: +20%	T17: +35%	T23: +15%
Deadline (ns)	100,000	100,000	100,000	100,000	100,000	100,000
Self-adaptability factor	Time without rescheduling					
Previous work	99,846	103,955	100,765	100,259	102,479	104,724
0	99,368	102,665	101,668	99,884	102,659	104,246
1.5	97,652	101,671	99,952	98,168	100,943	102,530
3	83,308	86,595	85,147	83,721	85,941	87,211
	Time after rescheduling					
Previous work	–	99,772	99,846	96,211	98,431	98,822
0	–	98,607	99,368	99,884	98,611	98,344
1.5	–	99,907	97,136	98,168	99,089	99,993
3	–	83,308	85,147	83,721	85,941	87,211
	Time-out (%)					
Previous work	–	3.95%	0.77%	0.26%	2.48%	4.72%
0	–	2.67%	1.67%	0.00%	2.66%	4.25%
1.5	–	1.67%	0.00%	0.00%	0.94%	2.53%
3	–	0.00%	0.00%	0.00%	0.00%	0.00%
	Energy (mJ)					
Previous work	990	1319	1238	1071	1071	1163
0	992	1203	1122	992	1073	1165
1.5	1026	1253	1036	1026	1253	1267
3	1426	1426	1426	1426	1426	1426

In hard real-time systems, usually a pessimistic estimation of execution time is applied (worst case execution time). It is because that the exceeding the deadline means that the system failed. But in most cases the time of execution occurs shorter than expected. Thus, it is important to verify the behaviour of system in such cases. The self-adaptive system should reschedule tasks and resource assignments to obtain more energy-efficient solution. In Tables 7, 8 and 9 the results showing the impact of Self-Adaptability Factor on the self-optimization capability is shown.

Also in this case, the influence of Self-Adaptability Factor is significant. For small values of this parameter, obtained systems are able to reduce the energy usage. We may notice some interesting observation: if the value of SAF increases, the solutions for long deadlines have smaller energy usage than ones, which were obtained for shorter deadlines. Further increasing Self-Adaptability Factor drives to obtaining more energy intensive results. Therefore, a value of Self-Adaptability Factor should not be high, to obtain the best makespans.

Table 7 Time decrease for deadline 90,000 ns

Case	0	1	2	3	4	5	6	7
Time decrease	(None)	T17 −43%	T23 −34%	T27 −15%	T30 −25%	T36 −38%	All −30%	All −50%
Deadline (ns)	90,000	90,000	90,000	90,000	90,000	90,000	90,000	90,000
Self-adaptability factor	Time (ns)							
Previous work	89,756	89,987	89,649	89,756	89,756	80,785	88,916	83,398
0	88,675	89,282	89,980	88,675	88,675	89,869	70,759	50,545
1.5	88,407	89,014	89,756	88,407	88,407	89,601	70,571	50,411
3	83,168	79,933	74,322	83,168	83,168	74,197	58,216	41,589
	Energy (mJ)							
Previous work	1494	1464	1376	1494	1494	1494	1116	998
0	1291	1253	1044	1291	1291	1043	1043	1043
1.5	1295	1257	1047	1295	1295	1047	1047	1047
3	1563	1563	1563	1563	1563	1563	1563	1563
	Energy without rescheduling (mJ)							
Previous work	1494	1494	1494	1494	1494	1494	1494	1494
0	1291	1291	1291	1291	1291	1291	1291	1291
1.5	1295	1295	1295	1295	1295	1295	1295	1295
3	1563	1563	1563	1563	1563	1563	1563	1563

Table 8 Time decrease for deadline 95,000 ns

Case	0	1	2	3	4	5
Time decrease	(None)	T7 −21%	T16 −53%	T30 −48%	All −30%	All −50%
Deadline (ns)	95,000	95,000	95,000	95,000	95,000	95,000
Self-adaptability factor	Time (ns)					
Previous work	94,463	94,463	91,810	94,463	94,543	69,690
0	94,717	94,717	94,717	94,717	81,961	73,301
1.5	91,810	91,810	91,810	91,810	94,764	68,362
3	83,168	83,168	83,168	83,168	58,216	41,589
	Energy (mJ)					
Previous work	1275	1281	1275	1275	1275	1275
0	1086	1086	1086	1086	1085	1085
1.5	1150	1150	1150	1150	1150	1150
3	1517	1517	1517	1517	1517	1517
	Energy without rescheduling (mJ)					
Previous work	1275	1275	1275	1275	1275	1275
0	1086	1086	1086	1086	1086	1086
1.5	1150	1150	1150	1150	1150	1150
3	1517	1517	1517	1517	1517	1517

Table 9 Time decrease for deadline 100,000 ns

Case	0	1	2	3	4	5	6	7
Time decrease	(None)	T7: −30%	T14: −50%	T15: −20%	T17: −35%	T36: −35%	All −30%	All −50%
Deadline (ns)	100,000	100,000	100,000	100,000	100,000	100,000	100,000	100,000
Self-adaptability factor	Time (ns)							
Previous work	99,846	99,846	99,846	99,433	99,324	91,371	81,970	89,570
0	99,368	97,934	99,368	98,852	97,931	90,894	70,857	50,615
1.5	97,652	97,048	97,652	97,136	98,409	93,226	72,489	51,781
3	83,308	83,308	83,308	83,308	80,675	75,046	58,314	41,659
	Energy (mJ)							
Previous work	990	990	990	990	953	953	953	953
0	992	986	992	992	955	955	955	955
1.5	1026	1085	1026	1026	945	908	908	908
3	1426	1426	1426	1426	1426	1426	1426	1426
	Energy without rescheduling (mJ)							
Previous work	990	990	990	990	990	990	990	990
0	992	992	992	992	992	992	992	992
1.5	1026	1026	1026	1026	1026	1026	1026	1026
3	1426	1426	1426	1426	1426	1426	1426	1426

6 Conclusions

In this paper the method of automatic synthesis of power-aware schedulers for real-time distributed embedded systems was presented. Starting from the system specification in the form of the task graph, we use developmental genetic programming to optimize the scheduling strategy that minimizes the power consumption. Finally, the best makespan as well as the optimized scheduler are generated. The capabilities of self-adaptation were evaluated for different values of the Self-Adaptability Factor parameter. It allows customizing the properties of self-adaptability. Depending on purpose of designed system, it is possible to synthesize it as more active one (reactive scheduling) or passive one (proactive scheduling).

The presented method is dedicated to ARM big.LITTLE technology, developed for low power systems. But, since we use general optimization method, it would be easily adapted to other energy-efficient architectures.

The experimental results confirmed that DGP method is able to generate efficient and flexible systems, optimized for low power as well as for self-adaptivity. These properties are also customizable during optimization of the solution. Based on them, the scheduler is able to quickly and effectively react to any changes of task execution times, by rescheduling remaining tasks.

The full impact of Self-Adaptability Factor has not been evaluated yet. Moreover some over types if multiobjective fitness functions may be developed and evaluated. In the future work, we will consider special types of adaptive genes that could support more possibilities for self-adaptation, we will also consider using other scheduling methods, alternative to list scheduling, e.g. based on mathematical/constrained programming [22].

References

1. big.LITTLE processing with $ARMCortex^{TM}$ - A15 & Cortex-A7, ARM holdings, September 2013. http://www.arm.com/files/downloads/big.LITTLE_Final.pdf
2. Deniziak, S., Ciopinski, L.: Synthesis of power aware adaptive schedulers for embedded systems using developmental genetic programming. In: Federated Conference on Computer Science and Information Systems (FedCSIS). IEEE (2015). http://dx.doi.org/10.15439/2015F313
3. Luo, J., Jha, N.K.: Low power distributed embedded systems: dynamic voltage scaling and synthesis. In: Proceedings of the 9th International Conference on High Performance Computing - HiPC 2002. Lecture Notes in Computer Science, vol. 2552, pp. 679–693 (2002). http://dx.doi.org/10.1007/3-540-36265-7_63
4. Hartmann, S., Briskorn, D.: A survey of variants and extensions of the resource-constrained project scheduling problem. Eur. J. Oper. Res.: EJOR. vol. 207, 1 (16.11.), pp. 1–15. Elsevier, Amsterdam (2010). http://dx.doi.org/10.1016/j.ejor.2009.11.005
5. Hartmann, S.: An competitive genetic algorithm for resource-constrained project scheduling. Nav. Res. Logist. 45(7), 733–750 (1998). http://dx.doi.org/10.1002/(SICI)1520-6750(199810)45:7%3C733::AID-NAV5%3E3.3.CO;2-7
6. Li, X., Kang, L., Tan, W.: Optimized research of resource constrained project scheduling problem based on genetic algorithms. Lecture Notes in Computer Science, vol. 4683, pp. 177–186 (2007). http://dx.doi.org/10.1007/978-3-540-74581-5_19
7. Zoulfaghari, H., Nematian, J., Mahmoudi, N., Khodabandeh, M.: A new genetic algorithm for the RCPSP in large scale. Int. J. Appl. Evol. Comput. 4(2), 29–40 (2013). http://dx.doi.org/10.4018/jaec.2013040103
8. Calhoun, K.M., Deckro, R.F., Moore, J.T., Chrissis, J.W., Hove, J.C.V.: Planning and re-planning in project and production scheduling, Omega Int. J. Manag. Sci. 30(3), 155–170 (2002). http://dx.doi.org/10.1016/S0305-0483(02)00024-5
9. Van de Vonder, S., Demeulemeester, E.L., Herroelen, W.S.: A classification of predictive-reactive project scheduling procedures. J. Sched. 10(3), 195–207 (2007). http://dx.doi.org/10.1007/s10951-007-0011-2
10. Sakkout, H., Wallace, M.: Probe backtrack search for minimal perturbation in dynamic scheduling. Constraints 5(4), 359–388 (2000). http://dx.doi.org/10.1023/A:1009856210543
11. Al-Fawzan, M., Haouari, M.: A bi-objective model for robust resourceconstrained project scheduling. Int. J. Prod. Econ. 96, 175–187 (2005). http://dx.doi.org/10.1016/j.ijpe.2004.04.002
12. Jeff, B.: Ten Things to Know About big.LITTLE. ARM Holdings (2013). http://community.arm.com/groups/processors/blog/2013/06/18/ten-things-to-know-about-biglittle
13. Michalewicz, Z.: Genetic Algorithms + Data Structures = Evolution Programs. Springer, Berlin (1996). http://dx.doi.org/10.1007/978-3-662-03315-9
14. Dick, R.P., Jha, N.K.: MOGAC: A multiobjective genetic algorithm for the cosynthesis of hardware-software embedded systems. IEEE Trans. Comput.Aided Des. Integr. Circuits Syst. 17(10), 920–935 (1998). http://dx.doi.org/10.1109/43.728914
15. Koza, J., Bennett III, F. H., Andre, D., Keane, M. A.: Evolutionary design of analog electrical circuits using genetic programming. In: Parmee, I.C. (ed.) Adaptive Computing in Design and Manufacture (1998). http://dx.doi.org/10.1007/978-1-4471-1589-2_14

16. Koza, J.R., Poli, R.: Genetic programming. In: Burke, E., Kendal, G. (eds.) Search Methodologies: Introductory Tutorials in Optimization and Decision Support Techniques. Springer, New York (2005). http://dx.doi.org/10.1007/0-387-28356-0_5
17. Deniziak, S., Górski, A.: Hardware/Software Co-Synthesis of Distributed Embedded Systems Using Genetic Programming. Lecture Notes in Computer Science, pp. 83–93. Springer, New York (2008). http://dx.doi.org/10.1007/978-3-540-85857-7_8
18. Deniziak, S., Ciopiński, L., Pawiński, G., Wieczorek, K., Bak, S.: Cost optimization of real-time cloud applications using developmental genetic programing. In: Proceedings of the 7th IEEE/ACM International Conference on Utility and Cloud Computing, pp. 774–779 (2014). http://dx.doi.org/10.1109/UCC.2014.126
19. Sapiecha, K., Ciopiński, L., Deniziak, S.: An application of developmental genetic programming for automatic creation of supervisors of multi-task real-time object-oriented systems. In: IEEE Federated Conference on Computer Science and Information Systems (FedCSIS) (2014). http://dx.doi.org/10.15439/2014F208
20. Hu, J., Marculescu, R.: Energy-and performance-aware mapping for regular NoC architectures. IEEE Trans. Comput.-Aided Des. Integr. Circuits Syst. **24**(4), 551–562 (2005). http://dx.doi.org/10.1109/TCAD.2005.844106
21. Han, S., Park, M.: Predictability of least laxity first scheduling algorithm on multiprocessor real-time systems. In: Proceedings of EUC Workshops. Lecture Notes in Computer Science, vol. 4097, pp. 755–764 (2006). http://dx.doi.org/10.1007/11807964_76
22. Sitek, P.: A hybrid CP/MP approach to supply chain modelling, optimization and analysis. In: Federated Conference on Computer Science and Information Systems (FedCSIS). IEEE (2014). http://dx.doi.org/10.15439/2014F89

Flow Design and Evaluation in Photonic Data Transport Network

Mateusz Dzida and Andrzej Bąk

Abstract Development of sophisticated photonic transmission systems enabled evolution of photonic data transport networks towards cost-efficient and energy-efficient platforms capable to carry enormous traffic. Given access to technologically advanced equipment, network operator faces a series of decision problems related to how to efficiently use this technology. In this paper, we propose a mathematical model of modern photonic network with wavelength division multiplexing (WDM). Proposed model is an instance of multi-commodity flow optimization formulation related to specific construction of network graph. Given optimal solution of the considered optimization problem can be evaluated through a series of indicators determining quality and performance of the solution.

1 Introduction

Recent advances in the photonic networking enabled rapid growth of the transmission rates in the modern photonic data transport networks. Thus, photonic data transport networks became considerable alternative for traditional electric-based transmission systems, and are more and more widely deployed in the Autonomous Systems composing the Internet.

Photonic data transmission exploits optical fibers to send photonic signals between transceivers (lasers) and receivers (photo-diodes). Photonic signal is modulated to represent values of consecutive bits, composing transmitted piece of digital data.

M. Dzida (✉) · A. Bąk
Optimax, ul. Wolbromska 19/A, 03-680, Warsaw, Poland
e-mail: mdzida@onet.eu; dzida.mateuszmd@orange.com

M. Dzida
Orange, ul. Sw. Barbary 2, 00-686, Warsaw, Poland

A. Bąk
Warsaw University of Technology, ul. Nowowiejska 15/19, 00-665, Warsaw, Poland
e-mail: abak@poczta.pl; bak@tele.pw.edu.pl

© Springer International Publishing Switzerland 2016
S. Fidanova (ed.), *Recent Advances in Computational Optimization*,
Studies in Computational Intelligence 655, DOI 10.1007/978-3-319-40132-4_8

Low interference with other electromagnetic signals makes photonic signals stable and robust to distortions. Thus, photonic signals can be successfully sent over long distances, reaching thousands of kilometers, and transmitted data can be still reproduced with small error rate.

Photonic signals can be easily multiplexed in the frequency domain. Frequency in the case of photonic networking is called wavelength, and the corresponding multiplexing is called Wavelength Division Multiplexing or WDM. WDM is also a term describing the related channel-oriented transport technology, exploiting regular channel widths and spacing (so-called optical grid). Each wavelength constitutes thus an isolated communication channel, in the following referred to as λ.

According to ITU-T recommendation G.709, signals transmitted through Optical Transport Network (OTN) compose the Optical Transport Hierarchy (OTH). OTH defines structure and bandwidth of tributary signals crossing the User-Network Interface (UNI) in the OTN network. As bandwidth of single WDM channel may be greater than bandwidth of typically used Layer 2 signals, multiple client signals may be concatenated into larger signals, better fitted to bandwidth of WDM channels. Signals are typically concatenated in the time domain through dedicated concentrator cards.

Tributary signals need to be further framed according to G.709 definition, coupled with error correction overhead, and transformed into colored optical signals to be transmitted through WDM network. Card responsible for these functions is called transponder. Optical signal transmitted by source transponder is again converted to the digital domain in the destination transponder. Between transponders, signal remains in the optical domain. Still, certain control information related to the transmitted signal is carried along signal path through dedicated administration channel.

In the balance of this paper, being an continuation of our work [6], we propose mathematical (optimization) model of the flow design problem related to photonic data transport network. Developed model is expressed in terms of integer programming. It refers to generic input data, including: network topology, infrastructure, and type/configuration of Optical Network Elements (ONEs). In particular, on one hand, input data must determine comprehensive cost characteristic of the considered equipment, usually in the form of cost of particular expansion cards. On the other hand, input data are supposed to include locations of client devices and their demand for data transport services. It is therefore assumed that knowledge possessed by a network operator about demand structure is certain. In practice, knowing exact demand for transport services can be difficult, and sometimes even impossible. Still, we assume that through appropriate statistical methodology, it is possible to determine demand structure and volume with reasonable degree of confidence.

Paper is organized as follows. In Sect. 2 we define mathematical model associated with designing flows in photonic data transport network. Assumptions and construction of network graph are discussed in Sects. 2.1 and 2.2, respectively. Considered flow design problem is formulated as mixed-integer programme in Sect. 2.3. Further, we investigate modeling specific aspects of the photonic data transport networks, related to: consistency of client flow at technology level (Sect. 2.4), redundancy

(Sect. 2.5), and network cost (Sect. 2.6). In Sect. 3 we investigate physical effects leading to degradation of photonic signal, and propose an extension of introduced mathematical model aimed at reducing some of these effects. Paper is summarized with estimation of formulation complexity in Sect. 5, and conclusions in Sect. 6.

2 Flow Design

In the balance of this section we consider flow design problem related to OTN/WDM photonic networks. Considered flow design problem is formulated in terms of mathematical programming. Having given basic WDM network topology and set of traffic demands to be realized, OTN/WDM flow design problem is aimed at identifying a flow distribution and composition of expandable WDM components (muxponders, transponders, multiplexers, etc.) leading to optimized value of certain objective function. In particular, feasible solution of the considered problem identifies design of the ONE nodes in terms of number, type, and configuration of expansion cards necessary to realize traffic demands, and associated cost.

Considered problem is described in the literature as Routing Wavelength Assignment (RWA) problem. It is commonly considered in combination with objective function maximizing the number of concurrent connections. Example integer linear programming formulation of this problem can be found in [23]. Independently in [4, 8] it was proved that RWA problem is $\mathcal{N}\mathcal{P}$-complete. Formulations proposed in the literature [3, 5, 10, 12, 14, 16, 17, 19, 21, 22, 24, 25] differ from formulation proposed in the following in terms of graph construction. Namely, in this paper it is assumed that each λ channel constitutes separate edge in the network graph. Such assumption is not common in other works, but it allows to simplify formulation, and increase problem flexibility through graph construction. Moreover, classical RWA problem is concerned with routing and λ selection only. Here, problem is extended with consideration of the access side. This extension is motivated by usage of objective function related to cost of elastic expansion cards.

2.1 Assumptions

Traffic demands are assumed to be known in advance, e.g., they can be sourced from some external business forecast and measurement tool. As demand variation is out of the scope at considered network design problem, demand volumes may be additionally adjusted with some security margin. Each traffic demand is defined by triple: source, destination, and bandwidth volume. Demand source and destination are external clients connected to local ONE nodes through intra-office or short-haul black&white fibers.

Client devices and ONE devices are installed within Points of Presence (PoPs) of a network operator, and each client device is connected to uniquely defined ONE, usually in the same PoP. Even if in some PoP, ONE device is not installed, client localized in such PoP must be unambiguously assigned and connected to one ONE in one of the other PoPs.

After installing full suite of channel multiplexers, ONE device is capable to handle N channels, equal to its maximum capacity. Each channel has precisely defined central frequency and width. Central frequencies of consecutive channels are supposed to be compatible with one of optical grids defined by ITU-T.

2.2 Network Graph

Network topology at the simplest level defines locations, configuration, and type of network elements, and arrangement of long-haul fibers connecting network elements. Depending on required level of granularity, network topology can be more or less detailed. At level of details required by flow optimization, this simple topology needs to be extended with deeper insight into composition of network elements. For this purpose, we define a directed graph $\mathcal{G}(\mathcal{V}, \mathcal{E})$ composed of set of nodes $\mathcal{V}$ and set of edges $\mathcal{E}$. Graph composition is used in the following as a basic modeling methodology. It allows to formulate considered flow design problem in terms of multi-commodity flow optimization.

2.2.1 Graph Nodes

In general, node set $\mathcal{V}$ can refer to four types of physical elements (cards or whole devices):

- optical network element—device responsible for multiplexing, switching, and (optionally) converting colorful λ signals ($\mathcal{O}$),
- transponder—expansion card responsible for adopting colorless tributary signals and modulating them as colorful λ signals ($\mathcal{T}$),
- muxponders—expansion card responsible for concatenating multiple colorless signals into higher-order colorful signals ($\mathcal{M}$),
- client—non-WDM device, consuming OTN services ($\mathcal{C}$).

In order to model switching and converting colorful λ signals, each ONE is represented in the network graph $\mathcal{G}$ by a set of graph nodes (referred to as *colorful nodes*), each associated with exactly one λ and one direction towards adjacent ONE. Accordingly, number of colorful graph nodes associated with single ONE is equal to $N \times D$, where D is the number of ONE neighbors. Similarly, basic graph of long-haul fiber connections is replicated, so there exists N (equal to number of λ's) parallel subgraphs, each topologically isomorphic with original network graph. If ONEs are capable to convert λ frequencies, all colorful nodes associated with single ONE needs to be interconnected. For example, if in the feasible solution, such artificial link is crossed on path between colorful nodes associated with λ_1 and λ_2, it means that signal incoming to the related ONE at λ_1 is transmitted out through λ_2. Example subgraph of colorful nodes associated with two neighboring 3-direction ONEs in ROADM configuration is presented in Fig. 1. T&ROADM counterpart extends the ROADM subgraph with full mesh connections between colorful nodes inside ONE, as presented in Fig. 2.

Transponder and concentrator are sometimes combined as one expansion card—muxponder. If not combined, clients can be connected to transponders two-fold: through direct connections or indirectly through hierarchy of compatible concentrators. In the network graph, stand-alone transponders are associated with a subset of graph nodes $\mathscr{T}$, where each graph node is associated with one transponder type and one ONE location. Associated subgraph is presented in Fig. 3. In the figure, there are three transponder types (say 10, 40, and 100 Gbps) and two clients. Transponder graph nodes representing each transponder type in one location are connected to all colorful nodes.

Muxponders form subset of graph nodes $\mathscr{M}$, where each muxponder type is replicated N times, so each colorful node can be connected to its own unique suite of muxponders. Despite graph contains all potential muxponder cards, only some subset of cards may be required in the optimal solution. All muxponder graph nodes are connected with graph nodes representing compatible interfaces in the client devices, as it is presented in Fig. 4. Whole muxponder hierarchy is represented by single graph node, which means that multiple different connection types (different transport

Fig. 1 ROADM subgraph

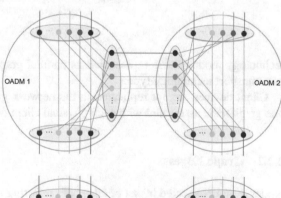

Fig. 2 T&ROADM subgraph

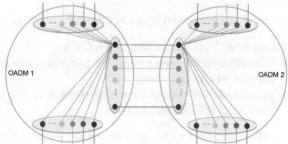

Fig. 3 Transponder
subgraph

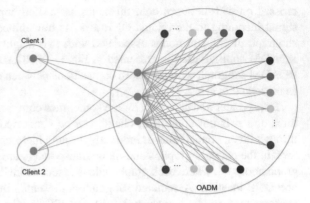

Fig. 4 Muxponder subgraph

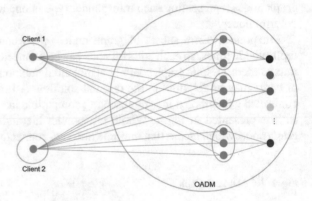

technology modules) are represented as parallel graph links, each associated with
one transport technology type.

Client devices are not replicated in the network graph, and there exists exactly
one graph node associated with each physical client device.

2.2.2 Graph Edges

Nodes $\mathcal{V}$ are connected by set of edges $\mathcal{F}$, referring to physical connections:

- long-haul fibers connecting ONEs ($\mathcal{H}$),
- intra-office fibers connecting line ports in client devices and tributary ports in
  muxponders/transponders,
- patch-cords and back-plane wiring connecting line ports in transponders and tributary ports in ONE multiplexers.

All enumerated types of connections: fibers, patch-cords, and wiring are further described by common term link. Set $\mathcal{F}$ is assumed to be a superset of the

link set defined by the basic network topology. In particular, it contains replicated edges between adjacent colorful nodes. Finally, not all edges contained in $\mathscr{F}$ will be deployed, because some graph nodes represent non-existent components and link deployment will depend on card installation. A subset of these potential links will be selected for deployment or activation.

2.3 *Mathematical Formulation*

Based on introduced network graph definition, the WDM flow design problem can be formulated as below mixed integer programme.

object sets

$\mathscr{E}$	(directed) demands (e.g., IP links)
$\mathscr{V}$	nodes
$\mathscr{O} \subset \mathscr{V}$	colorful ONE nodes
$\mathscr{T} \subset \mathscr{V}$	transponders
$\mathscr{M} \subset \mathscr{V}$	muxponders
$\mathscr{C} \subset \mathscr{V}$	clients (e.g., IP routers)
$\mathscr{F} = \mathscr{H} \cup \mathscr{L}$	(directed) edges (WDM links)
$\mathscr{G} \subset \mathscr{F}$	edges associated with transponder links
$\mathscr{H} \subset \mathscr{F}$	edges associated with long-haul links
$\mathscr{L} \subset \mathscr{F}$	edges associated with intra-office links
$\mathscr{A}_v \subset \mathscr{F}$	edges outgoing from node $v \in \mathscr{V}$
$\mathscr{B}_v \subset \mathscr{F}$	edges incoming to node $v \in \mathscr{V}$
$\mathscr{P}_v \subset \mathscr{F}$	edges associated with add-drop (tributary) ports in colorful ONE nodes $v \in \mathscr{O}$
$\mathscr{Q}$	data transmission technologies

predefined objects

$a(e) \in \mathscr{C}$	originating client node (source) of demand $e \in \mathscr{E}$
$b(e) \in \mathscr{C}$	terminating client node (sink) of demand $e \in \mathscr{E}$
$a(f) \in \mathscr{V}$	originating client node (source) of edge $f \in \mathscr{F}$
$b(f) \in \mathscr{V}$	terminating client node (sink) of edge $f \in \mathscr{F}$
$\alpha(fe) \in \mathscr{L}$	terminating line related to originating link $f \in \mathscr{L}$ with regard to demand $e \in \mathscr{E}$

constants

c_e	volume of demand $e \in \mathscr{E}$
l_f	capacity module of link $f \in \mathscr{F}$
t_v	equal to the maximum number of active tributary links of muxponder, if $v \in \mathscr{M}$; equal to 1, if $v \in \mathscr{O}$
n_f	equal to N, if $f \in \mathscr{G}$; equal to 1, if $f \in \mathscr{F} \backslash \mathscr{G}$

variables

$s_{fe} \in \{0, 1\}$ variable equal to 1 if demand $e \in \mathscr{E}$ is realized on link $f \in \mathscr{F}$, and 0 otherwise

$z_f \in \mathbb{Z}$ variable equal to the number of transport modules on link $f \in \mathscr{F}$

constraints

$$\sum_{f \in \mathscr{A}_v} l_f s_{fe} = c_e \qquad\qquad e \in \mathscr{E}, v = a(e) \in \mathscr{C} \qquad (1a)$$

$$\sum_{f \in \mathscr{B}_v} l_f s_{fe} = c_e \qquad\qquad e \in \mathscr{E}, v = b(e) \in \mathscr{C} \qquad (1b)$$

$$\sum_{f \in \mathscr{A}_v} s_{fe} = \sum_{f \in \mathscr{B}_v} s_{fe} \qquad e \in \mathscr{E}, v \in \mathscr{V} \setminus \{a(e), b(e)\} \qquad (1c)$$

$$\sum_{e \in \mathscr{E}} s_{fe} \le z_f \qquad\qquad f \in \mathscr{L} \qquad (1d)$$

$$\sum_{e \in \mathscr{E}} s_{fe} \le M z_f \qquad\qquad f \in \mathscr{H} \qquad (1e)$$

$$\sum_{f \in \mathscr{A}_v} z_f = \sum_{f \in \mathscr{B}_v} z_f \qquad v \in \mathscr{O} \qquad (1f)$$

$$\sum_{f \in \mathscr{B}_v} z_f \le t_v \qquad\qquad v \in \mathscr{M} \cup \mathscr{O} \qquad (1g)$$

$$\sum_{f \in \mathscr{A}_v} z_f \le 1 \qquad\qquad v \in \mathscr{M} \cup \mathscr{O} \qquad (1h)$$

$$z_f \le n_f \qquad\qquad f \in \mathscr{F}. \qquad (1i)$$

Presented formulation is a modified form of classical formulation of multi-commodity flow optimization problem (see [2, 11]). In this formulation, flow distribution is described by values of binary variables s representing flows on particular network links. Having given feasible values of s one can easily reconstruct particular paths selected to carry traffic.

Integer variables z determine in general number of transmission modules on particular links. However, in case of links associated with nodes $v \in \mathscr{M} \cup \mathscr{O}$ this number is strictly binary (due to constraints (1h) and (1i)). For the rest, variable z is integer (due to constraints (1g) and (1i)).

Due to classical flow conservation constraints (see [11]), in relation to specific demand, in all nodes, except end nodes of this demand, the total volume of incoming flows must be balanced by total volume of outgoing flows. Formulation (1) involves two groups of flow conservation constrains: constraints (1a)–(1c) related to variables s and constraints (1f) related to variables z.

Usage of particular network links, including all types of inter-card patch-cords and back-plane wiring, by flows determines consumption of transport modules (their number is expressed by variables z), according to constraints (1d) and (1e).

Constraints (1g) assure that only one transponder or muxponder can be coupled with each channel tributary port in ONE multiplexer. Similarly, number of active tributary and line links connected to muxponder ports are limited by constraints (1g) and (1h), respectively.

Formulation (1) gathers constraints related to using WDM transport to carry client traffic. Based on this formulation, in the following we consider a number of its extensions and composition of objective function related to the overall cost associated with WDM transport.

2.4 L2 Technology

To express that each demand can be realized using homogenous L2 technology, like GigabitEthernet, FC800, STM64, binary variable vector k was introduced. Non-zero value of variable k_{ge} enforces through constraints (2a) that demand $e \in \mathcal{E}$ can be realized using only links compliant with technology $g \in \mathcal{Q}$. If one technology (say $g \in \mathcal{Q}$) is selected (value k_{ge} is 1), links associated with other technologies cannot be used, what is assured by constrains (2b).

$$\sum_{f \in \mathcal{R}_g} s_{fe} \leq |\mathcal{R}_g| k_{ge} \quad g \in \mathcal{Q}, e \in \mathcal{E} \tag{2a}$$

$$\sum_{g \in \mathcal{Q}} k_{ge} \leq 1 \quad e \in \mathcal{E}. \tag{2b}$$

Above, $\mathcal{R}_g \subset \mathcal{F}$ denotes set of edges associated with technology $g \in \mathcal{Q}$.

2.5 Redundancy

To provide uninterrupted services, able to survive failures of optical network elements and fiber connections, client devices need additional bandwidth, allocated along paths not affected by considered failures. Additional bandwidth, required by protection, is associated with certain level of resource redundancy. Redundant resources are either not used in the nominal network state or can be used for transmitting low priority traffic, preempted in case of failure occurrence.

In case of the WDM networks, redundant resources can be provided either at client digital signal level (called client protection) or photonic signal level (called photonic protection). In the former case, client device is responsible for activating redundant resources. Redundant resources cover optical channels allocated along protection path, and transponder/muxponder cards and ports. In the latter case, specialized protection cards are required. Such protection cards split power of the protected optical signal between multiple (usually two) ports connected to different add-drop ports within multiplexer subsystem. In both cases, the nominal and protection paths should be topologically disjoint with regard to failure occurrence, so under any failure at least one of the paths survives.

Let set $\mathcal{F}_i$, $i \in \mathcal{I}$ represents an arbitrary set of links that share risk of failure. Such group is described in the literature as Shared Risk Link Group (SRLG) or in general Shared Risk Resource Group (SRRG) [20]. Each SRLG associated with single link $f \in \mathcal{F}$ failure contains exactly one element. Each SRLG associated with node $v \in \mathcal{O}$ failure contains all adjacent links, i.e., $\mathcal{A}_v \cup \mathcal{B}_v$. SRLG should be constructed case by case in relation to specific needs and requirements of a network operator. Specific composition of SRLG thus remains out of scope of this paper.

In order to determine capacity c_{ei} allocated to demand e and available during failure state i, constraints (3a)–(3b) should be added to the problem formulation (1):

$$0 \leq c_e - c_{ei} \leq Mr_{ei} \qquad e \in \mathcal{E}, i \in \mathcal{I} \qquad (3a)$$

$$0 \leq c_{ei} \leq M(1 - r_{ei}) \qquad e \in \mathcal{E}, i \in \mathcal{I} \qquad (3b)$$

$$0 \leq r_{ei} \leq \sum_{f \in \mathcal{F}_i} s_{fe} \leq Mr_{ei} \qquad e \in \mathcal{E}, i \in \mathcal{I}. \qquad (3c)$$

Above, each variable c_{ei} expresses volume of link flow associated with demand e in failure state i. Value of variable r_{ei} indicates if link e is available throughout failure state i. Consequently, if for some pair (e, i) $r_{ei} = 1$ then associated c_{ei} is equal to c_e, and c_{ei} is zero otherwise. In according to constraints (3c), value of r_{ei} is positive if and only if at least one link f realizing demand e is affected by failure i, i.e., when $\sum_{f \in \mathcal{F}_i} s_{fe} \geq 0$.

Redundancy required by protection mechanisms can be also modeled through multiplication of demand volume to be realized by the transport WDM/OTN network, and additional constraints assuring that only fraction of demand volume is transmitted through specific resources (network element or link). Protection method associated with described resource redundancy requirement is commonly described in the literature as path diversity [7]. To assure introduced requirement constraints (1a)–(1b) must be rewritten as:

$$\sum_{f \in \mathcal{A}_v} l_f s_{fe} = 2c_e \quad e \in \mathcal{E}, v = a(e) \qquad (4a)$$

$$\sum_{f \in \mathcal{B}_v} l_f s_{fe} = 2c_e \quad e \in \mathcal{E}, v = b(e) \qquad (4b)$$

$$\sum_{f \in \mathcal{F}_i} l_f s_{fe} \leq c_e \quad e \in \mathcal{E}, i \in \mathcal{I}. \qquad (4c)$$

Constraints (4c) assure that any link flow do not exceed demand volume. In result, each demand must be realized on at least two disjoint paths.

2.6   Network Cost Model

Adjacent optical devices are connected by optical fibers, attached to their line interfaces. Multiple fibers are further combined into optical cables, connecting network sites and wells. Cables are attached to optical distribution frames, where incoming fibers are interconnected in order to establish end-to-end optical spans between devices. Signaling between WDM devices is realized via proprietary out-of-band channel, outside optical grid.

Optical device is a complex device, equipped with variety of specialized functions required for photonic signal processing. Typically, those specialized functions are realized by expansion cards, fitted into device slots. Cards can be further interconnected by back-plane device wiring or through external patch-cords, fitted manually into their front panel interfaces. In practice, to transport client signals, optical device must be equipped with a minimal set of supervisory cards, responsible for device control and management. Those cards implement proprietary vendor-specific algorithms to enable proper transmission and reception of photonic signals between adjacent devices. In the following, cost related to those cards is treated as a part of device fixed cost, being in turn a part of CAPital EXpenditure (CAPEX) related to network deployment.

In practice, optical devices are sold in the form of racks (for example, one rack per direction), equipped with certain number of slots for shelves. Each shelf in turn may be equipped with a number of slots to host expansion cards. Tributary and linear cards may be combined with embedded or external Network Interface Controllers (NICs). In the latter case, external NICs are fitted into appropriate slots on cards. ONE slots must thus be filled with a set of required expansion cards providing specialized functionality. Expansion cards can be classified into two categories:

- fixed cards: switching matrices, band multiplexers, supervision cards, fans, power suppliers, etc.,
- elastic cards: channel multiplexers, transponders, and muxponders.

Number and type of fixed cards is predefined by system vendor. Cost of fixed cards, together with cost of racks and shelves, is accounted into system installation cost. In other words, cost of the fixed cards is independent of traffic amount handled by ONE. Contrary to fixed cards, number and type of elastic cards can vary according to specific usage of the WDM system. In particular, network operator may choose between different types of transponders providing different modulation types, capacity, and forward error correction methods. In some ONE designs, channel multiplexers can be not expandable, and full range of multiplexers must be installed in the form of fixed cards. However, in the following, we consider elastic channel multiplexer cards as more general case. Band multiplexers are typically fixed cards. Still, due to particular composition of the switching matrices, some ONE designs may require installation of one rack per each direction (representing long-haul connection with neighboring ONE). In that case, predefined set of fixed cards, including band multiplexers, must be installed in each rack.

Cost of WDM transport is mostly related to the number and type of used elastic expansion cards: channel multiplexers, transponders, and muxponders. Cost related to installation of transponder and muxponder cards can be expressed as follows:

$$\sum_{v \in \mathcal{O}} \sum_{f \in \mathcal{P}_v} \frac{1}{2} g_v z_f \qquad (5)$$

Above, unitary cost related to card associated with node $v \in \mathcal{M} \cup \mathcal{T}$ is given by constant g_v.

To calculate cost related to installation of multi-stage multiplexer expansion cards we need to introduce additional variables and constrains. Binary variable m_j associated with multiplexer $j \in \mathcal{J}$ states if card is installed or not. Variable is positive if at least one channel associated with this particular multiplexer is used. This relation is expressed by constrains (6). Set of colorful channel links and cost associated with multiplexer $j \in \mathcal{J}$ are given by $\mathcal{S}_j$ and h_j, respectively.

$$\sum_{f \in \mathcal{S}_j} z_f \leq |\mathcal{S}_j| m_j \quad j \in \mathcal{J}. \tag{6}$$

Finally, with respect to (6) and other constraints defined above, the objective function can be formulated as:

$$\min F(z, m) = \sum_{v \in \mathcal{O}} \sum_{f \in \mathcal{P}_v} \frac{1}{2} g_v z_f + \sum_{j \in \mathcal{J}} h_j m_j \tag{7}$$

Objective function (7) is related to minimization of number of expansion cards.

3   Optical Impairments

Fiber attenuation is the most fundamental impairment that affects optical signal propagation. Attenuation is a fiber property resulting from using various material, structural, and modular impairments. Still, fiber attenuation is an effect with intensity linearly proportional to fiber length and number of optical elements: connectors, splitters, etc. Thus, optical signal power lost due to fiber attenuation can be recovered by using so-called Linearized Optical Fiber Amplifier (LOFA) cards, containing amplification modules. Beyond amplifiers itself, LOFA cards may also contain a module called Variable Optical Attenuator (VOA), responsible for enforced attenuation of optical signal, so volume of power received by photo-diode is contained in strictly defined working window. Attenuation of VOA modules is typically dynamically set in a closed-loop feedback between adjacent ONE's. Unfortunately, amplifiers, as all active optical elements, introduce into transmitted signal some portion of noise. Fiber attenuation is well standardized for particular fiber types.

Further, shape of optical signal can be distorted due to dispersion. Dispersion cause that optical impulse is widened in time. In extreme case, two consecutive impulses can overlap, leading to potential reception errors. Two types of dispersion ca be categorized: chromatic dispersion and polarization mode dispersion.

Chromatic dispersion is related to velocity difference between different wavelengths in particular medium. Due to linear character of chromatic dispersion, impairments introduced by this effect can be eliminated by using fiber spans with reverse dispersion characteristics. Compensating modules, called Dispersion Compensating Unit (DCU), are typically attached after each fixed-length section of long-haul connection.

Polarization mode dispersion is caused by asymmetry of the fiber-optic strand. Polarization mode dispersion has thus completely random character. According to ITU-T recommendation G.652, unit dispersion coefficient of optical fiber G.652B should not cross $17\,ps/(nm \times km)$ for chromatic dispersion and $0.20\,ps/(\sqrt{km})$ for polarization mode dispersion.

Another serious source of impairments is light scattering. Light scattering results from localized non-uniformity in the fiber medium. It can be seen as a deflection of a ray from a straight path. Deviations from the law of reflection due to irregularities on a surface of optical connectors are also usually considered to be a form of scattering. Among light scattering effects, several effects can imply serious impairments in photonic networks: stimulated Brillouin scattering and stimulated Raman scattering. Scattering effects are non-linear, and they tend to manifest themselves when optical signal power is high.

Other serious non-linear impairments in photonic networks are related to: Four-Wave Mixing, Self-Phase Modulation, and Cross-Phase Modulation. These effects result from transmitting multiple wavelengths over single fiber. Non-linear effects increase level of noise in the optical signal.

Optical impairments affecting optical signals, transmitted through photonic network, can be modeled in the form of three limitations:

- fiber length limit,
- hop-count limit,
- noise accumulation limit.

Chromatic dispersion is responsible for spreading duration of optical signal peaks. As a linear effect, proportional to the total length of fiber, chromatic dispersion can be eliminated by using DCM modules. DCM modules are supposed to introduced chromatic dispersion in reverse direction than dispersion introduced by regular fiber. DCM modules compensate thus chromatic dispersion related to central frequency of the optical signal. Consequently, dispersion affecting frequencies far from the central frequency are not compensated completely. Amount of uncompensated chromatic dispersion is called residual dispersion. Residual dispersion is an important factor that limits the total length of fiber traversed by optical signal. Maximum admissible fiber length depends on transponder type and characteristics.

Optical signal propagating through fiber medium is attenuated. Fiber attenuation is proportional to the total length of fiber spans crossed by the signal. To restore signal power to the level required by photo-detector, optical signal is amplified by LOFA cards localized in selected points along the path. However, LOFA cards, beside signal amplification, introduce some portion of noise. To keep signal quality at high level, network operator should control the total amount of introduced noise. On one hand, low noise level can be assured by hop-count constraints. On the other hand, noise characteristic of active optical elements, as mentioned LOFA cards, can be expressed in terms of Optical Noise-to-Signal Ratio (ONSR). Total value of inverted OSNR is proportional to inverted partial OSNR of particular elements in the path.

Some of the considered limitations can be associated with additive metrics: length, hop-count, and inverse OSNR. Accordingly, all can be modeled similarly by set of so-called shortest path constraints. Formulation of shortest path constraints is based on path length variables $p = (p_v : v \in \mathcal{O})$. Each p_v represents the length of the shortest path from v with respect to weight system q. Then, for each link f outgoing from node v $(a(f) = v)$ contained in the shortest path crossed by edge the following shortest path condition must hold.

$$p_{a(f)} + q_f = p_{b(f)} \qquad (8)$$

Condition (8) is commonly used by the shortest path algorithms to validate if the path traversing edge f is shorter than the shortest path found so far. For our purposes we adopt condition (8) to formulate shortest path constraints:

$$p_{a(f)} + q_f - p_{b(f)} = 0 \text{ if value of } z_f \text{ is } 1 \quad f \in \mathcal{E} \qquad (9a)$$
$$p_{a(f)} + q_f - p_{b(f)} \geq m \text{ if value of } z_f \text{ is } 0 \quad f \in \mathcal{E}. \qquad (9b)$$

Conditions (9a)–(9b) state that if and only if edge f is contained in the shortest path to $b(f)$, length of this path must be equal to sum of the length of a shortest path to $a(f)$ and weight q_f. Otherwise; the value of the expression $p_{a(f)} + q_f - p_{b(f)}$ must be greater or equal to m, which is the smallest difference between lengths of two paths. Accordingly, length limitation constraints can be formulated as follows:

$$m(1 - z_f) \leq p_{a(f)} + q_f - p_{b(f)} \leq M z_f \quad f \in \mathcal{E} \qquad (10a)$$
$$p_v \leq p^* \qquad v \in \mathcal{V}. \qquad (10b)$$

Considered limitations require additional constraints (10b) to enforce that values of required parameters remain under maximum admissible level p^*. Weight system q is constant, and is supposed to express value of required parameter:

- link length,
- number of active elements associated with link (usually one),
- inverse OSNR associated with link.

4 Flow Evaluation

In order to assess how well or poor given flow design fits into the needs of a network operator, in this chapter we propose a flow design evaluation method. Proposed optimization model refers to a single criteria objective function minimizing network cost. In order to allow network operator to make business decision in relation to other factors than cost, in the following we propose a set of auxiliary indicators determining quality and performance of obtained flow design.

Flow in the WDM network represents photonic signal encoded at specific frequency within standardized WDM photonic spectrum, transmitted between Optical Network Elements composing WDM network. Such flow is used to represent a digital signal encoded and decoded at the edge of WDM transport network.

Signal attenuation and deformation lead to transmission errors in carried digital signal. Such errors usually imply retransmission at higher network layers what decreases efficiency of the WDM transport network. For this reason it may appear that selection of WDM flow design based on cost only is not enough in some cases, and further evaluation is needed. In Sect. 3 it is shown how some of the physical impairments can be taken into account in optimization model. However, not all of these effects, especially those of non-linear character, can be modeled in mixed-integer programs.

Let us consider end-to-end composition of a WDM path depicted in Fig. 5. Depicted WDM path begins and ends in active photonic components–transponders. Transponder is responsible for modulating and demodulating digital signal, provided through black&white tributary interface, as a photonic signal. Photonic signal, modulated at specific frequency of photonic spectrum, is further multiplexed at the path begin (and demultiplexed at the path end) by passive component called multiplexer (and demultiplexer respectively). Optical multiplexer outputs an aggregate photonic signal being a composition of particular tributary signals encoded at frequencies associated with particular WDM channels. Further, aggregate photonic signal traverses a series of active components responsible for adjusting power and shape of the aggregate photonic signal through attenuating, amplifying, and optionally compensating some physical effects (like dispersion).

On the one hand, power of photonic signal transmitted through WDM path is attenuated by each passive component in the path: multiplexer, a series of photonic switches, intra-office patchcord connectors, and finally long-haul fibers. On the other hand, photonic signal is amplified by active amplifiers in order to compensate accumulated power loss.

Fig. 5 WDM path structure

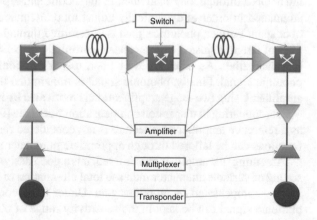

Fig. 6 Generic amplifier structure and example power characteristics of the photonic signal

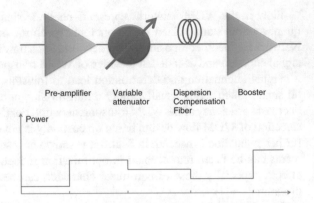

Amplifiers through Amplified Spontaneous Emission (ASE) produce photonic noise degrading quality of the photonic signal. Spectral density of ASE noise generated by optical amplifier is given by formula (see [1, 15]):

$$S_{ASE} = F * (G - 1) * E_{phot},$$ (11)

where F and G represent amplifier noise figure and gain, respectively, and E_{phot} represents photon energy defined by $E_{phot} = h * v = \frac{h*c}{\lambda}$ ($h = 6.626068 * 10^{-34} \frac{m^2 * kg}{s}$ is Plank's constant and $c = 299792458 \frac{m}{s}$ is light speed, v is optical frequency, and λ is optical wavelength).

Amplifier cards are commonly associated with other component used to compensate chromatic dispersion and other physical impairments. Thus, in the following we consider generic composition of an amplifier, depicted in Fig. 6. Lower part of the figure presents power levels of the photonic signal after crossing particular stages.

In our studies we consider generic amplifier composed of four stages (see [13]). First stage, described as pre-amplifier, amplifies power of the photonic signal incoming to ONE through long-haul fiber. In the second stage, photonic signal is artificially attenuated in order to dynamically adjust total attenuation related to the long-haul fiber span. Further, photonic signal is transmitted through DCU—inverse dispersion fiber of length adjusted to compensate chromatic dispersion accumulated along the long-haul fiber. As other types of fiber, dispersion compensation fiber attenuates photonic signal. Finally, photonic signal is transmitted through booster—last stage amplifier. Using two-stage amplification is motivated by mitigation of the non-linear effects appearing in dispersion compensation fiber (see [9]). Using generic amplifier as a reference amplifier architecture is not considered restrictive in sense that stage functions can be tailored through appropriate parameter setting.

Assuming that amplifier gain is fixed, it is a good design practice to through proper setting of variable attenuator increase total attenuation of each fiber span at the same value—compensated by amplifier gain. Under these conditions power level of the photonic signal can be kept in the sensitivity range of photonic receiver, even after crossing a number of fiber spans.

In our evaluation approach we consider simplified model of power management, based on the following assumptions:

1. each LOFA generates fixed amount of noise, noise figure of LOFA $OSNR_{LOFA}$ is set by system user (e.g. 5 dB),
2. LOFA gain G_{LOFA}, determined by difference of power levels between LOFA input and output, is either fixed or can be deterministically calculated in according to specific LOFA model (e.g. 20 dB),
3. each passive component attenuates power of the photonic signal by fixed factor $A_{mux|demux}$ with flat characteristic (e.g. 5 dB),
4. transponder output power P_{trans} is fixed (e.g. 5 dB),
5. fiber attenuation depends linearly with fiber length (attenuation coefficient denoted by A_{fiber}) (e.g. $0.2 \frac{dB}{km}$),
6. fiber chromatic dispersion $D_{chromatic}$ depends linearly with fiber length (e.g. $17 \frac{ps}{nm\,km}$),

Based on the above assumptions, we can construct simple emulator of power characteristic of WDM path, describing power levels at specific points of the path depicted in Fig. 5.

4.1 Optical Signal-to-Noise Ratio

Each fiber span connecting two adjacent ONEs is associated with two of amplifiers (one at each end of the span). Fiber power loss accumulated along the fiber span needs to be compensated by power amplification. However, as an active photonic component, amplifier generates photonic noise which affects quality of the proper photonic signal. Thus, noise accumulated along WDM paths traversing certain number of amplifiers can degraded signal power in degree which makes in practice reception of the encoded bit stream erroneous or impossible. Thus, after [9] we consider noise accumulation as basic quality factor in WDM transmission.

In general, signal quality is characterized by Optical Signal-to-Noise Ratio (OSNR), defined as:

$$OSNR = \frac{P}{N} \tag{12}$$

where P represents power of photonic signal and N represents power of unwanted noise.

Let $OSNR_j$ be OSNR associated with amplifier $j \in \mathscr{J}$. It can be calculated as:

$$OSNR_j = \frac{P_{out}}{P_{ASE}} = \frac{P_{out}}{S_{ASE} * B_o} = \frac{P_{out}}{F * G * E_{phot} * B_o} = \frac{P_{in}}{F * (G - 1) * E_{phot} * B_o} \tag{13}$$

where P_{in} and P_{out} represent input and output power of the amplifier, and B_o represents optical bandwidth of the amplifier.

Finally, fractional $OSNR_j$, $j \in \mathcal{J}$, characteristic for all components included in a WDM path depicted in Fig. 5, can be used to determine total $OSNR_p$ at the end of the path, according to formula:

$$\frac{1}{OSNR_p} = \sum_{j \in \mathcal{J}_p} \frac{1}{OSNR_j} \tag{14}$$

Two approaches are considered for calculating $OSNR_p$ (see [9]). First approach is based on assumption that OSNR of all N amplifiers along the path are the same (denoted $OSNR_{amp}$) and formula (14) can be rewritten as:

$$OSNR_p = \frac{OSNR_{amp}}{N} = \frac{P_{out}}{N * F * G * E_{phot} * B_o} \tag{15}$$

Further, assuming that loss L of each span is the same and is completely compensated by amplifier gain, we get $G = L$. Taking logarithm of Eq. (15) and putting values of constants we get approximate formula:

$$OSNR_p^{[dB]} = 58 + P_{out}^{[dBm]} - 10 \lg N - F^{[dB]} - L^{[dB]} \tag{16}$$

To enable power tuning in practical designs output power of transponder can be usually adjusted in certain range. Each product is thus characterized by applicable power settings. For the evaluation purposes we assume that certain values can be used in Eq. (16): min, max, avg.

Q-factor is a synthetic parameter which value is determined as a function of OSNR at the input of photodetector, according to formula:

$$Q = 20 \lg \sqrt{OSNR} \sqrt{\frac{B_o}{B_c}} \tag{17}$$

where B_o represents optical bandwidth of photodetector, and B_c represents electrical bandwidth of the receiver filter. After [9], for practical designs we consider estimation:

$$Q^{[dBm]} = OSNR^{[dBm]} - 2dB \tag{18}$$

Q-factor is related to *Bit Error Rate* (BER) through the following relation:

$$BER = \frac{1}{2} erfc(\frac{Q}{\sqrt{2}}) \tag{19}$$

BER is a valuable indicator which determines how often errors in transmitted signal occur. Large number of errors (and associated high value of BER) leads to numerous retransmissions further reducing efficient bandwidth of the WDM network. Thus, optimized WDM flows should be characterized by possibly smallest value of BER.

4.2 Residual Dispersion

Chromatic dispersion is an effect associated with velocity difference between particular wavelengths propagating in transmission medium. As a linear effect, chromatic dispersion can be compensated by attaching to regular transmission medium precisely determined section of fiber characterized by reverse chromatic dispersion. Nevertheless, length of compensating fiber DCU is usually adjusted to compensate chromatic dispersion at central frequency of the WDM channel. Therefore, chromatic dispersion associated with wavelengths at extreme frequencies in this channel are not fully compensated. Remaining portion of chromatic dispersion, called residual dispersion, is accumulating along the WDM path. Residual dispersion results in broadening encoded symbols in time, and leads to transmission errors. Described effect is illustrated by Fig. 7.

4.3 Path Length

As shown above, length of photonic path (expressed in terms of intermediate optical network elements) determines the quality of carried photonic signal. However, it also determines how much networking resources is consumed to realize specific transport service. For these two reasons photonic paths in general should be as short as possible. Using possibly shortest paths, it will be possible to maximize service

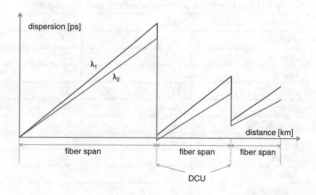

Fig. 7 Compensation of chromatic dispersion

quality and optimize resource usage. However, flow design optimization driven by cost minimization may produce relatively long paths (meaning length longer than length of the shortest possible path). Therefore, in this section we consider evaluation of the length of photonic path.

Having given photonic path $\bar{p}$ originated in node o_p and terminated in node d_p and network graph $G(\mathcal{V}, \mathcal{F})$ we denote the shortest path between o_p and d_p by $\widehat{p}$. Let $L(p)$ determine the length of path p with respect to uniform edge weights (normalized value of $L(p)$ determines the number of path hops). We define *relative path length* of path $\bar{p}$ as a relation of $L(\bar{p})$ to $L(\widehat{p})$:

$$RL_p = \frac{L(\bar{p})}{L(\widehat{p})} \tag{20}$$

Having given feasible solution of the flow design problem we define another set of performance indicators:

- value of relative path length in relation to the longest photonic path,
- mean value of relative path lengths in relation to all photonic paths,
- variance of relative path lengths in relation to all photonic paths.

5 Complexity

Throughout this section we estimate complexity of the considered formulation of the WDM flow design problem. Complexity estimation is based on calculation of the numbers of variables and constrains necessary to formulate the considered WDM flow design problem in relation to the selected instances of network instances defined in the SNDLib library [18]. Referenced network instances are characterized in Table 1, where particular columns contain the numbers of network nodes, network links, and traffic demands, respectively.

Table 1 Characteristics of the selected network instances

Network instance	Nodes	Links	Demands
abilene	12	15	132
atlanta	15	22	210
brain	161	332	14311
cost266	37	57	1332
geant	22	36	462
germany50	50	88	662
giul39	39	172	1471
france	25	45	300
janos-us	26	84	650
janos-us-ca	39	122	1482

Table 2 Characteristics of the network graphs

Network instance	$|\mathcal{O}|$	$|\mathcal{C}|$	$|\mathcal{L}|$	$|\mathcal{H}|$	$|\mathcal{F}|$
abilene	2400	12	1200	36	1236
atlanta	3520	15	1760	45	1805
brain	53120	161	26560	483	27043
cost266	9120	37	4560	111	4671
geant	5760	22	2880	66	2946
germany50	14080	50	7040	150	7190
giul39	27520	39	13760	117	13877
france	7200	25	3600	75	3675
janos-us	13440	26	6720	78	6798
janos-us-ca	19520	39	9760	117	9877

Table 3 Characteristics of the formulations

Network instance	Constraints	Variables
abilene	19704	164388
atlanta	38335	380855
brain	26151647	387039416
cost266	484005	6226443
geant	114852	1363998
germany50	372608	4766970
giul39	1263603	20426944
france	96825	1106175
janos-us	307484	4425498
janos-us-ca	956135	14647591

Further, assuming 80-channel WDM technology and three types of transponders (10, 40, 100 Gbps) we calculate the numbers of particular types of graph elements related to the considered network instances. Calculation results are presented in Table 2.

Finally, Table 3 contains the numbers of constrains and variables necessary to formulate the considered WDM flow design problem in relation to the selected SNDLib network instances. Number of constraints is contained in a range from 19 thousands to 26 millions. Number of variables is even larger and is contained in a range from 164 thousands to 387 millions. Such enormous numbers of constraints and variables make in practice the considered formulations numerically intractable for resolving with exact optimization methods.

6 Conclusion

Paper investigates mathematical modeling of photonic networks applying wavelength division multiplexing. Based on standardization efforts, research work, and commercial offerings, a mathematical model of WDM network is proposed. Proposed network model involves a series of specific aspects of photonic transmission, like fiber attenuation, dispersion, and noise accumulation. Paper also defines a series of indicators that can be used to determine quality and performance of flow designs being solutions of the considered optimization problem.

Number of integer variables used in a mathematical model in high degree determines computational complexity of optimization formulations based on this model. In case of the proposed model, this number is proportional to the squared number of WDM devices and number of WDM channels. For even small network instances (composed of several devices), this number can be at level of thousands. Thus, in the future work, in order to reduce complexity of the proposed model to numerically tractable level, authors will try to decompose it. In particular, future work will focus on adopting general decompositions methods proposed in context of large-scale linear programming, like Dantzig–Wolfe decomposition, Lagrangean relaxation, and Benders decomposition, for the case of proposed model.

Acknowledgments This work presents results of realization of ongoing project "Optymalizacja przepływów w wielowarstwowych sieciach dostępu do Internetu w technologii DWDM/IP/MPLS" (ODIN). Project is co-founded by The National Centre for Research and Development, Poland.

References

1. Agrawal, G.: Fiber-Optic Communication Systems, 4td edn. Wiley, New York (2010)
2. Ahuja, R.K., Magnanti, T.L., Orlin, J.B.: Network Flows: Theory, Algorithms, and Applications. Prentice Hall, Upper Saddle River (1993)
3. Aparicio-Pardo, R., Klinkowski, M., Garcia-Manrubia, B., Pavon-Marino, P., Careglio, D.: Offine impairment-aware rwa and regenerator placement in translucent optical networks. J. Light. Technol. **29**, 3. doi:10.1109/JLT.2010.2098393. (2011)
4. Chlamtac, I., Ganz, A., Karmi, G.: Lightpath communications: an approach to high bandwidth optical WAN's. IEEE Trans. Commun. **40**(7), 1171–1182. doi:10.1109/26.153361. (1992)
5. Christodoulopoulos, K., Manousakis, K., Varvarigos, E.: Considering physical layer impairments in offine RWA. IEEE Netw. **23**. doi:10.1109/MNET.2009.4939260. (2009)
6. Dzida, M., Bak, A.: Flow design in photonic data transport network. In: M.P. M. Ganzha L. Maciaszek (ed.) Proceedings of the 2015 Federated Conference on Computer Science and Information Systems. Annals of Computer Science and Information Systems, vol. 5, pp. 471–482. IEEE. doi:10.15439/2015F148. (2015)
7. Dzida, M., Sliwinski, T., Zagozdzon, M., Ogryczak, W., Pioro, M.: Path generation for a class survivable network design problems. In: NGI 2008 Conference on Next Generation Internet Networks, Cracow, Poland. doi:10.1109/NGI.2008.11. (2008)
8. Evan, S., Itai, A., Shamir, A.: On the complexity of timetable and multicommodity flow problems. SIAM J. Comput. **5**, 691–703. doi:10.1137/0205048. (1976)
9. Gumaste, A., Antony, T.: Dwdm Network Designs and Engineering Solutions, 1st edn. Cisco Press, Indianapolis (2002)

10. Manousakis, K., Christodoulopoulos, K., Kamitsas, E., Tomkos, I., Varvarigos, E.: Offine impairment-aware routing and wavelength assignment algorithms in translucent WDM optical networks. J. Light. Technol. **27**, 12. doi:10.1109/JLT.2009.2021534. (2009)
11. Minoux, M.: Mathematical Programming: Theory and Algorithms. Wiley, New York (1986)
12. Pavon-Marino, P., Azodolmolky, S., Aparicio-Pardo, R., Garcia-Manrubia, B., Pointurier, Y., Angelou, M., Sole-Pareta, J., Garcia-Haro, J., Tomkos, I.: Offine impairment aware RWA algorithms for cross-layer planning of optical networks. J. Light. Technol. **27**, 12. doi:10.1109/JLT.2009.2018291. (2009)
13. Poti, L.: Deliverable OCF-DS1.1. Technical report, GÉANT Open Call: Coherent Optical system Field trial For spectral Efficiency Enhancement (COFFEE) (2015)
14. Saradhi, C., Subramaniam, S.: Physical layer impairment aware routing (PLIAR) in WDM optical networks: issues and challenges. IEEE Commun. Surv. Tutor. **11**, 4. doi:10.1109/SURV.2009.090407. (2009)
15. Sckinger, E.: Broadband Circuits for Optical Fiber Communication. Wiley, New York(2005)
16. Sengezer, N., Karasan, E.: Static lightpath establishment in multilayer traffc engineering under physical layer impairments. IEEE/OSA J. Opt. Commun. Netw. **2**, 9. doi:10.1364/JOCN.2.000662. (2010)
17. Sengezer, N., Karasan, E.: Multi-layer virtual topology design in optical networks under physical layer impairments and multi-hour traffc demand. EEE/OSA J. Opt. Commun. Netw. **4**, 2. doi:10.1364/JOCN.4.000078. (2012)
18. SNDlib 1.0—Survivable network design data library. http://sndlib.zib.de (2005)
19. Sole, J., Subramaniam, S., Careglio, D., Spadaro, S.: Cross-layer approaches for planning and operating impairment-aware optical networks. In: Proceedings of the IEEE, vol. 100. doi:10.1109/JPROC.2012.2185669. (2012)
20. Strand, J., Chiu, A., Tkach, R.: Issues for routing in the optical layer. IEEE Commun. Mag. doi:10.1109/35.900635. (2001)
21. Varvarigos, E., Manousakis, K., Christodoulopoulos, K.: Cross layer optimization of static lightpath demands in transparent WDM optical networks. In: IEEE Information Theory Workshop on Networking and Information Theory. doi:10.1109/ITWNIT.2009.5158553. (2009)
22. Varvarigos, E., Manousakis, K., Christodoulopoulos, K.: Offline routing and wavelength assignment in transparent WDM networks. IEEE/ACM Trans. Netw. **18**, 5. doi:10.1109/TNET.2010.2044585. (2010)
23. Zang, H., Jue, J., Mukherjee, B.: A review of routing and wavelength assignment approaches for wavelength routed optical WDM networks. Opt. Netw. Mag. **1**(1), 47–60 (2000)
24. Zhai, Y., Askarian, A., Subramaniam, S., Pointurier, Y., Brandt-pearce, M.: Cross-layer approach to survivable DWDM network design. IEEE/OSA J. Opt. Commun. Netw. **2**, 6. doi:10.1364/JOCN.2.000319. (2010)
25. Zhang, W., Tang, J., Nygard, K., Wang, C.: REPARE: Regenerator placement and routing establishment in translucent networks. In: IEEE Global Telecommunications Conference GLOBECOM. doi:10.1109/GLOCOM.2009.5425649. (2009)

Introducing the Environment in Ant Colony Optimization

Antonio Mucherino, Stefka Fidanova and Maria Ganzha

Abstract Meta-heuristics are general-purpose methods for global optimization, which take generally inspiration from natural behaviors and phenomena. Among the others, Ant Colony Optimization (ACO) received particular interest in the last years. In this work, we introduce the environment in ACO, for the meta-heuristic to perform a more realistic simulation of the ants' behavior. Computational experiments on instances of the GPS Surveying Problem (GSP) show that the introduction of the environment in ACO allows us to improve the quality of obtained solutions.

1 Introduction

Meta-heuristics are general-purpose methods for global optimization. They are usually based on the simulation of animal behaviors and physical phenomena, that are believed to work optimally in nature. They are conceived for providing approximations of problem solutions when no deterministic algorithms, having a less than exponential complexity, can be developed. The meta-heuristics generally consist in a list of repetitive actions to be performed for finding an approximation of the global optimum of a given optimization problem.

In recent years, several meta-heuristic approaches have been proposed in the scientific literature. A classical example is the Simulating Annealing (SA), which was proposed in the 80s, but it is still considered for solving some particular problems. Other well-known examples of meta-heuristics are the Genetic Algorithms (GAs),

A. Mucherino
IRISA, University of Rennes 1, Rennes, France
e-mail: antonio.mucherino@irisa.fr

S. Fidanova (✉)
BAS, University of Sofia, Sofia, Bulgaria
e-mail: stefka@parallel.bas.bg

M. Ganzha
SRI, Polish Academy of Science, Warsaw, Poland
e-mail: maria.ganzha@ibspan.waw.pl

© Springer International Publishing Switzerland 2016
S. Fidanova (ed.), *Recent Advances in Computational Optimization*,
Studies in Computational Intelligence 655, DOI 10.1007/978-3-319-40132-4_9

Differential Evolution (DE), the Tabu Search (TS), the Variable Neighborhood Search (VNS), the Monkey Search (MS), and Ant Colony Optimization (ACO). For a quick survey on meta-heuristics and some references, the reader can refer to [16, 26].

Every meta-heuristic is developed in order to find the best trade-off between two main concepts: *diversification* and *intensification* [23]. While the former tries to widely extend the search in the domain of the optimization problem, the latter improves candidate solutions by focusing on local neighbors of the current best known solutions. We propose to include a third concept in meta-heuristics: the environment. The main idea comes from the observation for which the behaviors and the phenomena that are simulated in meta-heuristics have generally to deal with a variable environment, that should therefore be taken into consideration. As an immediate example, we can mention to the Darwinian theory that is at the basis of GAs: it is evident how various environments were able to influence the human genome (e.g. the color of the eyes and of the skin in humans is strongly conditioned by the exposition to the sun). Similar remarks apply for other natural and animal behaviors that gave inspiration to other meta-heuristics.

In this work, we introduce the concept of environment in Ant Colony Optimization (ACO, see Sect. 2). ACO takes inspiration from the behavior of a colony of ants foraging for food. In ACO, diversification is guaranteed by the simulation of the typical ant behavior, while intensification is performed by applying a local search to a set of candidate solutions. In our *environmental* ACO (*e*ACO), pheromone updates are not supposed to be performed only on the basis of found solutions, but also on the basis of the current environment "surrounding" the ants. As an example, particular real-life environment conditions, such as strong wind and rain, may alter the perception of the deposited pheromone. We will simulate environment changes in our *e*ACO by employing the Logistic map [25]. We warn the reader that the Logistic map has already been employed in optimization for performing chaotic searches [4, 8, 27]. However, its use is different from the one considered in this work for the simulation of environment changes. This work extends a previous publication on conference proceedings [15].

The paper is organized as follows. In Sect. 2, we will briefly describe the basic idea behind ACO, and we will provide a sketch of the meta-heuristic. In Sect. 3, we will focus our attention on the Logistic map, and we will explain how to exploit its features for performing ACO with variable environments. In Sect. 4, we will describe the problem that we will consider in our computational experiments: the GPS Surveying Problem (GSP). The experiments will be reported in Sect. 5, and conclusions will be given in Sect. 6.

2 Ant Colony Optimization

Ants foraging for food deposit a substance named pheromone on the paths they follow. An isolated ant would just move randomly. Ants encountering previously laid pheromone marks are however stimulated to follow the same paths. This way, the pheromone trails are reinforced around the optimal ones, so that the probability for the other ants to follow optimal paths increases with time. The repetition of this mechanism represents the auto-catalytic behavior of ant colonies in nature [1, 7].

Ant Colony Optimization (ACO) is inspired by this ant behavior. A colony of artificial ants working into a mathematical space is simulated. These ants search for candidate solutions of a given optimization problem, while possible paths are marked by artificial pheromone for guiding other ants in the regions of the search space where good-quality solutions were already found. In ACO, therefore, the artificial ants generally create a sort of environment by themselves, by depositing the pheromone on marked paths: we will perturb this environment by means of the Logistic map (see Sect. 3).

The ants' search space is represented by the so-called *construction graph*, which is a weighted undirected graph $G_C = (S_C, E_C, \eta)$ where vertices in S_C are solution components and edges in E_C indicate the possibility to combine the two connected components for obtaining a partial solution. The weight $\eta : (u, v) \in E_C \longrightarrow \mathbb{R}$ associated to each edge (u, v) is named heuristic information, which is generally defined on the basis of the problem at hand. A path on G_C allows to combine several partial solutions and to construct one complete solution.

Algorithm 1 is a sketch of the ACO-based meta-heuristic. The transition probability p_{uv}, necessary in ACO when the ants need to decide on which edge (u, v) to walk, is based on the heuristic information η_{uv} and on the current pheromone level τ_{uv}. Suppose that an ant has already constructed a partial solution X, and that it is standing right now on the vertex $u \in S_C$. Any edge belonging to the star of u can be the next candidate for pursuing the search. The probability to choose one edge or another in the star of u is given by

---

Algorithm 1 Ant Colony Optimization

1: **ACO** (*in*: N, G_C, α, β; *out*: X_{best})
2: **let** $X_{best} = \emptyset$;
3: **while** (stopping criteria not satisfied) **do**
4:   **for** ($k = 1, N$) **do**
5:     **place** k^{th} ant on a random vertex $u \in S_C$;
6:     **let** $X = \{u\}$;
7:     **while** (X is incomplete) **do**
8:       **select** the vertex v in the star of u having higher probability p_{uv} (*see Eq.* (1));
9:       **let** $X = X \cup \{v\}$;
10:      **let** $u = v$;
11:    **end while**
12:    **update** pheromone (*see Eq.* (2));
13:    **apply** local search starting from X (*optional*);
14:    **if** (X is better than X_{best}) **then**
15:      **let** $X_{best} = X$;
16:    **end if**
17: **end for**
18: **end while**

$$p_{uv} = \tau_{uv}^{\alpha} \eta_{uv}^{\beta} \left[\sum_{(u,w) \in E_S : w \not\subset X} \left(\tau_{uw}^{\alpha} \eta_{uw}^{\beta} \right)^{-1} \right], \tag{1}$$

where α and β are transition probability parameters. When the algorithm starts, small positive values are given to every τ_{uv}, which represent the current pheromone values on the edges $(u, v) \in E_C$. Then, every time a new solution is identified, the edges considered by the ants are marked with a new level of pheromone. In ACO, one possible updating rule (for a minimization problem) is the following:

$$\tau_{uv} = \tau_{uv} + \frac{1}{f(X)}, \tag{2}$$

where X is the current solution, and f is the objective function of the considered problem.

In ACO, the general ant behavior allows to perform a wide search on the search domain (diversification), while the local search (see line 13 in Algorithm 1) from constructed solutions X allows to focus on promising neighbors (intensification). Our implementation of ACO makes use of MaxMin Ant System (MMAS). The reader is referred to [22] for a wider explanation of the ACO implementation considered in this work.

3 ACO with Variable Environment

The Logistic map is a quadratic dynamical equation proposed in 1938 as a demographic model [25]. It is a rather simple quadratic polynomial

$$x_{n+1} = r x_n (1 - x_n), \quad n > 0, \tag{3}$$

where x_n represents the population size at time n and r is a constant, named growth coefficient. Given $x_0 \in [0, 1]$ and a value for $r \in [0, 4]$, we can define a sequence of values $\{x_n\}_{n \in \mathbb{N}}$. This sequence can either converge to an attraction domain, or be chaotic. There are actually three different possible situations: (i) the Logistic map converges to a singleton: from a given $k \in \mathbb{N}$, all sequence values $\{x_n\}_{n > k}$ correspond to a constant value; (ii) the sequence $\{x_n\}_{n \in \mathbb{N}}$ is periodic (from a certain k), and the attraction domain is given by the set of values that are periodically repeated in the sequence; (iii) the Logistic map is chaotic: $\{x_n\}_{n \in \mathbb{N}}$ is a sequence of values having no apparent order.

Figure 1 shows the behavior of the Logistic map for different values of r in the range $[2, 4]$ (its behavior is linear in the range $[0, 2]$). On the x-axis, we consider a discrete subset of 3000 equidistant values for r between 2 and 4; on the y-axis, for every considered value for r, we report the set of points obtained as follows: we take 1500 equidistant points in the interval $[0, 1]$, we apply Eq. (3) 1000 times for

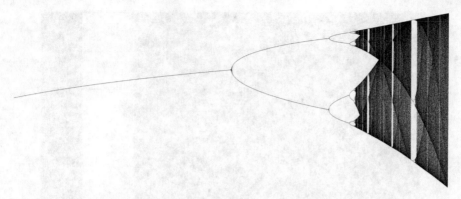

Fig. 1 The behavior of the Logistic map for different values of r (on the x-axis, for $r = 2$ to 4 in the figure). It can either be regular (it converges to an attraction domain) or be chaotic

each of them, and we report the element x_{1000} of the sequence. For small values of r, the Logistic map always converges to one single point, i.e. the attraction domain consists of one point only. The first bifurcation appears when $r = 3$, where the attraction domain consists of 2 points x' and x''. In this case, part of the original 1500 points in [0, 1] generated a sequence $\{x_n\}_{n \in \mathbb{N}}$ for which x_{1000} is x'; all other initial points defined instead a sequence such that x_{1000} is x''. All these sequences are in fact periodic, and with period 2. Another bifurcation in the Logistic map appears when $r = 1 + \sqrt{6}$, where the attraction domain consists of 4 points.

For larger values for r, the Logistic map experiences other bifurcations, and it can be chaotic for some subintervals of r. However, in these chaotic regions, it is still possible to identify regular attraction domains. For example, in Fig. 2, the same graphic reported in Fig. 1 is zoomed in the region $r = [3.8852, 3.8863]$, where this phenomenon is clearly shown. Regular regions, that can be glimpsed in Fig. 1, still contain bifurcations. Moreover, we can notice that the whole graphic in Fig. 1 reappears in our zoomed region. Other regular attraction domains can be identified by looking at tighter subintervals of r, as well as at other copies of the entire graphic.

The graphic in Fig. 1 is a fractal, because of its self-similarity [9, 19]. Fractals are beautiful geometrical objects that have found various applications, such as DNA analysis [17] and music [3], to name a few. For a given set of points normalized in [0, 1], the Logistic map provides us with a mapping of such points into a bounded subinterval of [0, 1], which can be either regular or chaotic. The modification of such points by the Logistic map represents, in our simulations, the influence of a given environment (a regular or a chaotic one) on the original points. The co-domain of the Logistic map is bounded in [0, 1] and, more precisely, for a given value of $r \in [0, 4]$, we can compute the bounds for the values of x as follows. One side of the graphic of the Logistic map is bounded by the quadratic equation:

$$x = \frac{r^2}{4} \left(1 - \frac{r}{4} \right),$$

Fig. 2 The behavior of the Logistic map for values of r in the interval [3.8852, 3.8863]. This region of the Logistic map is mostly chaotic, but regular attraction domains (with the typical bifurcations) can still be identified. The entire graphic in Fig. 1 reappears several times in the selected interval

while the other side of the graphic is bounded by the linear equation:

$$x = \frac{r}{4}.$$

As a consequence, Logistic map can have narrower attraction domains for smaller values of r.

In the practice, we simulate regular and chaotic changes of environment in ACO by introducing the Logistic map in Eq. (2), which is used in ACO for updating the pheromone trails. In the hypothesis the objective function of the considered problem is positive and greater than 1, the term $1/f(X)$ in Eq. (2) has always values ranging between 0 and 1. It can therefore take the place of x_0 in the Logistic map, so that a perturbed value x_1 can be computed, for a given value of r in $[0, 4]$. The equation for updating the pheromone therefore becomes:

$$\tau_{uv} = \tau_{uv} + r \cdot \frac{1}{f(X)} \cdot \left(1 - \frac{1}{f(X)}\right). \tag{4}$$

With this simple change in the rule for updating the pheromone, we artificially perturb the environment of the ants, which would otherwise only depend on the function values. Different values for r can produce different environment changes, depending on the Logistic map.

We refer to ACO with environment changes as *environmental* ACO (*e*ACO). In this work, we present some computational experiments (see Sect. 5) where *e*ACO is employed for solving the GSP (see next section).

4   GPS Surveying

The Global Positioning System (GPS) consists of a certain number of satellites that constantly orbit around earth and that are provided with sensors able to communicate with machines located on earth [13]. Given a ground sensor network, the GPS technology can provide very accurate locations for all sensors forming the network. The related costs can however be too high when it is necessary to deal with large networks. For this reason, over the last years, researchers have been working for designing and installing local ground networks having the task of recording satellite signals with the aim of decreasing the overall network functioning cost [6, 14, 20].

A ground network is composed by a certain number of *receivers* working at different *stations* in different time slots. A *session* represents the temporarily assignment of a given number of receivers to a set of distinct stations. The main problem is to find a suitable *order* for such sessions for reducing the overall cost: the expenses are in fact caused by the need of moving receivers from one station to another, when stepping from one session to the next one. The session order is also named *session schedule*. The distance between two involved stations is an important factor for the computation of the expenses. However, there are additional costs that we need to consider: if the number of working days necessary to perform the operation is more than one, for example, then the need of planning an over-night stop at a company office can make the cost of the operation increase. For this reason, the problem of finding an *optimal session schedule* can be seen as the Multiple Asymmetric Traveling Salesman Problem (MATSP) [2], which is an NP-hard problem [18]. In order to alleviate the impact of measurement errors in the data, at least two receivers per session are generally considered [24].

Let $S = \{s_1, s_2, \ldots, s_n\}$ be a set of stations, and let $R = \{r_1, r_2, \ldots, r_m\}$ be a set of receivers, with $m < n$. Sessions can be defined by a function $\sigma : R \longrightarrow S$ that associates one receiver to one station. Considering that no more than one receiver should be assigned to the same station, σ can be represented by an m-vector $(\varsigma_1, \varsigma_2, \ldots, \varsigma_m)$ containing, for each of the m receivers, the labels of the chosen stations. Let C be an $n \times n$ matrix providing the costs $c(\varsigma_u, \varsigma_v)$ for moving one receiver from the station ς_u to the station ς_v. This matrix can be symmetric when moving between ς_u and ς_v is independent from the orientation of the movement; the non-symmetric case is however more realistic.

We introduce a weighted directed graph $G = (V, E, c)$, where vertices represent sessions σ_v and arcs (σ_u, σ_v) indicate the possibility to switch from session σ_u to session σ_v. The upper bound on the cardinality of V is $n!/(n - m)!$, which corresponds to the maximum number of possible sessions. The weight associated to the arcs provides the cost $c(\sigma_u, \sigma_v)$ for moving every receiver from the station $\varsigma_{u,i}$ to the station $\varsigma_{v,i}$, for each i:

$$c(\sigma_u, \sigma_v) = \sum_{i=1}^{m} c(\varsigma_{u,i}, \varsigma_{v,i}).$$

Notice that the weight associated to (σ_u, σ_v) can be different from the weight on (σ_v, σ_u). The GPS Surveying Problem (GSP) consists in finding an optimal path on graph G, i.e. a path for which the selected set of arcs gives the minimal total cost, while covering the entire vertex set V [6].

First attempts for solving the GSP were based on the idea of transforming GSP instances into instances of the class of TSP-like problems, and to employ existing methods and algorithms. In [5], a branch-and-bound approach was employed, which is actually able to find the optimal solutions for small GSP instances. As the size of the networks increases, the complexity grows and the time necessary for a branch-and-bound to converge becomes prohibitive. On the other side, in real-life applications, there is generally the need to obtain GSP solutions as fast as possible, even if only approximated ones. Therefore, heuristic approaches particularly developed for this application have been proposed over the last years for identifying optimal or near-optimal session orders [10, 11, 20]. In this work, we consider a set of instances of the GSP for testing our ACO approach with environment changes described in Sect. 3.

5 Computational Experiments

This section will present some computational experiments where the standard ACO and our eACO are employed for solving instances of the GSP. ACO's construction graph G_C (see Sect. 2) corresponds to a weighted directed graph G representing an instance of the GSP. In this graph, vertices are sessions σ_u and edges indicate the possibility to switch from one session to another (see Sect. 4). As test cases, we consider data from two real networks: Malta [20], composed by 38 sessions, and Seychelles [21], composed by 71 sessions. We also consider larger instances designed for testing the Asymmetric Traveling Salesman Problem (ATSP), which are freely available on the Internet.[1]

Our eACO implementation is based on Algorithm 1, where the transition probability p_{uv} is computed by Eq. (1), and the pheromone update is performed by applying Eq. (4) (see Sect. 3). The heuristic information η_{uv} is given by the formula:

$$\eta_{uv} = \frac{1}{c(\sigma_u, \sigma_v)},$$

where $c(\sigma_u, \sigma_v)$ is the total cost for switching from session σ_u to session σ_v. Finally, the values for the transition probability parameters α and β are fixed, respectively, to 1 and 2 [12].

In the following experiments, we will focus on the quality of the found solutions, rather than on the algorithms' performances. In fact, the increase in complexity in using Eq. (4), rather than Eq. (2), can be neglected when considering the overall algorithms' complexity. We do not compare our results to the best results currently known for the considered instances. Our aim is not in fact to improve those results, but only to give a validation of the simulation of a variable environment in ACO.

We remind the reader that r is the growth coefficient in the Logistic map, which we use in eACO for simulating a variable environment. For values of r in [0, 3], the Logistic map has either a constant attraction domain, or a periodic one with period 2. Therefore, more dynamical environments can be simulated by fixing r in the interval [3, 4], where the Logistic map can also be chaotic.

Table 1 shows some computational experiments with the standard ACO and our eACO, and, in the latter case, for different values of r. For every instance of the GSP, we report its size in terms of sessions. For every experiment, we report the average result over 30 runs (the average function value in the found solutions), and, between parentheses, we give the best found result over the 30 runs.

It is evident from the experiments that eACO is always able to provide better results, for all considered values of r. The only exception is given by the small instance "Malta", where three experiments with eACO gave, in average over the 30 runs, three slightly worse results. It is moreover interesting to remark that all simulated environments (given by different values for r) are able to guarantee an improvement on the performances. As Table 1 shows, all simulated environments were able, at least once, to provide better results, with respect to the standard ACO, as well as with respect to the other simulated environments in eACO.

---

[1]http://www.informatik.uni-heidelberg.de/groups/comopt/software/TSLIB95/ATSP.html.

Table 1 Comparison between ACO and eACO on a set of GSP instances, and for different values of r

		ACO	eACO				
Instances	$|V|$		$r = 3.00$	$r = 3.25$	$r = 3.50$	$r = 3.75$	$r = 4.00$
Malta	38	899.50 (895)	897.00 (895)	897.00 (895)	900.16 (895)	900.16 (895)	900.33 (895)
Seyshels	71	922.50 (865)	887.33 (865)	909.10 (865)	907.63 (865)	906.03 (865)	906.73 (865)
kro124p	100	40910.60 (39096)	40753.00 (39899)	40742.50 (39899)	40697.83 (39798)	40653.66 (39511)	40803.76 (39572)
ftv170	171	3341.93 (3115)	3319.83 (3163)	3345.93 (3163)	3336.16 (3158)	3324.46 (3130)	3338.53 (3180)
rgb323	323	1665.90 (1615)	1649.43 (1612)	1649.36 (1612)	1649.36 (1612)	1649.16 (1615)	1649.55 (1604)
rgb358	358	1692.66 (1648)	1682.80 (1642)	1679.13 (1672)	1626.70 (1627)	1681.53 (1628)	1685.95 (1636)
rgb403	403	3428.56 (3382)	3393.76 (3280)	3392.90 (3303)	3395.43 (3320)	3392.15 (3329)	3386.10 (3290)
rgb443	443	3765.80 (3701)	3742.43 (3660)	3736.16 (3660)	3747.90 (3662)	3756.88 (3672)	3754.50 (3658)

We report the average function values in the found solutions over 30 runs. In parentheses, we give the function value of the best found solution

6 Conclusions

We believe this is the first research contribution where an "environment" is introduced in a meta-heuristic search. This idea appears natural for us when considering that most meta-heuristics are inspired by natural phenomena and behaviors, where a variable environment is actually present to influence them. It is our opinion therefore that it is more appropriate to perform simulations where a variable environment is as well included in the simulation. Our first experiences with ACO and the Logistic map are quite promising. Our eACO is able, in all proposed experiments, to provide better-quality results, when compared to the standard ACO.

Future works will be aimed at extending the idea of simulating a variable environment to other meta-heuristic searches. We will also consider a wider set of test problems to be used for validating this idea. Our conjecture is that, together with intensification and diversification, the environment is another important concept to take into consideration in the development of meta-heuristics.

Acknowledgments This work was partially supported by two grants of the Bulgarian National Scientific Fund: "Efficient Parallel Algorithms for Large Scale Computational Problems" and "Inter-Criteria Analysis. A New Approach to Decision Making".

References

1. Atanassova, V., Fidanova, S., Popchev, I., Chountas, P.: Generalized nets, ACO-algorithms and genetic algorithm. In: Sabelfeld, K.K., Dimov, I. (eds.) Monte Carlo Methods and Applications, pp. 39–46. De Gruyter, Berlin (2012)
2. Bektas, T.: The multiple traveling salesman problem: an overview of formulations and solution procedures. Omega **34**(3), 209–219 (2006)
3. Busiello, S.: Fractals and Music (in Italian), Aracne, p. 464 (2000)
4. Chen, L., Aihara, K.: Chaotic simulated annealing by a neural network model with transient chaos. Neural Netw. **8**(6), 915–930 (1995)
5. Dare, P.: Optimal design of GPS networks: operational procedures. Ph.D. Thesis, School of Surveying, University of East London, UK (1995)
6. Dare, P., Saleh, H.A.: GPS network design: logistics solution using optimal and near-optimal methods. J. Geod. **74**, 467–478 (2000)
7. Dorigo, M., Birattari, M.: Ant colony optimization. In: Sammut, C., Webb, G.I. (eds.) Encyclopedia of Machine Learning, pp. 36–39. Springer, Heidelberg (2010)
8. El-Shorbagy, M.A., Mousa, A.A., Nasr, S.M.: A chaos-based evolutionary algorithm for general nonlinear programming problems. Chaos, Solitons Fractals **85**, 8–21 (2016)
9. Falconer, K.: Fractal Geometry: Mathematical Foundations and Applications, 400 p. Wiley, New York (2013)
10. Fidanova, S.: Hybrid heuristics algorithms for GPS surveying problem. In: Boyanov, T., Dimova, S., Georgiev, K., Nikolov, G. (eds.) Proceedings of the 6th International Conference on Numerical Methods and Applications. Lecture Notes in Computer Science, vol. 4310, pp. 239–248 (2007)
11. Fidanova, S., Alba, E., Molina, G.: Memetic simulated annealing for GPS surveying problem. In: Margenov, S., Vulkov, L.G., Waśniewski, J. (eds.) Proceedings of the 4th International Conference on Numerical Analysis and Its Applications. Lecture Notes in Computer Science, vol. 5434, pp. 281–288 (2009)
12. Fidanova, S., Alba, E., Molina, G.: Hybrid ACO algorithm for the GPS surveying problem. In: Lirkov, I., Margenov, S., Waśniewski, J. (eds.) Proceedings of Large Scale Scientific Computing. Lecture Notes in Computer Science, vol. 5910, pp. 318–325 (2010)
13. Hofmann-Wellenhof, B., Lichtenegger, H., Collins, J.: Global Positioning System: Theory and Practice, 326 p. Springer, Heidelberg (1993)
14. Leick, A.: GPS Satellite Surveying, 3rd edn, 464 p. Wiley, New York (2004)
15. Mucherino, A., Fidanova, S., Ganzha, M.: Ant colony optimization with environment changes: an application to GPS surveying. IEEE Conference Proceedings, Federated Conference on Computer Science and Information Systems (FedCSIS15), Workshop on Computational Optimization (WCO15), Lodz, Poland, 495–500 (2015)
16. Mucherino, A., Seref, O.: Modeling and solving real life global optimization problems with meta-heuristic methods. In: Papajorgji, P.J., Pardalos, P.M. (eds.) Advances in Modeling Agricultural Systems. Springer Optimization and Its Applications, vol. 25, pp. 403–420 (2008)
17. Pandit, A., Dasanna, A.K., Sinha, S.: Multifractal analysis of HIV-1 genomes. Mol. Phylogenetics Evol. **62**(2), 756–763 (2012)
18. Papadimitriou, C.H.: The Euclidean travelling salesman problem is NP-complete. Theor. Comput. Sci. **4**(3), 237–244 (1977)
19. Rani, M., Agarwal, R.: Generation of fractals from complex logistic map. Chaos, Solitions Fractals **42**, 447–452 (2009)
20. Saleh, H.A., Dare, P.: Effective heuristics for the GPS survey network of Malta: simulated annealing and Tabu search techniques. J. Heuristics **7**, 533–549 (2001)
21. Saleh, H.A., Dare, P.: Heuristic methods for designing a global positioning system surveying network in the republic of Seychelles. Arab. J. Sci. Eng. **26**(1B), 74–93 (2002)
22. Stutzle, T., Hoos, H.H.: MAX-MIN ant system. In: Dorigo, M., Stutzle, T., Di Caro, G. (eds.) Future Generation Computer Systems, vol. 16, pp. 889–914 (2000)

23. Talbi, E-G.: Metaheuristics: From Design to Implementation, 624 p. Wiley, New York (2009)
24. Teunissen, P., Kleusberg, A., GPS for Geodesy, 2nd edn., 650 p. Springer, Heidelberg (1998)
25. Verhulst, P.F.: A note on the law of population growth. Correspondence Mathematiques et Physiques **10**, 113–121 (1938) (in French)
26. Yang, X.S.: Nature-Inspired Optimization Algorithms, 300 p. Elsevier Insights, Amsterdam (2014)
27. Yang, D., Li, G., Cheng, G.: On the efficiency of chaos optimization algorithms for global optimization. Chaos, Solitions Fractals **34**, 1366–1375 (2007)

Fast Preconditioned Solver for Truncated Saddle Point Problem in Nonsmooth Cahn–Hilliard Model

Pawan Kumar

Abstract The discretization of Cahn–Hilliard equation with obstacle potential leads to a block 2×2 non-linear system, where the $(1, 1)$ block has a non-linear and non-smooth term. Recently a globally convergent Newton Schur method was proposed for the non-linear Schur complement corresponding to this non-linear system. The proposed method is similar to an inexact active set method in the sense that the active sets are first approximately identified by solving a quadratic obstacle problem corresponding to the $(1, 1)$ block of the block 2×2 system, and later solving a reduced linear system by annihilating the rows and columns corresponding to identified active sets. For solving the quadratic obstacle problem, various optimal multigrid like methods have been proposed. In this paper, we study a block tridiagonal Schur complement preconditioner for solving the reduced linear system. The preconditioner shows robustness with respect to problem parameters and truncations of domain.

1 Introduction

We are interested in the problem of solving the linear system

$$\begin{pmatrix} \hat{A} & \hat{B}^T \\ \hat{B} & -C \end{pmatrix} \begin{pmatrix} \hat{u} \\ \hat{w} \end{pmatrix} = \begin{pmatrix} \hat{f} \\ \hat{g} \end{pmatrix}, \quad \hat{A}, \hat{B}, C \in \mathbb{R}^{n \times n}, \quad \hat{u}, \hat{w} \in K^n \quad (1)$$

that arises as a subproblem of a nonsmooth nonlinear solver [9] for set-valued saddle point problem such as the Cahn–Hilliard problem with obstacle potential. Solving such systems constitute significant part of the total simulation time. In brief, the Cahn–Hilliard equation was first proposed in 1958 by Cahn and Hilliard [7] to study the phase separation process in a binary alloy. The obstacle potential was proposed by Oono and Puri [18] in 1987 to model the deep quench phenomena, thus leading

P. Kumar (✉)
Department of Mathematics and Computer Science, Freie Universität Berlin,
Arnimallee 6, 14195 Berlin, Germany
e-mail: kumar@zedat.fu-berlin.de

© Springer International Publishing Switzerland 2016
S. Fidanova (ed.), *Recent Advances in Computational Optimization*,
Studies in Computational Intelligence 655, DOI 10.1007/978-3-319-40132-4_10

to a nonlinear and nonsmooth equations. For a motivation behind such model, an interested reader is referred to the excellent introductory sections of [1, 6].

The non-linear problem corresponding to Cahn–Hilliard problem with obstacle potential could be written as a non-linear system in block 2×2 matrix form as follows:

$$\begin{pmatrix} F & B^T \\ B & -C \end{pmatrix} \begin{pmatrix} u^* \\ w^* \end{pmatrix} \ni \begin{pmatrix} f \\ g \end{pmatrix}, \quad f, g, u^*, w^* \in \mathbb{R}^n \tag{2}$$

where u^*, w^* are unknowns, $F = A + \partial I_\mathcal{K}$, where $I_\mathcal{K}$ denotes the indicator functional of the admissible set $\mathcal{K}$. The matrices A, C are essentially Laplacian where A is augmented by a rank one term. By nonlinear Gaussian elimination of the u^* variables, the system above could be reduced to a nonlinear Schur complement system in w^* variables [8], where the nonlinear Schur complement is given by $-(C + BF^{-1}B^T)$. In [8], a globally convergent Newton method is proposed for this nonlinear Schur complement system which is interpreted as a preconditioned Uzawa iteration. Note that $F(x)$ is a set valued mapping due to the presence of set-valued operator $\partial I_\mathcal{K}$, which denote the subdifferential of $I_\mathcal{K}$. To solve the inclusion $F(x) \ni y$, or equivalently, $x \in F^{-1}y$ corresponding to the quadratic obstacle problem, many methods have been proposed: projected block Gauss–Seidel [2], monotone multigrid method [11, 12, 16], truncated monotone multigrid [10], and recently introduced truncated Newton multigrid [10]. See the excellent review article [10] that compares all these methods. Solving the quadratic obstacle problem corresponds to identifying the active sets. By annihilating the corresponding rows and columns that belong to the identified active sets, we obtain a reduced linear system (1). The overall nonlinear iteration is performed in the sense of inexact Uzawa, and the preconditioners are updated during each time step.

In this paper our goal is to design effective preconditioner for (1). Various preconditioners for saddle point problems have been proposed [3, 13]. In this paper, we consider the Schur complement preconditioner proposed in [6, 15] that was proposed for similar system, and seems to be optimal. It exploits the structure of the matrix and requires only two elliptic type solves. We adapt this preconditioner to our linear system. Our linear system differs from [6]; moreover, we have nontrivial kernels due to truncations, and the $(2, 2)$ block being a stiffness matrix corresponding to Neumann boundary has non trivial kernel too. We study the effectiveness of the preconditioner for various active set configurations, thus verifying the robustness of the linear solver.

The rest of this paper is organized as follows. In Sect. 3, we describe the Cahn–Hilliard model with obstacle potential, we discuss the time and space discretizations and variational formulations. In Sect. 4, we discuss the recent Nonsmooth Newton Schur method; we describe how the linear system appears. The preconditioner for the reduced linear system is discussed in Sect. 4.5. Finally, in Sect. 5, we show numerical experiments with the proposed preconditioner.

2 Notation

Let SPD and SPSD denote symmetric positive definite and symmetric positive semi-definite respectively. Let $|x|$ denote the absolute value of x, whereas, for a set $\mathcal{K}$, $|\mathcal{K}|$ denotes the number of elements in $\mathcal{K}$. Let $Id \in \mathbb{R}^{n \times n}$ denote the identity matrix. For a vector u let $u(i)$ denote the ith entry of vector u. Similarly for a matrix we use the notation $K(i, j)$ to denote the (i, j)th entry of K. For any matrix K, K^+ shall denote a pseudoinverse of K.

3 Cahn–Hilliard Problem with Obstacle Potential

3.1 The Model

The Cahn–Hilliard equation in PDE form with inequality constraints reads:

$$\partial_t u = \Delta w, \tag{3}$$
$$w = -\epsilon \Delta u + \psi'_0(u) + \mu,$$
$$\mu \in \partial \beta_{[-1,1]}(u),$$
$$|u| \leq 1,$$
$$\frac{\partial u}{\partial n} = \frac{\partial w}{\partial n} = 0 \text{ on } \partial \Omega, \tag{4}$$

where $\partial \beta_{[-1,1]}(u)$ is the subdifferential of $\beta_{[-1,1]}(u) := \int_\Omega I_{[-1,1]}(u)$. The obstacle potential ψ is given as follows:

$$\psi(u) = \psi_0(u) + I_{[-1,1]}(u), \quad \text{where } \psi_0(u) = \frac{1}{2}(1 - u^2).$$

Here the indicator function $I_{[-1,1]}(u)$ is defined as follows:

$$I_{[-1,1]} = \begin{cases} 0, & \text{if } u(i) \in [-1, 1], \\ \infty, & \text{otherwise.} \end{cases}$$

The subscript $[-1, 1]$ correspond to the fact that u is allowed to take values only between -1 and $+1$, which we refer to as upper and lower obstacles respectively.

In (3)–(4) the unknowns u and w are called order parameter and chemical potential respectively. For a given $\epsilon > 0$, final time $T > 0$ and initial condition $u_0 \in \mathcal{K}$ where

$$\mathcal{K} = \{v \in H^1(\Omega) : |v| \leq 1\},$$

the equivalent initial value problem for Cahn–Hilliard equation with obstacle potential interpreted as variational inequality reads:

$$\left\langle \frac{du}{dt}, v \right\rangle + (\nabla w, \nabla v) = 0, \ \forall v \in H^1(\Omega), \tag{5}$$

$$\epsilon(\nabla u, \nabla(v - u)) - (u, v - u) \geq (w, v - u), \ \forall v \in \mathcal{K}, \tag{6}$$

where we use the notation $\langle \cdot, \cdot \rangle$ to denote the duality pairing of $H^1(\Omega)$ and $H^1(\Omega)'$. Note that we used the fact that $\psi_0'(u) = -u$ in the second term on the left of inequality (6) above. The existence and uniqueness of the solution of (5), (6) above has been established in Blowey and Elliot [4]. We next consider discretization in time and space for the Eqs. (5) and (6) above.

3.2   Time and Space Discretizations

We consider a fixed non-adaptive grid in time $(0, T)$ and in space $\Omega = (0, 1) \times (0, 1)$. The time step $\tau = T/N$ is kept uniform, N being the number of time steps. We consider the semi-implicit Euler discretization in time and finite element discretization as in Barrett et. al. [2] with triangulation $\mathcal{T}_h$ with the following spaces:

$$S_h = \{v \in C(\overline{\Omega}) : v|_T \text{ is linear } \forall T \in T_h\},$$
$$\mathcal{P}_h = \{v \in L^2(\Omega) : v_T \text{ is constant } \forall T \in \mathcal{T} \in \mathcal{T}_h\},$$
$$\mathcal{K}_h = \{v \in \mathcal{P}_h : |v_T| \leq 1 \forall T \in \mathcal{T}_h\} = \mathcal{K} \cap S_h \subset \mathcal{K},$$

which leads to the following discrete Cahn–Hilliard problem with obstacle potential: Find $u_h^k \in \mathcal{K}_h, w_h^k \in S_h$ such that

$$\langle u_h^k, v_h \rangle + \tau(\nabla w_h^k, \nabla v_h) = \langle u_h^{k-1}, v_h \rangle, \quad \forall v_h \in S_h, \tag{7}$$

$$\epsilon(\nabla u_h^k, \nabla(v_h - u_h^k)) - \langle w_h^k, v_h - u_h^k \rangle \geq \langle u_h^{k-1}, v_h - u_h^k \rangle, \qquad \forall v_h \in \mathcal{K}_h.$$

holds for each $k = 1, \ldots, N$. The initial solution $u_h^0 \in \mathcal{K}_h$ is taken to be the discrete L^2 projection $\langle u_h^0, v_h \rangle = (u_0, v_h), \ \forall v_h \in S_h$. Existence and uniqueness of the discrete Cahn–Hilliard equations has been established in [5]. The discrete Cahn–Hilliard equation is equivalent to the set valued saddle point block 2×2 nonlinear system (2) with $F = A + \partial I_{\mathcal{K}_h}$, and

$$A = \epsilon((\langle \lambda_p, 1 \rangle \langle \lambda_p, 1 \rangle + (\nabla \lambda_p, \nabla \lambda_q))_{p,q \in \mathcal{N}_h}, \tag{8}$$
$$B = -((\langle \lambda_p, \lambda_q \rangle)_{p,q \in \mathcal{N}_h}, \ C = \tau((\nabla \lambda_p, \nabla \lambda_q))_{p,q \in \mathcal{N}_h}, \tag{9}$$

where $\mathcal{N}_h$ stands for the set of vertices in $\mathcal{T}_h$, and λ_p, $p \in \mathcal{N}_h$ denote the standard nodal basis. We write the above in more compact notation as follows

$$A = \epsilon(K + mm^T), \quad B = -M, \quad C = \tau K, \tag{10}$$

where K and M are stiffness and mass matrices respectively.

4 Iterative Solver for Cahn–Hilliard with Obstacle Potential

In [8], a nonsmooth Newton Schur method is proposed which is also interpreted as a preconditioned Uzawa iteration. For a given time step k, the Uzawa iteration reads:

$$u^{i,k} = F^{-1}(f^k - B^T w^{i,k}) \tag{11}$$

$$w^{i+1,k} = w^{i,k} + \rho^{i,k} \hat{S}_{i,k}^{-1}(Bu^{i,k} - Cw^{i,k} - g^k) \tag{12}$$

for the saddle point problem (2). Here i denotes the ith Uzawa step and k denotes the kth time step. Here f^k and g^k are defined in (7) and (8). The time loop starts with an initial value for $w^{0,0}$ which is taken arbitrary, and with the initial value $u^{0,0}$. The Uzawa iteration requires three main computations that we describe below.

4.1 Computing $u^{i,k}$

The first step (11) corresponds to a quadratic obstacle problem:

$$u^{i,k} = \arg\min_{v \in \mathcal{K}} \left(\frac{1}{2}\langle Av, v \rangle - \langle f^k - B^T w^{i,k}, v \rangle \right).$$

As mentioned in the introduction, this problem has been extensively studied during last decades [2, 10–12].

4.2 Computing $\hat{S}_{i,k}^{-1}(Bu^{i,k} - Cw^{i,k} - g^k)$

The descent direction $d^{i,k+1} = \hat{S}_{i,k}^{-1}(Bu^{i,k} - Cw^{i,k} - g)$ in (12) is obtained as a solution of the following reduced linear block 2×2 system:

$$\begin{pmatrix} \hat{A} & \hat{B}^T \\ \hat{B} & -C \end{pmatrix} \begin{pmatrix} \tilde{u}^{i,k} \\ d^{i,k} \end{pmatrix} = \begin{pmatrix} 0 \\ g + Cw^{i,k} - Bu^{i,k} \end{pmatrix}, \tag{13}$$

where

$$\hat{A} = TAT + \hat{T}, \quad \hat{B} = TB. \tag{14}$$

Here T and $\hat{T}$ are defined as follows:

$$T = \mathrm{diag}\begin{pmatrix} 0, \text{ if } u^{i,k}(j) \in \{-1, 1\} \\ 1, \text{ otherwise} \end{pmatrix}, \quad j = 1, \dots, |\mathcal{N}_h|, \tag{15}$$

$$\hat{T} = Id - T, \quad Id \in \mathbb{R}^{|\mathcal{N}_h| \times |\mathcal{N}_h|},$$

where $u^{i,k}(j)$ is the jth component of $u^{i,k}$. In other words, in (14), $\hat{A}$ is the matrix obtained from A by replacing the ith row and the ith column by the unit vector e_i corresponding to the active sets identified by diagonal entries of T. Similarly, $\hat{B}$ is the matrix obtained from B by annihilating columns, and $\hat{B}^T$ is the matrix obtained from B^T by annihilating rows.

4.3 Computing Step Length $\rho^{i,k}$

The step length $\rho^{i,k}$ is computed using a bisection method. We refer the reader to [9, p. 88]. The correct step length lends global convergence for the Uzawa method.

4.4 Algebraic Monotone Multigrid for Obstacle Problem

To solve the quadratic obstacle problem (11), we may use the truncated monotone multigrid method proposed in [11]. However, here we may use algebraic coarsening to create initial set of interpolation operators. We describe it briefly as in [14].

4.4.1 Aggregation Based Coarsening

We first discuss the coarsening for two-grid, the multilevel interpolations are applied recursively. In classical two-grid, a set of coarse grid unknowns is selected and the matrix entries are used to build interpolation rules that define the prolongation matrix P, and the coarse grid matrix A_c is computed from the following Galerkin formula

$$A_c = P^T A P.$$

In contrast to the classical two-grid approach, in aggregation based multigrid, first a set of aggregates G_i is defined. Let $|\mathcal{N}_{h,c}|$ be the total number of such aggregates, then the interpolation matrix P is defined as follows

$$P_{ij} = \begin{cases} 1, & \text{if } i \in G_j, \\ 0, & \text{otherwise,} \end{cases}$$

Here, $1 \leq i \leq |\mathcal{N}_h|$, $1 \leq j \leq |\mathcal{N}_{h,c}|$. Further, we assume that the aggregates G_i are such that

$$G_i \bigcap G_j = \phi, \text{ for } i \neq j \text{ and } \bigcup_i G_i = \{i \in \mathbb{N} : 1 \leq i \leq |\mathcal{N}_h|\}.$$

The matrix P defined above is a $|\mathcal{N}_h| \times |\mathcal{N}_{h,c}|$ matrix, but since it has only one non-zero entry (which are "one") per row, the matrix is compactly represented by a single array of length $\mathcal{N}_{h,c}$ storing the location of the non-zero entry on each row. The coarse grid matrix A_c may be computed as follows

$$(A_c)(i, j) = \sum_{k \in G_i} \sum_{l \in G_j} A(k, l),$$

where $1 \leq i, j \leq |\mathcal{N}_{h,c}|$, and $A(k, l)$ is the (k, l)th entry of A.

Numerous aggregation schemes have been proposed in the literature, but in this paper we consider the standard aggregation based on strength of connection [21, Appendix A, p. 413] where one first defines a set of nodes S_i to which i is strongly negatively coupled, using the Strong/Weak coupling threshold β:

$$S_i = \{ j \neq i \mid A(i, j) < -\beta \max |A(i, k)| \}.$$

Then an unmarked node i is chosen such that priority is given to a node with minimal M_i, here M_i being the number of unmarked nodes that are strongly negatively coupled to i. For a complete algorithm of aggregation, the reader is referred to Notay [14, 17].

4.5 Preconditioner for Reduced Linear System

In Bosch et. al. [6], a preconditioner is proposed in the framework of semi-smooth Newton method combined with Moreau–Yosida regularization for the same problem. However, the preconditioner was constructed for a linear system which is different from the one we consider in (13). For convenience of notation, we rewrite the system matrix in (13) as follows

$$\mathcal{A}x = b,$$

where scripted $\mathcal{A}$ above is

$$\mathcal{A} = \begin{pmatrix} \hat{A} & \hat{B}^T \\ \hat{B} & -C \end{pmatrix}, \quad x = \begin{pmatrix} x_1 \\ x_2 \end{pmatrix}, \quad b = \begin{pmatrix} b_1 \\ b_2 \end{pmatrix}, \tag{17}$$

where

$$x_1 = \tilde{u}^{i,k}, \quad x_2 = d^{i,k}, \quad b_1 = 0, \quad b_2 = g + Cw^{i,k} - Bu^{i,k}.$$

The preconditioner proposed in [6] has the following block lower triangular form

$$\mathcal{B} = \begin{pmatrix} \hat{A} & 0 \\ \hat{B} & -S \end{pmatrix},$$

where $S = C + \hat{B}\hat{A}^{-1}\hat{B}^T$ is the negative Schur complement. Before we define preconditioner, it is essential to know whether S is nonsingular.

In the following, we denote the set of truncated nodes by

$$\mathcal{N}_h^{\bullet} = \{i : T(i, i) = 0\}.$$

That is, $\mathcal{N}_h^{\bullet}$ is the set of truncated nodes.

Theorem 1 *The negative Schur complement $S = C + \hat{B}\hat{A}^{-1}\hat{B}^T$ is non-singular, in particular, SPD if and only if $|\mathcal{N}_h^{\bullet}| < |\mathcal{N}|_h$.*

Proof If $|\mathcal{N}_h^{\bullet}| = |\mathcal{N}|_h$, then $\hat{B}$ is the zero matrix, consequently $S = C = \tau K$ is singular since K correspond to stiffness matrix with pure Neumann boundary condition. For other implication, we recall that $\hat{B}^T = \hat{M}^T = -TM$, where T is defined in (15). The (i, j)th entry of element mass matrix is given as follows

$$M_{ij}^K = \int_K \phi_i \phi_j dx = \frac{1}{12}(1 + \delta_{ij}|K|) \quad i, j = 1, 2, 3, \tag{18}$$

where δ_{ij} is the Kronecker symbol, that is, it is equal to 1 if $i = j$, and 0 if $i \neq j$. Here ϕ_1, ϕ_2, and ϕ_3 are hat functions on triangular element K with local numbering, and $|K|$ is the area of triangle element K. From (19), it is easy to see that

$$M^K = \frac{1}{12}\begin{pmatrix} 2 & 1 & 1 \\ 1 & 2 & 1 \\ 1 & 1 & 2 \end{pmatrix}. \tag{19}$$

Evidently, entries of global mass matrix $M = \sum_K M^K$ are also all positive, hence all entries of truncated mass matrix $\hat{M}$ remain non-negative. In particular, due to our hypothesis $|\mathcal{N}^\bullet| > 0$, there is atleast one untruncated column, hence, atleast few positive entries. Consequently, $M\mathbf{1} \neq 0$, i.e., $\mathbf{1}$ or span$\{\mathbf{1}\}$ is neither in kernel of M, nor in the kernel of $\hat{M}$, in particular, $\mathbf{1}^T \hat{M}^T \mathbf{1} > 0$. The proof of the theorem then follows since C is SPD except on $\mathbf{1}$ for which $\hat{B}^T \mathbf{1}$ is non-zero, and the fact that $\hat{A}$ is SPD yields

$$\left\langle \hat{B}\hat{A}^{-1}\hat{B}^T \mathbf{1}, \mathbf{1} \right\rangle = \left\langle \hat{A}^{-1}(\hat{B}^T \mathbf{1}), (\hat{B}^T \mathbf{1}) \right\rangle = \left\langle \hat{A}^{-1}(-\hat{M}^T \mathbf{1}), (-\hat{M}^T \mathbf{1}) \right\rangle > 0. \quad \square$$

Note that such preconditioners are also called inexact or preconditioned Uzawa preconditioners for the *linear* saddle point problems. By block 2×2 inversion formula we have

$$\mathcal{B}^{-1} = \begin{pmatrix} \hat{A} & 0 \\ \hat{B} & -S \end{pmatrix}^{-1} = \begin{pmatrix} \hat{A}^{-1} & 0 \\ S^{-1}\hat{B}^T \hat{A}^{-1} & -S^{-1} \end{pmatrix}.$$

Let $\hat{S}$ be an approximation of Schur complement S in $\mathcal{B}$. The new preconditioner $\hat{\mathcal{B}}$, and the corresponding preconditioned operator $\hat{\mathcal{B}}^{-1}\mathcal{A}$ are given as follows

$$\hat{\mathcal{B}} = \begin{pmatrix} \hat{A} & 0 \\ \hat{B} & -\hat{S} \end{pmatrix}, \quad \hat{\mathcal{B}}^{-1}\mathcal{A} = \begin{pmatrix} I & \hat{A}^{-1}\hat{B}^T \\ 0 & \hat{S}^{-1}S \end{pmatrix}. \tag{20}$$

Using (20) above, we can note the following trivial result that justifies the need for a good preconditioner for the Schur complement.

Theorem 2 *Let $\mathcal{B}$ defined in (20) be a preconditioner for $\mathcal{A}$ defined in (17), then there are $|\mathcal{N}_h|$ eigenvalues of $\mathcal{B}^{-1}\mathcal{A}$ equal to one, and the rest are the eigenvalues of the preconditioned Schur complement $\hat{S}^{-1}S$.*

Remark 1 When using GMRES [20], right preconditioning is preferred. Similar result as for the left preconditioner above Theorem 2 holds.

The preconditioned system $\mathcal{B}^{-1}\mathcal{A}x = \mathcal{B}^{-1}b$ is given as follows

$$\begin{pmatrix} I & \hat{A}^{-1}\hat{B}^T \\ 0 & \hat{S}^{-1}S \end{pmatrix} \begin{pmatrix} x_1 \\ x_2 \end{pmatrix} = \begin{pmatrix} \hat{A}^{-1} & 0 \\ S^{-1}\hat{B}^T \hat{A}^{-1} & -S^{-1} \end{pmatrix} \begin{pmatrix} b_1 \\ b_2 \end{pmatrix}$$

from which we obtain the following set of equations

$$x_1 + \hat{A}^{-1}\hat{B}^T x_2 = \hat{A}^{-1}b_1,$$
$$\hat{S}^{-1}Sx_2 = S^{-1}(\hat{B}^T \hat{A}^{-1}b_1 - b_2).$$

Algorithm 1 Objective: Solve $\mathcal{B}^{-1}\mathcal{A}x = \mathcal{B}^{-1}b$

1. Solve for $x_2 : \hat{S}^{-1}Sx_2 = \hat{S}^{-1}(\hat{B}^T\hat{A}^{-1}b_1 - b_2)$
2. Set $x_1 = \hat{A}^{-1}(b_1 - \hat{B}^T x_2)$

Here if Krylov subspace method is used to solve for x_2, then a matrix vector product with S and a solve with $\hat{S}$ is needed. However, when the problem size, it won't be feasible to do exact solve with $\hat{A}$, and we need to solve it inexactly, for example, using algebraic multigrid methods. In the later case, the decoupling of x_1 and x_2 as in Algorithm 1 is not possible, and in this case, we shall need matrix vector product with $\mathcal{A}$ (17) and a solve (forward sweep) with $\hat{\mathcal{B}}$. We discuss at the end of this subsection on how to take advantage of the special structure of $\hat{A}$.

As a preconditioner $\tilde{S}$ of S, we choose the preconditioner first proposed in [6]. The preconditioner is given as follows:

$$\tilde{S} = S_1\hat{A}^{-1}S_2 = -(\hat{B} - \tau^{1/2}K)\hat{A}^{-1}(\hat{B}^T - \tau^{1/2}\hat{A}),$$

where K is the stiffness matrix from (9). We observe that the preconditioned Schur complement $\tilde{S}^{-1}S$ is not symmetric, in particular, it is not symmetric positive (semi)definite w.r.t. $\langle \cdot, \cdot \rangle_S$ or w.r.t. $\langle \cdot, \cdot \rangle_{\tilde{S}}$, thus, we may not use preconditioned conjugate gradient method [20, p. 262]. Consequently, we shall use GMRES in Saad [20, p. 269] that allows nonsymmetric preconditioners. However, it is easy to see that S_1, S_2 are SPD provided we use mass lumping which is possible since we use non adaptive uniform grid in space, that keeps the mass matrices consistent.

4.5.1 Solve with $\hat{A}$

In step 1 of Algorithm 1, we need to solve with $\hat{A}$ when constructing right hand side, and also in step 2. Let P be a permutation matrix, then solving a system of the form $\hat{A}h = g$ is equivalent to solving $P^T\hat{A}h = P^Tg$ as P^T is nonsingular. With a change of variable $Py := h$, we then solve for y in

$$P^T\hat{A}Py = Pg, \tag{21}$$

and we set $h = Py$ to obtain the desired solution. By choosing P that renumbers the nodes corresponding to the coincidence set, we obtain

$$P^T\hat{A}P = \begin{pmatrix} I & \\ & R^T P^T \hat{A} P R \end{pmatrix},$$

where R is the restriction operator that compresses the matrix $P^T\hat{A}P$ to untruncated nodes $\mathcal{N}_h$. Here R is given as follows:

$$R = \left(\begin{pmatrix} 0 & 0 & \dots & 0 \\ \vdots & & \dots & 0 \\ 0 & 0 & \dots & 0 \\ 1 & 0 & \dots & 0 \\ 0 & 1 & \dots & 0 \\ \vdots & \ddots & \dots & 0 \\ 0 & 0 & \dots & 1 \end{pmatrix} \right), \quad R \in \mathbb{R}^{|\mathcal{N}_h| \times |\mathcal{N}_h \setminus \mathcal{N}_h^\bullet|}$$

Let $\hat{K} = TKT$, we have

$$R^T P^T \hat{A} P R = R^T P^T \epsilon (\hat{K} + \hat{m}\hat{m}^T) P R$$
$$= \epsilon (R^T P^T \hat{K} P R + R^T P^T \hat{m}\hat{m}^T P R),$$

where

$$R^T P^T \hat{K} P R = (P^T \hat{K} P)|_{\mathcal{N}_h \setminus \mathcal{N}_h^\bullet}, \quad \hat{K} = TKT, \quad \hat{m} = Tm,$$

where m is the rank-one term defined in (10). For convenience of notation, we write

$$R^T P^T \hat{A} P R = \epsilon(\widetilde{K} + \tilde{z}\tilde{z}^T),$$

where $\widetilde{K} = R^T P^T \hat{K} P R$ and $\tilde{m} = R^T P^T \hat{m}$. In the new notation, we have

$$P^T \hat{A} P = \begin{pmatrix} I & \\ & \epsilon(\widetilde{K} + \tilde{m}\tilde{m}^T) \end{pmatrix}. \tag{22}$$

Thus (21) now reads

$$\begin{pmatrix} I & \\ & \epsilon(\widetilde{K} + \tilde{m}\tilde{m}^T) \end{pmatrix} \begin{pmatrix} y_1 \\ y_2 \end{pmatrix} = Pg =: \begin{pmatrix} g_1 \\ g_2 \end{pmatrix},$$

which reduces to two-set of equations

$$y_1 = g_1,$$
$$\epsilon(\widetilde{K} + \tilde{m}\tilde{m}^T)y_2 = g_2.$$

To solve the latter, we use the following Sherman–Woodbury formula

$$(\widetilde{K} + \tilde{m}\tilde{m}^T)^+ = \widetilde{K}^+ - \frac{\widetilde{K}^+ \tilde{m}\tilde{m}^T \widetilde{K}^+}{1 + \tilde{m}^T \widetilde{K}^+ \tilde{m}}.$$

The aggregation based AMG discussed before can be used to solve with $\tilde{K}$, thus, we avoid constructing the prohibitively dense matrix which would be the case when rank one term is explicitly added. We stress here that, in case, there are no truncations, i.e., when the set of truncated nodes $\mathcal{N}_h^\bullet = \phi$, then $\tilde{K} = K$ is singular, then some AMG may fail. This is the reason why we use pseudo-inverse notation. However, $\tilde{K}^+$ may be replaced by $\tilde{K}^{-1}$ in case $|\mathcal{N}_h| > 1$, because of the following result that follows later.

Lemma 1 (Poincaré separation theorem for eigenvalues) *Let $Z \in \mathbb{R}^{n \times n}$ be a symmetric matrix with eigenvalues $\lambda_1 \leq \lambda_2 \leq \cdots \lambda_n$, and let P be a semi-orthogonal $n \times k$ matrix such that $P^T P = Id \in \mathbb{R}^{k \times k}$. Then the eigenvalues $\mu_1 \leq \mu_2 \cdots \mu_{n-k+i}$ of $P^T Z P$ are separated by the eigenvalues of Z as follows*

$$\lambda_i \leq \mu_i \leq \lambda_{n-k+i}.$$

Proof The theorem is proved in [19, p. 337].   □

Lemma 2 (Eigenvalues of the truncated matrix) *Let $\lambda_1 \leq \lambda_2 \cdots \leq \lambda_n$ be the eigenvalues of A, and let $\hat{\lambda}_1 \leq \hat{\lambda}_2 \cdots \leq \hat{\lambda}_n$ be the eigenvalues of truncated matrix $\hat{A}$. Let $k = \sum_{i=1}^n T(i, i)$ be the number of untruncated rows in $\hat{A}$. Let $\hat{\lambda}_{n_1} \leq \hat{\lambda}_{n_2} \ldots \hat{\lambda}_{n_k}$ be the eigenvalues of $R^T P^T \hat{A} P R$. Then the following holds*

$$\lambda_i \leq \hat{\lambda}_{n_i} \leq \lambda_{n-k+i}.$$

Proof The proof shall follow by application of Poincare separation theorem, to this end, we need to reformulate our problem. Let P be a permutation matrix that renumbers the rows such that the truncated rows are numbered first, then we have

$$P^T \hat{A} P = \begin{pmatrix} I & \\ & R^T P^T \hat{A} P R \end{pmatrix},$$

where $R \in \mathbb{R}^{n \times k}$ is the restriction operator defined as follows

$$R = \begin{pmatrix} \begin{pmatrix} 0 & 0 & \ldots & 0 \\ \vdots & \ldots & \ldots & 0 \\ 0 & 0 & \ldots & 0 \end{pmatrix}_{n-k \times k} \\ \begin{pmatrix} 1 & 0 & \ldots & 0 \\ 0 & 1 & \ldots & 0 \\ \vdots & \ddots & \ldots & 0 \\ 0 & 0 & \ldots & 1 \end{pmatrix}_{k \times k} \end{pmatrix}.$$

Clearly, $R^T R = Id \in \mathbb{R}^{k \times k}$. Since $P^T \hat{A} P$ and $\hat{A}$ are similar, because P being a permutation matrix $P P^T = Id$, consequently, from Lemma 1, theorem follows.   □

Corollary 1 *From Theorem 2 and from (22), since* $\lambda_{\min}(\epsilon(\widetilde{K} + \tilde{m}\tilde{m}^T)) = \lambda_{\min}$ $(P^T \hat{A} P) = \lambda_{\min}(\hat{A}) \geq \lambda_{\min}(A), \epsilon > 0$, *hence,* $\widetilde{K} + \tilde{m}\tilde{m}^T$ *is SPD, since* $A = K + mm^T$ *is SPD.*

Let $\| \cdot \|$ denote $\| \cdot \|_2$ which is a submultiplicative norm, and determines the maximum eigenvalue. Although, S is symmetric, the preconditioner $\tilde{S}$ is not symmetric, but for the error $\|S - \tilde{S}\|$ we have the following bound.

Theorem 3 (Error in preconditioner) *Let* $\|E\| = \|S - \tilde{S}\|$. *There holds*

$$\|E\| \leq \sqrt{\tau} \left(\lambda_{\max}(\hat{M}) + \epsilon^{-1} \lambda_{\max}(K)(\lambda_{\max}(K) + mm^T)\lambda_{\max}(\hat{M}) \right).$$

In particular, we have

$$\|E\| \leq \sqrt{\tau} \cdot Ch^2 + \frac{\sqrt{\tau}}{\epsilon}(C_3 \cdot h^2) \left(\lambda_{n-k+1}(K) + \frac{(\lambda_{n-k+1}(K))^2(\hat{m}^T\hat{m})}{|1/(\hat{m}^T\hat{m}) + \lambda_k(K)|} \right).$$

Proof We have

$$\|E\| = \|S - \tilde{S}\| = \sqrt{\tau}\|\hat{B} + K\hat{A}^{-1}\hat{B}^T\|$$
$$\leq \sqrt{\tau}(\|\hat{B}\| + \|K\|\|\hat{A}^{-1}\|\|\hat{B}^T\|). \tag{23}$$

Recalling that T being a truncation matrix, $\|T\| = \lambda_{\max}(T) = 1$, we observe that

$$\|\hat{B}\| = \|TM\| \leq \|T\|\|M\| = 1 \cdot \lambda_{\max}(M) = \lambda_{\max}(M).$$

Same estimate holds for $\|\hat{B}^T\|$. To estimate $\|\hat{A}^{-1}\|$, we first estimate $\|\hat{A}\|$. We write $\hat{A} = \epsilon(\hat{K} + \hat{m}\hat{m}^T)$, where $\hat{K} = TKT$ and $\hat{m} = Tm$. From Theorem 2, we have $\lambda_{\min}(\hat{A}) \geq \lambda_{\min}(A)$ and $\lambda_{\max}(\hat{A}) \leq \lambda_{\max}(A)$. Consequently, $\lambda_{\min}(\hat{A}^{-1}) \geq \lambda_{\min}(A^{-1})$ and $\lambda_{\max}(\hat{A}^{-1}) \leq \lambda_{\min}(A^{-1}) = \lambda_{\max}(A) \leq \|K\| + \|mm^T\|)$. We have

$$\|E\| \leq \sqrt{\tau} \left(\lambda_{\max}(\hat{M}) + \epsilon^{-1} \lambda_{\max}(K)(\lambda_{\max}(K) + mm^T)\lambda_{\max}(\hat{M}) \right).$$

More insight is obtained by using Sherman–Woodbury inversion

$$\|\hat{A}^{-1}\| = \frac{1}{\epsilon} \left\| (\hat{K} + \hat{m}\hat{m}^T)^{-1} \right\| = \frac{1}{\epsilon} \left\| \hat{K}^{-1} - \frac{\hat{K}^{-1}\hat{m}\hat{m}^T\hat{K}^{-1}}{1 + \hat{m}^T\hat{K}^{-1}\hat{m}} \right\|$$
$$\leq \frac{1}{\epsilon} \left(\|\hat{K}^{-1}\| + \left\| \frac{(\hat{K}^{-1}\hat{m})(\hat{K}^{-1}\hat{m})^T}{1 + \hat{m}^T\hat{m}\lambda_{\min}(\hat{K}^{-1})} \right\| \right)$$
$$= \frac{1}{\epsilon} \left(\lambda_{\max}(\hat{K}^{-1}) + \frac{\left\| (\hat{K}^{-1}\hat{m})(\hat{K}^{-1}\hat{m})^T \right\|}{1 + (\hat{m}^T\hat{m})\lambda_{\max}(\hat{K})} \right)$$

$$= \frac{1}{\epsilon} \left(\lambda_{\max}(\hat{K}^{-1}) + \frac{\lambda_{\max}\left((\hat{K}^{-1}\hat{m})(\hat{K}^{-1}\hat{m})^T\right)}{1 + (\hat{m}^T\hat{m})\lambda_{\max}(\hat{K})} \right)$$

$$= \frac{1}{\epsilon} \left(\lambda_{\min}(\hat{K}) + \frac{\hat{m}^T(\hat{K}^{-1})^2\hat{m}}{\left| 1 + (\hat{m}^T\hat{m})\lambda_{\max}(\hat{K}) \right|} \right)$$

$$\leq \frac{1}{\epsilon} \left(\lambda_{\max}(\hat{K}) + \frac{(\lambda_{\max}(\hat{K}^{-1}))^2(\hat{m}^T\hat{m})}{\left| 1 + (\hat{m}^T\hat{m})\lambda_{\max}(\hat{K}) \right|} \right), \quad \left(\frac{\hat{m}^T\hat{K}^{-2}\hat{m}}{\hat{m}^T\hat{m}} \leq \lambda_{\max}(\hat{K}^{-2}) \right)$$

$$= \frac{1}{\epsilon} \left(\lambda_{\min}(\hat{K}) + \frac{(\lambda_{\min}(\hat{K}))^2(\hat{m}^T\hat{m})}{\left| 1 + (\hat{m}^T\hat{m})\lambda_{\max}(\hat{K}) \right|} \right).$$

For convenience of notation, let number of untruncated rows be denoted by $k = |\mathcal{N}_h \setminus \mathcal{N}_h^\bullet|$. From Lemma 1, $\lambda_{\min}(\hat{K}) = \lambda_1(\hat{K}) \leq \lambda_{n-k+1}(K)$, estimate above becomes

$$\|\hat{A}^{-1}\| \leq \frac{1}{\epsilon} \left(\lambda_{n-k+1}(K) + \frac{(\lambda_{n-k+1}(K))^2(\hat{m}^T\hat{m})}{\left| 1 + (\hat{m}^T\hat{m})\lambda_{\max}(\hat{K}) \right|} \right).$$

Again from Lemma 1, we have $\lambda_{\max}(\hat{K}) = \lambda_k(\hat{K}) \geq \lambda_k(K)$. The error (23) now becomes

$$\|E\| \leq \sqrt{\tau} \cdot \lambda_{\max}(M)$$

$$+ \frac{\sqrt{\tau}}{\epsilon}(\lambda_{\max}(K) \cdot \lambda_{\max}(M)) \left(\lambda_{n-k+1}(K) + \frac{(\lambda_{n-k+1}(K))^2(\hat{m}^T\hat{m})}{\left| 1 + (\hat{m}^T\hat{m})\lambda_k(K) \right|} \right).$$

To observe the dependence on the mesh size h, we recall that on (quasi)uniform grid, the eigenvalues of M and K are known. On each element K, the element mass matrix is given by (19). Let $\xi \in \mathbb{R}^n$. We write $\xi^T M \xi$ as the sum $\sum_{K \in \mathcal{K}} \xi^T|_K M^K \xi|_K$, where M^K is proportional to $|K|$ which is proportional to h_K^2. We then have $ch^2(Id) \leq M^K \leq Ch^2(Id)$, which provides bound for $\lambda_{\max}(M)$ as follows

$$\lambda_{\max}(M) \leq Ch^2.$$

Also, we have

$$\frac{\xi^T K \xi}{\xi^T \xi} \leq C_2 h^{-2} \frac{\xi^T M \xi}{\xi^T \xi} \leq C_2 C = C_3.$$

The error now becomes

$$\|E\| \leq \sqrt{\tau} \cdot Ch^2 + \frac{\sqrt{\tau}}{\epsilon}(C_3 \cdot h^2)\left(\lambda_{n-k+1}(K) + \frac{(\lambda_{n-k+1}(K))^2 \hat{m}^T \hat{m}}{|1/(\hat{m}^T \hat{m}) + \lambda_k(K)|}\right). \qquad \square$$

5   Numerical Experiments

All the experiments were performed in double precision arithmetic in MATLAB.

The Krylov solver used was GMRES with subspace dimension of 200, and maximum number of iterations allowed was 300. The GMRES iteration was stopped as soon as the relative residual was below the tolerance of 10^{-7}, and for AMG, the tolerance was set to be 10^{-8}. The maximum number of iterations for AMG was 100. The system matrices were scaled (Fig. 1).

The initial random active set is similar to the one considered in [6], where the initial value is set to be random taking values between -0.3 and 0.5. The subfigures in Fig. 2 show the Cahn–Hilliard evolution for $\tau = 1, 20, 40, \ldots, 200$ steps. For the evolution of random initial configuration as in Fig. 2, we set $\epsilon = 2 * 10^{-2}, \tau = 10^{-5}$ as in [6]. In Table 1, we compare the iteration count and times for $h = 1/256, 1/400$. We note that the iteration count more or less remain comparable except for initial time steps where it takes roughly double the number of iterations than at later time steps. We also observe that the number of truncations increase with time, and this helps in bringing down the iteration count especially during first few time steps. For time steps until $\tau = 80$, the time for $h = 1/400$ is relatively larger because of AMG needing more iterations (Table 2).

Next we consider larger problems with two samples of active set configurations as shown in Fig. 2, and study the effect of parameter and mesh sizes on the iteration count for these fixed configurations. The region between the two squares and the circles is the interface between two bulk phases taking values $+1$ and -1; we set random values between -0.3 and 0.5 in this interface region. The width of the interface is kept to be 10 times the chosen mesh size. The time step τ is chosen to be equal to ϵ. We compare various mesh sizes leading to total number of grid points upto just above 1 million. We observe that the number of iterations remain independent of the mesh size, however it depends on ϵ. But we observe that for a fixed epsilon, with finer mesh, the number of iterations actually decrease significantly. For example the number of iterations for $h = 2^{-7}, \epsilon = 10^{-6}$ is 84 but the number of iterations for $h = 2^{-10}, \epsilon = 10^{-6}$ is 38, a reduction of 46 iterations! It seems that finer mesh size makes the preconditioner more efficient as also suggested by the error bound in Theorem 3. We also observe that the time to solve is proportional to number of iterations; the inexact solve for the $(1, 1)$ block remains optimal because the $(1, 1)$ block is essentially Laplacian for which AMG remains very efficient.

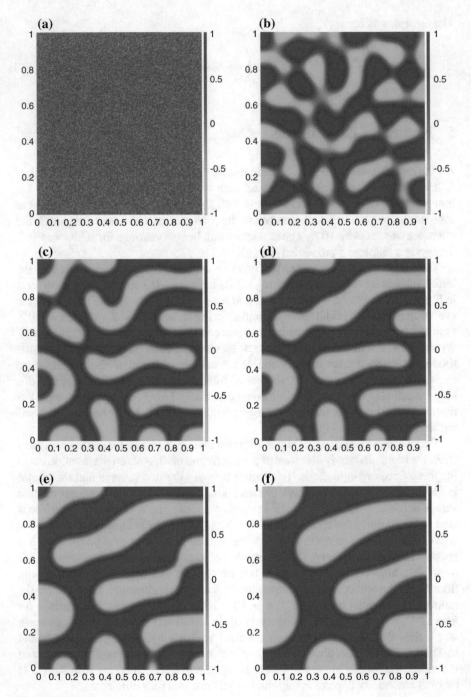

Fig. 1 Evolution of random initial active set configuration

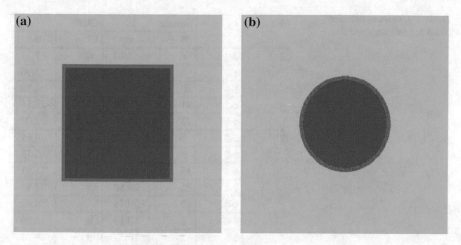

Fig. 2 Square and circle sample active set configurations

Table 1 Compare number of iterations for various mesh size h for various time steps τ for random initial configuration

τ	h	Random active set		
		Its	Time	#trunc
1	1/256	40	109.76	2
	1/400	41	272.42	2
20	1/256	35	65.6	13865
	1/400	41	233.81	16136
40	1/256	25	29.86	25696
	1/400	26	103.05	55886
80	1/256	23	26.25	31109
	1/400	28	112.93	72514
100	1/256	25	29.88	37336
	1/400	20	55.88	92787
120	1/256	21	20.14	39922
	1/400	20	50.96	95995
140	1/256	21	21.85	40357
	1/400	20	50.75	98593
160	1/256	21	21.26	40861
	1/400	20	54.78	100733
180	1/256	21	20.95	41215
	1/400	23	67.88	102625
200	1/256	21	21.05	41490
	1/400	20	53.44	104522

Here #trunc stands for the number of rows/columns truncated

Table 2 Compare iterations
count for various ϵ and h

h	ϵ	Square		Circle	
		Its	Time	Its	Time
2^{-7}	e-2	6	1.40	6	1.37
	e-3	8	2.00	9	2.16
	e-4	20	4.50	22	4.76
	e-5	41	9.23	45	10.14
	e-6	77	17.83	83	22.02
2^{-8}	e-2	6	6.22	6	4.61
	e-3	5	4.88	5	5.12
	e-4	13	10.56	15	11.91
	e-5	31	24.39	35	27.31
	e-6	60	50.44	67	54.09
2^{-9}	e-2	5	18.13	6	18.50
	e-3	5	16.57	5	18.92
	e-4	8	28.21	9	32.09
	e-5	22	72.59	25	80.03
	e-6	47	166.46	52	180.58
2^{-10}	e-2	5	81.86	6	89.16
	e-3	5	81.69	5	85.23
	e-4	5	98.62	5	97.76
	e-5	14	218.52	15	254.48
	e-6	34	527.49	38	612.41

6 Conclusion

For the solution of large scale optimization problem corresponding to Cahn–Hilliard
problem with obstacle problem, we studied an efficient preconditioning strategy for
a saddle point problem on truncated domains. It requires two elliptic solves. In our
initial experiments upto over million unknowns, the preconditioner remains mesh
independent. Although, for coarser mesh, there seems to be strong dependence on
epsilon, but as the mesh becomes finer, we observe a significant reduction in iteration
count, thus making the preconditioner effective and useful on finer meshes. It is likely
that the iteration count continues to decrease on finer meshes. Moreover, when the
parameters related to the interface, and the one corresponding to the time step are
kept fixed, we observe that the iteration counts decrease with time as more and more
truncations occur, which, in principle, makes the preconditioner more efficient.

Acknowledgments This research was carried out in the framework of MATHEON supported by
Einstein Foundation Berlin.

References

1. Banas, L., Nurnberg, R.: A multigrid method for the Cahn-Hilliard equation with obstacle potential. Appl. Math. Comput. **213**(2), 290–303 (2009)
2. Barrett, J.W., Nurnberg, R., Styles, V.: Finite element approximation of a phase field model for void electromigration. SIAM J. Numer. Anal. **42**(2), 738–772 (2004)
3. Benzi, M.: Numerical solution of saddle point problems. Acta Numer. **14**, 1–137 (2005)
4. Blowey, J.F., Elliott, C.M.: The Cahn-Hilliard gradient theory for phase separation with non-smooth free energy Part I: numerical analysis. Eur. J. Appl. Math. **2**, 233–280 (1991)
5. Blowey, J.F., Elliott, C.M.: The Cahn-Hilliard gradient theory for phase separation with non-smooth free energy Part II: numerical analysis. Eur. J. Appl. Math. **3** (1992)
6. Bosch, J., Stoll, M., Benner, P.: Fast solution of Cahn-Hilliard variational inequalities using implicit time discretization and finite elements. J. Comput. Phys. **262**, 38–57 (2014)
7. Cahn, J.W., Hilliard, J.E.: Free energy of a nonuniform system. I. Interfacial free energy. J. Chem. Phys. **28**(2) (1958)
8. Graeser, C., Kornhuber, R.: Nonsmooth newton methods for set-valued saddle point problems. SIAM J. Numer. Anal. **47**(2), 1251–1273 (2009)
9. Graser, C.: Convex minimization and phase field models. Ph.D. thesis, FU Berlin (2011)
10. Graser, C., Kornhuber, R.: Multigrid methods for obstacle problems. J. Comput. Math. **27**(1), 1–44 (2009)
11. Kornhuber, R.: Monotone multigrid methods for elliptic variational inequalities I. Numerische Mathematik **69**(2), 167–184 (1994)
12. Kornhuber, R.: Monotone multigrid methods for elliptic variational inequalities II. Numerische Mathematik **72**(4), 481–499 (1996)
13. Kumar, P.: Purely Algebraic Domain Decomposition Methods for the Incompressible Navier-Stokes Equations (2011). arXiv:1104.3349
14. Kumar, P.: Aggregation based on graph matching and inexact coarse grid solve for algebraic two grid. Int. J. Comput. Math. **91**(5), 1061–1081 (2014)
15. Kumar, P.: Fast solvers for nonsmooth optimization problems in phase separation. In: 8th International Workshop on Computational Optimization, FedCSIS 2015, IEEE, pp. 589–594 (2015)
16. Mandel, J.: A multilevel iterative method for symmetric, positive definite linear complementarity problems. Appl. Math. Optim. **11**, 77–95 (1984)
17. Notay, Y.: An aggregation-based algebraic multigrid method. Electron. Trans. Numer. Anal. **37**, 123–146 (2010)
18. Oono, Y., Puri, S.: Study of phase-separation dynamics by use of cell dynamical systems. I. Modeling. Phys. Rev. A, **38**(1) (1987)
19. Rao, C.R., Rao, M.B.: Matrix Algebra and Its Applications to Statistics and Econometrics. World Scientific, Singapore (1998)
20. Saad, Y.: Iterative Methods for Sparse Linear Systems, 2nd edn. SIAM, Philadelphia (2003)
21. Trottenberg, U., Oosterlee, C., Schuller, A.: Multigrid. Academic, Cambridge (2001)

The Constraints Aggregation Technique for Control of Ethanol Production

Paweł Drąg and Krystyn Styczeń

Abstract In the article a new method for control and optimization of the ethanol production process was presented. The fed batch reactor for ethanol production can be described by the system of the nonlinear differential-algebraic equations. The new control approach base on the direct shooting method, which leads to a medium- or large-scale nonlinear optimization problem. In the presented research, three main aims of the optimization with the differential-algebraic constraints are indicated. The first one is the minimization of the cost function, which represents a process performance index. The second and third aims are to provide consistent initial conditions for the system to be solved and to ensure the continuity of the differential state trajectories. For these reasons, the large number of decision variables and equality constraints, connected with both consistent initial conditions for the differential-algebraic system and continuity of the differential state trajectories, was treated using the constraints aggregation technique. The direct shooting approach together with constraints aggregation technique enables us to obtain a new form of the nonlinear optimization task.

Keywords Optimal control · Constraints aggregation · Differential-algebraic equations · Multiple shooting method · Ethanol production

1 Introduction

Control and optimization of chemical batch reactors has received major attention and play a key role in many real-life complex industrial systems. The value of the predefined performance index, such as profitability, product quality or productivity often

P. Drąg (✉) · K. Styczeń
Department of Control Systems and Mechatronics, Wrocław University of Technology, Janiszewskiego 11/17, 50-372 Wroclaw, Poland
e-mail: pawel.drag@pwr.edu.pl

K. Styczeń
e-mail: krystyn.styczen@pwr.edu.pl

© Springer International Publishing Switzerland 2016
S. Fidanova (ed.), *Recent Advances in Computational Optimization*,
Studies in Computational Intelligence 655, DOI 10.1007/978-3-319-40132-4_11

can be measured using the industrial cameras. This approach, which combines effective control methods with fast industrial cameras and image processing algorithms, is the subject of intensive research [2, 23, 27, 30, 34].

In the article a mathematical model of the fed batch reactor for ethanol production with both differential and algebraic relations is considered [3, 5, 8]. The modeling of the complex chemical processes generally leads to systems of nonlinear differential-algebraic equations (*DAEs*). After some elimination steps, *DAE* system can be rewritten in the form of the ordinary differential equations (*ODEs*), which do not have to present the same process properties like *DAEs* [10].

Modeling of the technological processes using the differential-algebraic equations exhibits a few advantages. The most important one is, that the considered variables keep their original physical interpretation. The algebraic equations typically describe physical laws or explicit equality constraints and they should be kept invariant. Moreover, while modeling, simulation and optimization of the process, it is easier to vary design parameters in an implicit model. Also, the system structure can be exploited by problem-specific solvers [4].

The implicit models do not require the modeling simplifications, which is often necessary to get an *ODE* model. Because of this reason, the differential-algebraic systems do not need an advanced mathematical knowledge from user to obtain an *ODE* model. Then, it may be difficult to reformulate the problem as an *ODE* when nonlinearities are present [6].

Methods for dynamic optimization of differential-algebraic processes can be classified into simultaneous and sequential approaches. In the simultaneous approach, both the control and state trajectories are discretized as a set of piecewise polynomials and the *DAE* model is represented by a set of nonlinear algebraic equations. The simultaneous approach solves the dynamic model while the problem is optimized. In this way, the simultaneous approach leads to the large-scale optimization problems and nonlinear programming strategies designed for such problems have to be used [7].

When the sequential approach is applied, only the control profile is approximated by the piecewise polynomial elements. The properties of these elements are treated as the decision variables in nonlinear optimization problem. The sequential methods allow us to solve the models with standard *DAE solvers*. Then, the gradients are calculated for those decision variables on the basis of the objective function and the constraints. This optimization strategy is motivated by using the existing *DAE solvers* and enables us to solve large, detailed and nonlinear models [19].

Direct methods reformulate the original infinite dimensional optimal control problem into a finite dimensional nonlinear programming problem. It is possible by parametrization of both the state variables and the control function [12]. One of the most popular parametrization approach is the direct multiple shooting, which offers some computational advantages in the control of the technological processes:

(a) efficient well-known nonlinear programming procedures can be used,
(b) procedures for solving the differential-algebraic systems can be easily incorporated into the solution method,

(c) the method is well suited for parallel computations, because the integrators are decoupled on different intervals,
(d) the direct shooting approach, in the opposite to the simultaneous approach, leads to the medium-scale nonlinear programming problem,
(e) if the actual solution is sufficiently close to the optimum, then the computations can be prematurely terminated.

The computational complexity of the control problem solved by using the direct shooting approach is dominated by the cost of computing the derivative matrices of the both cost function and continuity constraints. One way to reduce the computational time is to reduce the number of constraints by using the constraints aggregation technique [14, 29].

This paper is organized as follows. In the next section the nonlinear differential-algebraic model of the fed-batch reactor for the ethanol production was presented. In the Sect. 3, the dynamic optimization strategy for single-stage differential-algebraic systems was considered. Extension of the dynamic optimization strategy for control and optimization of nonlinear and interconnected technological processes with differential-algebraic constraints was discussed in the Sect. 4. Then, the Kreisselmeier–Steinhauser (KS) aggregation function was introduced into the presented solving procedure. The results of numerical simulation were presented in the Sect. 6.

2 Model of the Process

In this study the ethanol production from the anaerobic glucose fermentation by *Saccharomyces cerevisiae* is considered [11, 20, 25, 35].

The optimal control problem is to maximize the yield of ethanol using the feed rate as the control variable. From the computational purposes, the maximization of the yield ethanol was transformed into the minimization problem. The mathematical statement with the known terminal time is as follows.

Find the feed flowrate $\mathbf{u}(t)$ over $t \in [t_0, t_F]$ to minimize

$$\min_{\mathbf{u}(t)} \mathcal{J} = -y_3(t_F) \cdot y_4(t_F), \tag{1}$$

subject to

$$\frac{dy_1}{dt} = z_1 y_1 - \mathbf{u}\frac{y_1}{y_4}, \tag{2}$$

$$\frac{dy_2}{dt} = -10z_1 y_1 + \mathbf{u}\frac{150 - y_2}{y_4}, \tag{3}$$

$$\frac{dy_3}{dt} = z_2 y_1 - \mathbf{u}\frac{y_3}{y_4}, \tag{4}$$

$$\frac{dy_4}{dt} = \mathbf{u}, \tag{5}$$

$$0 = z_1 - \left(\frac{0.408}{1 + y_3/16}\right)\left(\frac{y_2}{0.22 + y_2}\right), \tag{6}$$

$$0 = z_2 - \left(\frac{1}{1 + y_3/71.5}\right)\left(\frac{y_2}{0.44 + y_2}\right) \tag{7}$$

with

$$\mathbf{y}(t) = \begin{bmatrix} y_1(t) \\ y_2(t) \\ y_3(t) \\ y_4(t) \end{bmatrix}, \tag{8}$$

where y_1, y_2 and y_3 are the cell mass, substrate and product concentrations (g/L), respectively, and y_4 is the volume (L). The initial values of differential states are

$$\mathbf{y}(0) = [1 \quad 150 \quad 0 \quad 10]^T. \tag{9}$$

The constraints on the control variable are

$$0 \leq \mathbf{u}(t) \leq 12. \tag{10}$$

There is an end-point constraint on the volume

$$0 \leq y_4(t_F) \leq 200 \tag{11}$$

for $t_F = 54$ h.

3 Dynamic Optimization with Differential-Algebraic Constraints

Dynamic optimization methods are aimed at obtaining the optimal operating policies, which optimize a scalar-valued objective function, known as the process performance index. The performance index is of a great practical importance because it reflects the productivity or any other parameters characterizing the production process [16, 32].

In the our researches, the optimal control methods are aimed at the minimization of the following performance cost function

$$\min_{(\mathbf{y}(t),\mathbf{z}(t),\mathbf{u}(t),\mathbf{p})} \int_{t_0}^{t_F} \mathcal{L}(\mathbf{y}(t), \mathbf{z}(t), \mathbf{u}(t), \mathbf{p}, t)dt + \mathcal{E}(\mathbf{y}(t_F), \mathbf{z}(t_F)), \tag{12}$$

with two scalar functions, which are continuously differentiable with respect to all of their arguments:

the Lagrange term

$$\mathcal{L} : \mathcal{R}^{n_y} \times \mathcal{R}^{n_z} \times \mathcal{R}^{n_u} \times \mathcal{R}^{n_p} \times \mathcal{R} \to \mathcal{R} \tag{13}$$

and the Mayer term

$$\mathcal{E} : \mathcal{R}^{n_y} \times \mathcal{R}^{n_z} \to \mathcal{R}. \tag{14}$$

Let us consider the process equations. The relationships, which describes the real-life systems can take a general *fully-implicit* form

$$F(\mathbf{y}(t), \mathbf{z}(t), \mathbf{u}(t), \mathbf{p}, t) = 0, \tag{15}$$

where $\mathbf{y}(t) \in \mathcal{R}^{n_y}$ is a differential state, $\mathbf{z}(t) \in \mathcal{R}^{n_z}$ is an algebraic state, $\mathbf{u}(t) \in \mathcal{R}^{n_u}$ is the control function and $\mathbf{p} \in \mathcal{R}^{n_p}$ is a vector of global parameters constant in the time. The vector-valued nonlinear function

$$F : \mathcal{R}^{n_y} \times \mathcal{R}^{n_z} \times \mathcal{R}^{n_u} \times \mathcal{R}^{n_p} \times \mathcal{R} \to \mathcal{R}^{n_y} \times \mathcal{R}^{n_z} \tag{16}$$

is considered. The description of the process, presented like in Eq. (15) is very general and can be difficult to analyze in all possible situations.

The technological processes presented in a descriptor form are more convenient to the analysis than presented like in Eq. (15). They enable us a compact description of the systems with complicated structure nad process relationships.

The descriptor models are applicable in wide range of industry and science. They can be applied to model many processes and systems in electrical engineering [18], mechanical engineering [26], chemistry [13, 33], environmental engineering [22], biotechnology [15] and glaciology [1]. Moreover, in many practical applications the *DAE* systems can have some highly nonlinear components.

In the practical applications the following assumption was made

Assumption 3.1 The process can be modeled by a system of the index-one differential-algebraic equations (DAEs) of the form

$$\begin{aligned} B(\cdot)\mathbf{y}(t) &= F_1(\mathbf{y}(t), \mathbf{z}(t), \mathbf{u}(t), \mathbf{p}, t) \\ 0 &= F_2(\mathbf{y}(t), \mathbf{z}(t), \mathbf{u}(t), \mathbf{p}, t), \end{aligned} \tag{17}$$

with the non-singular matrix B.

The consistent initial values of the differential and algebraic states, as well as the other system parameters are denoted as follows

$$\mathbf{y}(t_0) = \mathbf{y}_0, \tag{18}$$

$$\mathbf{z}(t_0) = \mathbf{z}_0, \tag{19}$$

$$\mathbf{p}(t_0) = \mathbf{p}(t) = \mathbf{p}_0. \tag{20}$$

Remark 3.2 The consistent initial values of the single-stage technological process, can be obtained by the physical interpretation of the differential-algebraic relations.

There is a quite different situation, when the multistage technological systems are considered.

Remark 3.3 Each stage of the multi-stage technological system can be described by different set of the nonlinear differential-algebraic equations.

4 Multiple Shooting Method

The multiple shooting approach is treated as a tool for decomposition and parametrization of the nonlinear differential-algebraic systems.

Let us assume, that N characteristic stages can be distinguished in the process. Moreover, there is an independent variable t (eg. time or length of the chemical reactor).

Let us to divide the considered time horizon $[t_0 \quad t_F]$ into N subintervals of the following form

$$t^i \in [t_0^i \quad t_F^i], \quad i = 1, \ldots, N. \tag{21}$$

It means, that

$$t_0 = t_0^1 < t_F^1 = t_0^2 < t_F^2 = \cdots < t_F^{N-1} = t_0^N < t_F^N = t_F. \tag{22}$$

At the next step, the control function $\mathbf{u}(t)$ can be discretized in an appropriate way. The most frequently encountered representations of the control function are a piecewise constant, a piecewise linear, as well as a polynomial approximation of the control [31].

If the control function is parametrized as a piecewise constant vector function, then

$$\mathbf{u}(t^i) = \mathbf{x}_{\mathbf{u}^i} \tag{23}$$

for $t^i \in [t_0^i \quad t_F^i], \quad i = 1, \ldots, N.$

By the multiple shooting method, the *DAEs* is parametrized and the solution of the *DAE* system is decoupled on the N intervals $[t_0^i \quad t_F^i]$.

In this way the initial values $\mathbf{x}_{\mathbf{y}^i}$ and $\mathbf{x}_{\mathbf{z}^i}$ of differential and algebraic states at times t_0^i are introduced as the additional optimization variables.

Remark 4.1 The consistent initial values of the multi-stage technological process are usually unknown and can be treated as the additional decision variables.

The trajectories $\mathbf{y}(t)$ and $\mathbf{z}(t)$ are obtained as a set of trajectories $\mathbf{y}^i(t^i)$ and $\mathbf{z}^i(t^i)$ on each interval $t^i \in [t_0^i \quad t_F^i]$. Moreover, the trajectories $\mathbf{y}^i(t^i)$ and $\mathbf{z}^i(t^i)$ are the solutions of an initial value problem

$$B^i(\cdot)\dot{\mathbf{y}}(t) = F_1^i(\mathbf{y}^i(t^i), \mathbf{z}^i(t^i), \mathbf{u}^i(t^i), \mathbf{p}, t^i)$$
$$0 = F_2^i(\mathbf{y}^i(t^i), \mathbf{z}^i(t^i), \mathbf{u}^i(t^i), \mathbf{p}, t^i) + \alpha^i(t_0^i) F_2^i(\mathbf{x}_{\mathbf{y}^i}, \mathbf{x}_{\mathbf{z}^i}, \mathbf{x}_{\mathbf{u}^i}, p, t^i)$$
$$t^i \in [t_0^i \quad t_F^i], \quad i = 1, \dots, N. \tag{24}$$

Remark 4.2 To solve the DAE system with inconsistent initial conditions, the relaxation parameter $\alpha^i(t_0^i)$ was introduced.

In this way, the trajectories $\mathbf{y}^i(t^i)$ and $\mathbf{z}^i(t^i)$ on the interval $t^i \in [t_0^i \quad t_F^i]$ are the functions of the initial values, controls, global parameters and independent variable $\mathbf{x}_{\mathbf{y}^i}, \mathbf{x}_{\mathbf{z}^i}, \mathbf{x}_{\mathbf{u}^i}, \mathbf{p}, t^i$, respectively.

The integral part of the performance cost function can be evaluated on each interval independently

$$\int_{t_0^1}^{t_F^1} \mathcal{L}^1(\mathbf{y}^1(t^1), \mathbf{z}^1(t^1), \mathbf{u}^1(t^1), \mathbf{p}, t^1)dt + \cdots$$

$$+ \int_{t_0^N}^{t_F^N} \mathcal{L}^N(\mathbf{y}^N(t^N), \mathbf{z}^N(t^N), \mathbf{u}^N(t^N), p, t^N)dt + \mathcal{E}(\mathbf{y}(t_F^N), \mathbf{z}(t_F^N)) \tag{25}$$

$$= \sum_{i=1}^{N} \int_{t_0^i}^{t_F^i} \mathcal{L}^i(\mathbf{y}^i(t^i), \mathbf{z}^i(t^i), \mathbf{u}^i(t^i), \mathbf{p}, t^i)dt + \mathcal{E}(\mathbf{y}(t_F^N), \mathbf{z}(t_F^N)).$$

Remark 4.3 The multiple shooting method enables us to solve the DAE system parallel, independently for each time interval $t^i \in [t_0^i \quad t_F^i]$.

The parameterization of the optimal control problem with the multistage *DAE* constraints using the multiple shooting approach and a piecewise constant control representation leads to the following nonlinear programming problem

$$\min_{\mathbf{X}_{\mathbf{y}}, \mathbf{X}_{\mathbf{z}}, \mathbf{X}_{\mathbf{u}}, \mathbf{X}_{\mathbf{p}}} \sum_{i=1}^{N} \int_{t_0^i}^{t_F^i} \mathcal{L}^i(\mathbf{y}^i(t^i), \mathbf{z}^i(t^i), \mathbf{u}^i(t^i), \mathbf{p}, t^i)dt$$
$$+ \mathcal{E}(\mathbf{y}^N(t_F^N), \mathbf{z}^N(t_F^N)) = \min_{\mathbf{X}} \Phi(\mathbf{X}), \tag{26}$$

with

$$\mathbf{X} = [\mathbf{X}_{\mathbf{y}} \quad \mathbf{X}_{\mathbf{z}} \quad \mathbf{X}_{\mathbf{u}} \quad \mathbf{X}_{\mathbf{p}}], \tag{27}$$

$$X_y = [x_{y^1} \quad x_{y^2} \quad \cdots \quad x_{y^N}], \tag{28}$$

$$X_z = [x_{z^1} \quad x_{z^2} \quad \cdots \quad x_{z^N}], \tag{29}$$

$$X_u = [x_{u^1} \quad x_{u^2} \quad \cdots \quad x_{u^N}], \tag{30}$$

$$X_p = x_p \tag{31}$$

subject to the continuity conditions

$$c_{cont}^i = x_{y^i} - y^{i-1}(t_F^{i-1}) = 0, \quad i = 2, \ldots, N, \tag{32}$$

the consistency conditions

$$c_{cons}^i = F_2^i(x_{y^i}, x_{z^i}, x_{u^i}, p, t_0^i) = 0, \quad i = 1, \ldots, N, \tag{33}$$

control path constraints imposed pointwise at the multiple shooting nodes

$$c_u^i(x_{y^i}, x_{z^i}, x_{u^i}, p, t_0^i) \le 0, \quad i = 1, \ldots, N \tag{34}$$

and the lower and upper bounds on the decision variables

$$X^L \le X \le X^U \tag{35}$$

with the system of the differential-algebraic constraints in each interval

$$
\begin{aligned}
B^i(\cdot)\dot{y}(t) &= F_1^i(y^i(t^i), z^i(t^i), u^i(t^i), p, t^i) \\
0 &= F_2^i(y^i(t^i), z^i(t^i), u^i(t^i), p, t^i) + \alpha^i(t_0^i)F_2^i(x_{y^i}, x_{z^i}, x_{u^i}, p, t^i) \\
t^i &\in [t_0^i \quad t_F^i], \quad i = 1, \ldots, N.
\end{aligned}
\tag{36}
$$

5 The Kreisselmeier–Steinhauser Function for Constraints Aggregation

The Kreisselmeier–Steinhauser (*KS*) function was introduced as a tool, which transform optimization problem with multiple constraints into a scalar optimization function [21, 28].

It is known, that each equality constraint can be expressed as a pair of the inequality constraint in the following manner

$$c(X) = 0 \iff (c(X) \le 0 \land c(X) \ge 0). \tag{37}$$

Let us to consider an optimization problem with m inequality constraints

$$\mathbf{c}(\mathbf{X}) = \begin{bmatrix} c^1(\mathbf{X}) \\ c^2(\mathbf{X}) \\ \vdots \\ c^m(\mathbf{X}) \end{bmatrix} \leq 0. \tag{38}$$

The Kreisselmeier–Steinhauser function, which estimates the values of all the constraints, is defined in the following form

$$KS(\mathbf{X}, \rho) = \frac{1}{\rho} \ln \sum_{j=1}^{m} e^{\rho c^j(\mathbf{X})}, \tag{39}$$

where ρ is an approximation parameter with a limiting value of infinity.

It follows, that for $\rho > 0$,

$$KS(\mathbf{X}, \rho) \geq \max(c^j(\mathbf{X})) \tag{40}$$

and for $\rho_1 > \rho_2$,

$$KS(\mathbf{X}, \rho_2) \geq KS(\mathbf{X}, \rho_1) \geq \max(c^j(\mathbf{X})). \tag{41}$$

From this it follows, that as ρ increases, the KS function gives a better estimate of the feasible region. One of the sources of the numerical difficulties is, that $e^{c^j(\mathbf{X})}$ becomes very large for $c^j(\mathbf{X}) >> 0$ and $\rho \to \infty$.

To avoid this problem, the scalar $\bar{c}$ can be defined as an upper bound to $c^j(\mathbf{X})$

$$\bar{c} \geq \mathbf{c}_{\max} = \max_{j=1,...,m} \{c^j(\mathbf{X})\}. \tag{42}$$

Then, the equivalent expression can be presented as

$$KS(\mathbf{X}, \rho) = \bar{c} + \frac{1}{\rho} \ln \left(\sum_{j=1}^{m} e^{\rho(c^j(\mathbf{X}) - \bar{c})} \right). \tag{43}$$

In this way the expression $e^{\rho(c^j(\mathbf{X}) - \bar{c})} << 1$ and the computational difficulties can be avoided.

One can observe, that the constraints (43) underestimate the feasible region. One way to expand this region is to consider average violation of the constraints within a modified KS function [9]

$$\overline{KS}(\mathbf{X}, \rho) = \overline{c} + \frac{1}{\rho} \ln \left(\frac{1}{m} \sum_{j=1}^{m} e^{\rho(c^j(\mathbf{X}) - \overline{c})} \right). \tag{44}$$

It is known, that KS function allows us to aggregate a large number of both equality and inequality constraints into a scalar function. For active constraints, $c^j(\mathbf{X})$ becomes active, $c^j(\mathbf{X}) \to 0$ and $e^{\rho(c^j(\mathbf{X}))} \to 1$. Otherwise, as $\rho \to \infty$, then $e^{\rho(c^j(\mathbf{X}))} \approx 0$ for $c^j(\mathbf{X}) << 0$.

To apply this approach to the state constraints, at the shooting points the state variable constraints $c^j(\mathbf{x}_{y^i}) < 0$ were applied

$$KS(\mathbf{X}, \rho) = \overline{c} + \frac{1}{\rho} \ln \left(\sum_{j=1}^{n_y} \sum_{i=1}^{N} e^{\rho(c^j(\mathbf{x}_{y^i}) - \overline{c})} \right), \tag{45}$$

or

$$\overline{KS}(\mathbf{X}, \rho) = \overline{c} + \frac{1}{\rho} \ln \left(\frac{1}{m} \sum_{j=1}^{n_y} \sum_{i=1}^{N} e^{\rho(c^j(\mathbf{x}_{y^i}) - \overline{c})} \right). \tag{46}$$

6 Results

The control problem of the fed batch reactor for ethanol production was based on the presented KS function aggregation approach.

At the beginning, the model of the reactor was divided into 100 parts. The decision variables vector with the initial values of the differential states was as follows

$$\mathbf{x}_{y_1}^{1-99} = 1.0, \tag{47}$$

$$\mathbf{x}_{y_2}^{1-99} = 150.0, \tag{48}$$

$$\mathbf{x}_{y_3}^{1-99} = 0.0, \tag{49}$$

$$\mathbf{x}_{y_4}^{1-99} = 10.0, \tag{50}$$

$$\mathbf{x}_{z_1}^{1-100} = 0.0, \tag{51}$$

$$\mathbf{x}_{z_2}^{1-100} = 0.0, \tag{52}$$

$$\mathbf{x}_{u}^{1-100} = 0.0. \tag{53}$$

There are constraints of the lower and upper bounds type

$$1.0 \le \mathbf{x_{y_1}}^{1-99} \le 100, \tag{54}$$

$$0.0 \le \mathbf{x_{y_2}}^{1-99} \le 150, \tag{55}$$

$$0.0 \le \mathbf{x_{y_3}}^{1-99} \le 100, \tag{56}$$

$$10 \le \mathbf{x_{y_4}}^{1-99} \le 200, \tag{57}$$

$$0.0 \le \mathbf{x_{z_1}}^{1-100} \le 150, \tag{58}$$

$$0.0 \le \mathbf{x_{z_2}}^{1-100} \le 150, \tag{59}$$

$$0.0 \le \mathbf{x_u}^{1-100} \le 12. \tag{60}$$

As a result, there are 4×99 equality continuity constraints and 2×100 equality constraints for the consistent initial conditions.

The most of the computational effort is connected with large-scale matrix of the partial derivatives of the constraints.

The applied constraints function aggregation causes, that there is a smaller number of the considered constraints

$$\mathbf{c}_{cont}(\mathbf{X}) = \frac{1}{\rho} \ln \sum_{i=1}^{396} e^{c_{cont}^i(\mathbf{X})}, \tag{61}$$

$$\mathbf{c}_{cons}(\mathbf{X}) = \frac{1}{\rho} \ln \sum_{i=1}^{200} e^{c_{cons}^i(\mathbf{X})}. \tag{62}$$

The biggest advantage of using these functions is much smaller size of the matrix derivatives. The property is valid if the structure of the considered process cannot be determined in advance.

Results for nonlinear programming with constraints aggregation were presented in the Tables 1 and 2. The obtained state trajectories were plotted on the Fig. 1.

Table 1 Results of *fmincon* [17] procedure with KS constraints aggregation function		
	The final value	20699.80
	Number of iterations	187
	Number of objective function evaluations	18994
	Computation time	2454.7 s

Table 2 Results of *fmincon* procedure without constraints aggregation

The final value	20699.80
Number of iterations	201
Number of objective function evaluations	20426
Computation time	3061.5 s

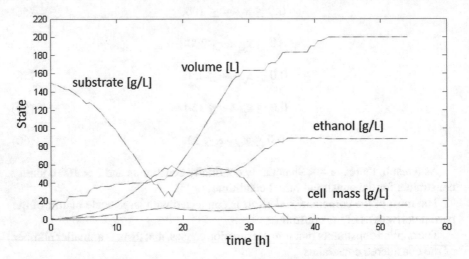

Fig. 1 The state trajectories obtained with KS constraints aggregation function

7 Conclusion

In the article the new algorithm for control of the ethanol production process was presented. The main advantages of the algorithm is connected with the reduction of computational effort in gradient matrices by aggregating continuity constraints by the KS function. Moreover, the KS function was explored as a way of handling the path constraints by aggregating each set of path constraints into a single endpoint constraints.

The general formulation of the presented approach enables us to apply the new algorithm in complex technological processes and other large-scale systems described by the highly nonlinear differential-algebraic systems with unknown feasible initial conditions.

Acknowledgments This work has been supported by the National Science Center under grant: 2012/07/B/ST7/012166.

References

1. Ahlkrona, J., Lötstedt, P., Kirchner, N., Zwinger, T.: Dynamically coupling the non-linear stokes equations with the shallow ice approximation in glaciology: Description and first applications of the ISCAL method. J. Comput. Phys. **308**, 1–19 (2016). doi:10.1016/j.jcp.2015.12.025
2. An, Y.-K., Yang, J., Hwang, S., Sohn, H.: Line laser lock-in thermography for instantaneous imaging of cracks in semiconductor chips. Opt. Lasers Eng. **73**, 128–136 (2015). doi:10.1016/j.optlaseng.2015.04.013
3. Bai, F.W., Anderson, W.A., Moo-Young, M.: Ethanol fermentation technology from sugar and starch feedstocks. Biotechnol. Adv. **26**, 89–105 (2008). doi:10.1016/j.biotechadv.2007.09.002
4. Balsa-Canto, E., Vassiliadis, V.S., Banga, J.R.: Dynamic optimization of single- and multi-stage systems using a hybrid stochastic-deterministic method. Ind. Eng. Chem. Res. **44**, 1514–1523 (2005). doi:10.1021/ie0493659
5. Banga, J.R., Alonso, A.A., Singh, R.P.: Stochastic dynamic optimization of batch and semi-continuous bioprocesses. Biotechnol. Prog. **13**, 326–335 (1997). doi:10.1021/bp970015+
6. Betts J.T.: Practical Methods for Optimal Control and Estimation Using Nonlinear Programming, 2nd edn. SIAM, Philadelphia (2010). doi:10.1137/1.9780898718577
7. Biegler L.T.: Nonlinear Programming. Concepts, Algorithms and Applications to Chemical Processes. SIAM, Philadelphia (2010). doi:10.1137/1.9780898719383
8. Birol, G., Doruker, P., Kirdar, B., Önsan, Z.I., Ülgen, K.: Mathematical description of ethanol fermentation by immobilized Saccharomyces cerevisiae. Process Biochem. **33**, 763–771 (1998). doi:10.1016/S0032-9592(98)00047-8
9. Bloss, K.F., Biegler, L.T., Schiesser, W.E.: Dynamic process optimization through adjoint formulations and constraint aggregation. Ind. Eng. Chem. Res. **38**, 421–432 (1999). doi:10.1021/ie9804733
10. Brenan K. E., Campbell S. L., Petzold L. R.: Numerical Solution of Initial-Value Problems in Differential Algebraic Equations. SIAM, Philadelphia (1996). doi:10.1137/1.9781611971224
11. Chen, C.T., Hwang, C.: Optimal on-off control for fed-batch fermentation processes. Ind. Eng. Chem. Res. **29**, 1869–1875 (1990). doi:10.1021/ie00105a019
12. Diehl, M., Bock, H.G., Schlder, J.P., Findeisen, R., Nagy, Z., Allgwer, F.: Real-time optimization and nonlinear model predictive control of processes governed by differential-algebraic equations. J. Process Control **12**, 577–585 (2002). doi:10.1016/S0959-1524(01)00023-3
13. Drąg P., Styczeń K.: A Two-Step Approach for Optimal Control of Kinetic Batch Reactor with Electroneutrality Condition. Przeglad Elektrotechniczny, **88**, 176-180 (2012)
14. Drąg P., Styczeń K.: Simulated annealing with constraints aggregation for control of the multi-stage processes. In: 2015 Federated Conference on Computer Science and Information Systems (FedCSIS), pp. 461-469. IEEE, New York (2015). doi:10.15439/2015F255
15. Fidanova S., Paprzycki M., Roeva O.: Hybrid GA-ACO Algorithm for a model parameters identification problem. In: 2014 Federated Conference on Computer Science and Information Systems (FedCSIS), pp. 413-420. IEEE, New York (2014). doi:10.15439/2014F373
16. Flores-Tlacuahuac, A., Moreno, S.T., Biegler, L.T.: Global optimization of highly nonlinear dynamic systems. Ind. Eng. Chem. Res. **47**, 2643–2655 (2008). doi:10.1021/ie070379z
17. Global Optimization Toolbox: Users Guide. Mathworks, Inc, Natick (2010)
18. Günther, M.: A joint DAE/PDE model for interconnected electrical networks. Math. Comput. Model. Dyn. Syst. **6**, 114–128 (2000). doi:10.1076/1387-3954(200006)6:2;1-M;FT114
19. Hartwich, A., Stockmann, K., Terboven, C., Feuerriegel, S., Marquardt, W.: Parallel sensitivity analysis for efficient large-scale dynamic optimization. Optim. Eng. **12**, 489–508 (2011). doi:10.1007/s11081-010-9104-4
20. Hong, J.: Optimal substrate feeding policy for a fed batch fermentation with substrate and product inhibition kinetics. Biotechnol. Bioeng. **28**, 1421–1431 (1986). doi:10.1002/bit.260280916
21. Jeon, M.: Parallel optimal control with multiple shooting constraints aggregation and adjoint methods. J. Appl. Math. Comput. **19**, 215–229 (2005). doi:10.1007/BF02935800

22. Kwiatkowska M.: Badanie zmienności w czasie parametrów powietrza wewnętrznego z zastosowaniem metody DAEs. Interdyscyplinarne zagadnienia w inżynierii i ochronie środowiska T. 6. Wrocław, Oficyna Wydawnicza Politechniki Wrocławskiej, 214-220 (2015)
23. Liu, W., Shen, H., Xu, Y., Song, Y., Li, H., Jia, J., Ding, Y.: Developing a thermal control strategy with the method of integrated analysis and experimental verification. Optik—Int. J. Light Electron Opt. **126**, 2378–2382 (2015). doi:10.1016/j.ijleo.2015.05.138
24. Liu, C.-Z., Wang, F., Ou-Yang, F.: Ethanol fermentation in a magnetically fluidized bed reactor with immobilized Saccharomyces cerevisiae in magnetic particles. Bioresour. Technol. **100**, 878–882 (2009). doi:10.1016/j.biortech.2008.07.016
25. Najafpour, G., Younesi, H., Ismail, K.S.K.: Ethanol fermentation in an immobilized cell reactor using Saccharomyces cerevisiae. Bioresour. Technol. **92**, 251–260 (2004). doi:10.1016/j.biortech.2003.09.009
26. Nugroho L.: Comparison of classical and modern landing control system for a small unmanned aerial vehicle. In: 2014 International Conference on Computer, Control, Informatics and Its Applications (IC3INA), pp. 187-192. IEEE, New York (2014). doi:10.1109/IC3INA.2014.7042625
27. Ojha S., Sakhare S.: Image processing techniques for object tracking in video surveillance - A survey. In: 2015 International Conference on Pervasive Computing: Advance Communication Technology and Application for Society, ICPC 2015, article number 7087180 (2015). doi:10.1109/PERVASIVE.2015.7087180
28. Poon, N.M.K., Martins, J.R.R.A.: An adaptive approach to constraint aggregation using adjoint sensitivity analysis. Struct. Multidisc. Optim. **34**, 61–73 (2007). doi:10.1007/s00158-006-0061-7
29. Rogers, D.F., Plante, R.D., Wong, R.T., Evans, J.R.: Aggregation and disaggregation techniques and methodology in optimization. Oper. Res. **39**, 553–582 (1991). doi:10.1287/opre.39.4.553
30. Schmitt R.L., Leclair T.T., Hedderich J.O.: Infrared thermography technologies for thermal measurement and control of HMA pavement construction. In: Airfield and Highway Pavements 2015: Innovative and Cost-Effective Pavements for a Sustainable Future—Proceedings of the 2015 International Airfield and Highway Pavements Conference, 236-247 (2015). doi:10.1061/9780784479216.022
31. Vassiliadis, V.S., Sargent, R.W.H., Pantelides, C.C.: Solution of a class of multistage dynamic optimization problems. 1. problems without path constraints. Ind. Eng. Chem. Res. **33**, 2111–2122 (1994). doi:10.1021/ie00033a014
32. Vassiliadis, V.S., Sargent, R.W.H., Pantelides, C.C.: Solution of a class of multistage dynamic optimization problems. 2. problems with path constraints. Ind. Eng. Chem. Res. **33**, 2122–2123 (1994). doi:10.1021/ie00033a015
33. Vazquez-Castillo, J.A., Segovia-Hernandez, J.G., Ponce-Ortega, J.M.: Multiobjective optimization approach for integrating design and control in multicomponent distillation sequences. Ind. Eng. Chem. Res. **54**, 12320–12330 (2015). doi:10.1021/acs.iecr.5b01611
34. Wong, A., Guo, Y., Park, C.B., Zhou, N.Q.: A polymer visualization system with accurate heating and cooling control and high-speed imaging. Int. J. Mol. Sci. **16**, 9196–9216 (2015). doi:10.3390/ijms16059196
35. Zaldivar, J., Nielsen, J., Olsson, L.: Fuel ethanol production from lignocellulose: a challenge for metabolic engineering and process integration. Appl. Microbiol. Biotechnol. **56**, 17–34 (2001). doi:10.1007/s002530100624

InterCriteria Analysis by Pairs and Triples of Genetic Algorithms Application for Models Identification

Olympia Roeva, Tania Pencheva, Maria Angelova and Peter Vassilev

Abstract In this investigation the InterCriteria Analysis (ICrA) approach is applied. The apparatuses of index matrices and intuitionistic fuzzy sets are at the core of ICrA. They are used to examine the influences of two main genetic algorithms (GA) parameters—the rates of crossover ($xovr$) and mutation ($mutr$). A series of parameter identification procedures for *S. cerevisiae* and *E. coli* fermentation process models is fulfilled. Twenty GA with different $xovr$ and $mutr$ values are applied. Relations between ICrA criteria—GA parameters and outcomes, on the one hand, and fermentation process model parameters, on the other hand, are investigated. The ICrA approach is applied by pairs, as well as by triples. The obtained results are thoroughly analysed towards computation time and model accuracy and some conclusions about the derived criteria interactions are reported.

Keywords InterCriteria analysis · Index matrices · Intuitionistic fuzzy sets · Genetic algorithm · Parameter identification · Algorithm performance · *S. cerevisiae* · *E. coli* · Fermentation

1 Introduction

InterCriteria Analysis (ICrA) [11] is a contemporary approach for multi-criteria decision making. ICrA implements the apparatuses of index matrices (IM) and intuitionistic fuzzy sets (IFS) in order to compare some criteria reflecting the behaviour of

O. Roeva (✉) · T. Pencheva · M. Angelova · P. Vassilev
Institute of Biophysics and Biomedical Engineering, Bulgarian Academy of Sciences,
Sofia, Bulgaria
e-mail: olympia@biomed.bas.bg

T. Pencheva
e-mail: tania.pencheva@biomed.bas.bg

M. Angelova
e-mail: maria.angelova@biomed.bas.bg

P. Vassilev
e-mail: peter.vassilev@gmail.com

© Springer International Publishing Switzerland 2016   193
S. Fidanova (ed.), *Recent Advances in Computational Optimization*,
Studies in Computational Intelligence 655, DOI 10.1007/978-3-319-40132-4_12

considered objects. Recently, ICrA has been successfully applied to model parameter identification of fermentation processes (FP). In [20] ICrA is applied to establish relations between two of the main genetic algorithms (GA) parameters—number of individuals and number of generations, on the one hand, and computation time T, model accuracy J and model parameters, on the other hand. This study provokes the further applications of ICrA in FP model identification.

FP are objects of increased research interest because of their widespread use in different branches of industry. The modeling and optimization of FP are a real challenge to the investigators. FP models have complex structures based on systems of non-linear differential equations with several specific growth rates [13]. The choice of an appropriate model parameter identification procedure is the most important problem for adequate modeling of FP. Among biologically inspired optimization techniques, GA [16] has been proved as a global search method [14] for solving different engineering and optimization problems [23], i.e. for parameter identification of FP [1, 2, 22]. GA efficiency strongly depends on the tuning of different operators, functions, and parameters. These settings are specifically implemented to different problems. The current investigation is focused on the evaluation of the impact of two of the main GA parameters, namely crossover ($xovr$) and mutation ($mutr$) rates. Simple GA (SGA) is applied for the purposes of model parameter identification of two fed-batch FP—*S. cerevisiae* and *E. coli*. Both yeast and bacteria have numerous applications in food and pharmaceutical industries. These microorganisms are widely used as model organisms in genetic engineering and cell biology due to their well known metabolic pathways [15, 18].

In this investigation the obtained results from SGA parameter identification of considered here FP models are used to determine some dependencies between preliminary defined as of significant importance criteria. The establishment of the relations between criteria—*S. cerevisiae* and *E. coli* FP models parameters, on the one hand, and SGA parameters ($xovr$ and $mutr$) and outcomes (T and J), on the other hand, is performed by ICrA implementation. The ICrA approach is applied by pairs [11] and by triples [25]. This is expected to lead to additional knowledge about considered ICrA criteria, which will be valuable especially in the case of modelling of living systems, such as FP.

2 Problem Formulation

2.1 Mathematical Models of Fermentation Processes

In order to illustrate ICrA by pairs and triples for the purposes of model parameter identification using SGA, two Case studies are going to be presented here: Case study 1—yeast *S. cerevisiae* fed-batch FP, and Case study 2—bacteria *E. coli* fed-batch FP.

Case study 1. *S. cerevisiae* fed-batch fermentation process

The mathematical model of *S. cerevisiae* fed-batch FP is presented by the following non-linear differential equations system [21]:

$$
\frac{dX}{dt} = \left(\mu_{2S} \frac{S}{S + k_S} + \mu_{2E} \frac{E}{E + k_E} \right) X - \frac{F_{in}}{V} X,
\tag{1}
$$

$$
\frac{dS}{dt} = -\frac{\mu_{2S}}{Y_{S/X}} \frac{S}{S + k_S} X + \frac{F_{in}}{V} (S_{in} - S),
\tag{2}
$$

$$
\frac{dV}{dt} = F_{in},
\tag{3}
$$

where X is the biomass concentration, [g/l]; S—substrate concentration, [g/l]; E—ethanol concentration, [g/l]; F_{in}—feeding rate, [l/h]; V—bioreactor volume, [l]; S_{in}—substrate concentration in the feeding solution, [g/l]; μ_{2S}, μ_{2E}—the maximum values of the specific growth rates, [1/h]; k_S, k_E—saturation constants, [g/l]; $Y_{S/X}$—yield coefficient, [-].

For the considered here model (Eqs. (1)–(3)), the vector of parameters to be identified is $p_1 = [\mu_{2S} \ \mu_{2E} \ k_S \ k_E \ Y_{S/X}]$.

Case study 2. *E. coli* fed-batch fermentation process

The mathematical model of *E. coli* fed-batch FP is presented by the following non-linear differential equations system [22]:

$$
\frac{dX}{dt} = \mu_{max} \frac{S}{k_S + S} X - \frac{F_{in}}{V} X,
\tag{4}
$$

$$
\frac{dS}{dt} = -\frac{\mu_{max}}{Y_{S/X}} \frac{S}{S + k_S} X + \frac{F_{in}}{V} (S_{in} - S),
\tag{5}
$$

$$
\frac{dV}{dt} = F_{in},
\tag{6}
$$

where all notations keep their meaning as described above, and, additionally, μ_{max} is the maximum value of the specific growth rate, [1/h].

For the considered here model (Eqs. (4)–(6)), the vector of parameters to be identified is $p_2 = [\mu_{max} \ k_S \ Y_{S/X}]$.

Model parameters identification of both fed-batch FP is performed based on experimental data for biomass, glucose and ethanol concentrations. The detailed description of the process conditions and experimental data is presented in [21, 22].

2.2 Optimization Criterion

The objective function is designed aiming at identification of parameter vectors p_1 and p_2 in order to obtain the best fit to a data set and it is defined as:

$$J = \sum_{i=1}^{m} \left(X_{\exp}(i) - X_{\mod}(i) \right)^2 + \sum_{i=1}^{n} \left(S_{\exp}(i) - S_{\mod}(i) \right)^2 \rightarrow \min, \qquad (7)$$

where m and n are the experimental data dimensions; $X_{\exp}$ and $S_{\exp}$—available experimental data for biomass and substrate; $X_{\mod}$ and $S_{\mod}$—model predictions for biomass and substrate with a given model parameter vector.

2.3 Simple Genetic Algorithms for Parameter Identification

Simple genetic algorithm, initially presented in Goldberg [16], searches a global optimal solution using three main genetic operators in a sequence selection, crossover and mutation. SGA starts with a creation of a randomly generated initial population. Each solution is then evaluated and assigned a fitness value. According to the fitness function, the most suitable solutions are selected. After that, crossover proceeds to form a new offspring. Mutation is then applied with determinate probability aiming to prevent falling of all solutions into a local optimum. The execution of SGA has been repeated until the termination criterion (i.e. reached number of generations, or found solution with a specified tolerance, etc.) is satisfied.

Crossover and mutation are among of the most important operators that can increase the efficiency of GA. The crossover operator is used to generate offspring by exchanging bits in a pair of parents chromosomes chosen from the population. Crossover occurs with a crossover probability (crossover rate, $xovr$), that indicates a ratio of how many couples will be picked for mating. The mutation operator changes some elements in selected chromosomes with a mutation probability (mutation rate, $mutr$). As such, the operator introduces genetic diversity and helps GA to escape the local optimum. It is well known that optimal $xovr$ and $mutr$ values vary for different problems and GA efficacy depends on their choice [17]. Usually, the determination what values of $xovr$ and $mutr$ should be used is processed on the trial-and-error basis. In the literature there exist a number of guidelines how $xovr$ and $mutr$ values to be tuned [16, 17, 19]. Recommended values of $xovr$ are high, usually in the range 0.5–1.0 [17, 19]. On the other hand, low mutation rate values for preventing search process to be turn into a simple random search are commonly adopted in GA. Typical values of $mutr$ are in the range 0.001–0.1 [17, 19].

In this investigation the impact of $xovr$ and $mutr$ values is going to be examined choosing different values of the both SGA parameters. In Case study 1 SGA is applied with the following values of crossover rate: $xovr = \{0.65; 0.75; 0.85; 0.95\}$, while in Case study 2—with $xovr = \{0.5; 0.6; 0.7; 0.8; 0.9; 1\}$. Due to the specific peculiarities of two fed-batch FP, again different strategies were applied for mutation rates in both Case studies. In Case study 1 SGA is applied with the following values of mutation rate: *mutr* = {0.02; 0.04; 0.06; 0.08; 0.1}, while in Case study 2—with *mutr* = {0.001; 0.01; 0.1; 0.5; 1}. The selected values of *xovr* and *mutr* are chosen based on the following prerequisites: (i) concerning the recommended by the literature values and trying to comprise different values in the ranges for both Case studies [16, 17, 19]; (ii) concerning the previous authors' experience of modelling of FP using GA [1, 2, 21–23]. All other SGA operators and parameters are tuned as presented in [1, 22].

3   InterCriteria Analysis

3.1   *InterCriteria Analysis by Pairs*

InterCriteria analysis, based on the apparatuses of IM [4, 7, 8] and IFS [6], is given in details in [11]. Here, for completeness, the proposed idea is briefly presented.

An intuitionistic fuzzy pair (IFP) [12] is an ordered pair of real non-negative numbers $\langle a, b \rangle$, where $a, b \in [0, 1]$ and $a + b \leq 1$, that is used as an evaluation of some object or process. According to [12], the components (a and b) of IFP might be interpreted as degrees of "membership" and "non-membership" to a given set, degrees of "agreement" and "disagreement", degrees of "validity" and "non-validity", degrees of "correctness" and "non-correctness", etc.

The apparatus of IM is presented initially in [4] and discussed in more details in [7, 8]. For the purposes of ICrA application, the initial index set consists of the criteria (for rows) and objects (for columns) with the IM elements assumed to be real numbers. Further, an IM with index sets consisting of the criteria (for rows and for columns) with IFP elements determining the degrees of correspondence between the respective criteria is constructed, as it is going to be briefly presented below.

Let the initial IM is presented in the form of Eq. (8), where, for every p, q, $(1 \leq p \leq m, 1 \leq q \leq n)$, C_p is a criterion, taking part in the evaluation; O_q—an object to be evaluated; $C_p(O_q)$—a real number (the value assigned by the p-th criteria to the q-th object).

$$A = \begin{array}{c|ccccccc}
 & O_1 & \cdots & O_k & \cdots & O_l & \cdots & O_n \\
\hline
C_1 & C_1(O_1) & \cdots & C_1(O_k) & \cdots & C_1(O_l) & \cdots & C_1(O_n) \\
\vdots & \vdots & \ddots & \vdots & \ddots & \vdots & \ddots & \vdots \\
C_i & C_i(O_1) & \cdots & C_i(O_k) & \cdots & C_i(O_l) & \cdots & C_i(O_n) \\
\vdots & \vdots & \ddots & \vdots & \ddots & \vdots & \ddots & \vdots \\
C_j & C_j(O_1) & \cdots & C_j(O_k) & \cdots & C_j(O_l) & \cdots & C_j(O_n) \\
\vdots & \vdots & \ddots & \vdots & \ddots & \vdots & \ddots & \vdots \\
C_m & C_m(O_1) & \cdots & C_m(O_k) & \cdots & C_m(O_l) & \cdots & C_m(O_n)
\end{array} \quad (8)$$

Let O denotes the set of all objects being evaluated, and $C(O)$ is the set of values assigned by a given criterion C (i.e., $C = C_p$ for some fixed p) to the objects, i.e.,

$$O \overset{\text{def}}{=} \{O_1, O_2, O_3, \ldots, O_n\},$$
$$C(O) \overset{\text{def}}{=} \{C(O_1), C(O_2), C(O_3), \ldots, C(O_n)\}.$$

Let $x_i = C(O_i)$. Then the following set can be defined:

$$C^*(O) \overset{\text{def}}{=} \{\langle x_i, x_j \rangle | i \neq j \,\&\, \langle x_i, x_j \rangle \in C(O) \times C(O)\}.$$

Further, if $x = C(O_i)$ and $y = C(O_j)$, $x \prec y$ iff $i < j$ will be written.

In order to find the agreement of different criteria, the vectors of all internal comparisons for each criterion are constructed, which elements fulfil one of the three relations R, $\overline{R}$ and $\tilde{R}$. The nature of the relations is chosen such that for a fixed criterion C and any ordered pair $\langle x, y \rangle \in C^*(O)$:

$$\langle x, y \rangle \in R \Leftrightarrow \langle y, x \rangle \in \overline{R}, \tag{9}$$
$$\langle x, y \rangle \in \tilde{R} \Leftrightarrow \langle x, y \rangle \notin (R \cup \overline{R}), \tag{10}$$
$$R \cup \overline{R} \cup \tilde{R} = C^*(O). \tag{11}$$

For example, if "R" is the relation "$<$", then $\overline{R}$ is the relation "$>$", and vice versa. Hence, for the effective calculation of the vector of internal comparisons (denoted further by $V(C)$) only the considering of a subset of $C(O) \times C(O)$ is needed, namely:

$$C^{\prec}(O) \overset{\text{def}}{=} \{\langle x, y \rangle | \, x \prec y \,\&\, \langle x, y \rangle \in C(O) \times C(O)\},$$

due to Eqs. (9)–(11). For brevity, $c^{i,j} = \langle C(O_i), C(O_j) \rangle$.

Then for a fixed criterion C the vector of lexicographically ordered pair elements is constructed:

$$V(C) = \{c^{1,2}, c^{1,3}, \ldots, c^{1,n}, c^{2,3}, c^{2,4}, \ldots, c^{2,n}, c^{3,4}, \ldots, c^{3,n}, \ldots, c^{n-1,n}\}. \tag{12}$$

In order to be more suitable for calculations, $V(C)$ is replaced by $\hat{V}(C)$, where its k-th component ($1 \leq k \leq \frac{n(n-1)}{2}$) is given by:

$$\hat{V}_k(C) = \begin{cases} 1, & \text{iff } V_k(C) \in R, \\ -1, & \text{iff } V_k(C) \in \overline{R}, \\ 0, & \text{otherwise.} \end{cases}$$

When comparing two criteria the degree of "agreement" is determined as the number of matching components of the respective vectors (divided by the length of the

---

Algorithm 1 Calculating $\mu_{C,C'}$ and $\nu_{C,C'}$ between two criteria

Require: Vectors $\hat{V}(C)$ and $\hat{V}(C')$

```
1:  function DEGREES OF AGREEMENT AND DISAGREEMENT(V̂(C), V̂(C'))
2:      V ← V̂(C) − V̂(C')
3:      μ ← 0
4:      ν ← 0
5:      for i ← 1 to n(n−1)/2 do
6:          if Vᵢ = 0 then
7:              μ ← μ + 1
8:          else if abs(Vᵢ) = 2 then          ▷ abs(Vᵢ): the absolute value of Vᵢ
9:              ν ← ν + 1
10:         end if
11:     end for
12:     μ ← 2/(n(n−1)) μ
13:     ν ← 2/(n(n−1)) ν
14:     return μ, ν
15: end function
```

---

vector for normalization purposes). This can be done in several ways, e.g. by counting the matches or by taking the complement of the Hamming distance. The degree of "disagreement" is the number of components of opposing signs in the two vectors (again normalized by the length). An example pseudocode for two criteria C and C' is presented below (Algorithm 1). If the respective degrees of "agreement" and "disagreement" are denoted by $\mu_{C,C'}$ and $\nu_{C,C'}$, it is obvious (from the way of computation) that $\mu_{C,C'} = \mu_{C',C}$ and $\nu_{C,C'} = \nu_{C',C}$. Also it is true that $\langle \mu_{C,C'}, \nu_{C,C'} \rangle$ is an IFP.

In the most of the obtained pairs $\langle \mu_{C,C'}, \nu_{C,C'} \rangle$, the sum $\mu_{C,C'} + \nu_{C,C'}$ is equal to 1. However, there may be some pairs, for which this sum is less than 1. The difference

$$\pi_{C,C'} = 1 - \mu_{C,C'} - \nu_{C,C'} \tag{13}$$

is considered as a degree of "uncertainty".

The following IM is constructed as a result of applying the Algorithm 1 to IM A (Eq. (8)):

$$
\begin{array}{c|ccc}
 & C_2 & \cdots & C_m \\
\hline
C_1 & \langle \mu_{C_1,C_2}, \nu_{C_1,C_2} \rangle & \cdots & \langle \mu_{C_1,C_m}, \nu_{C_1,C_m} \rangle \\
\vdots & \vdots & \ddots & \vdots \\
C_{m-1} & & \cdots & \langle \mu_{C_{m-1},C_m}, \nu_{C_{m-1},C_m} \rangle
\end{array} \, ,
$$

that determines the degrees of correspondence between criteria $C_1, \ldots, C_m$.

3.2 InterCriteria Analysis by Triples

Let an IM with real number elements (see [5, 7, 8]), with index sets consisting of
the names of the criteria (for rows) and of the objects (for columns) is given, as in
Eq. (8). It is possible to obtain a three dimensional IM (3D-IM) [9] with index sets
consisting of the names of the criteria and elements corresponding to the degrees
of "agreement" and "disagreement" between the respective triples of criteria in the
form of IFPs. Further, [25] is followed.

Using the same reasoning as outlined in Sect. 3.1 it is assumed that for all criteria
$V(C)$ from Eq. (12) is constructed. To make it more suitable for calculations, $V(C)$
is replaced by $\hat{V}(C)$, where its k-th component ($1 \leq k \leq \frac{n(n-1)}{2}$) is given by:

$$\hat{V}_k(C) = \begin{cases} 3 \text{ iff } V_k(C) \in R, \\ 2 \text{ iff } V_k(C) \in \tilde{R}, \\ 1 \text{ otherwise.} \end{cases}$$

When considering triples of criteria the degree of "agreement" is understood as the
number of matching components of the respective vectors (divided by the length of
the vector for normalization purposes). The degree of "disagreement" is considered
as the number of components with completely different values (again normalized
by the length). An example pseudocode for three criteria C, C' and C'' is presented
below (Algorithm 2).

For the respective degrees denoted by $\mu_{C,C',C''}$ and $\nu_{C,C',C''}$, it is obvious (from
the way of computation) that the order in which the criteria are taken has no impact
on their value. It is also evident that $\langle \mu_{C,C',C''}, \nu_{C,C',C''} \rangle$ is an IFP.

Algorithm 2 Calculating $\mu_{C,C',C''}$ and $\nu_{C,C',C''}$ between three criteria

---

Require: Vectors $\hat{V}(C)$, $\hat{V}(C')$ and $\hat{V}(C'')$

1: **function** DEGREES OF AGREEMENT AND DISAGREEMENT($\hat{V}(C)$, $\hat{V}(C')$, $\hat{V}(C'')$)
2:     $V \leftarrow \hat{V}(C) \odot \hat{V}(C') \odot \hat{V}(C'')$ ▷ $\odot$ denotes Hadamard (entrywise) product
3:     $\mu \leftarrow 0$
4:     $\nu \leftarrow 0$
5:     **for** $i \leftarrow 1$ to $\frac{n(n-1)}{2}$ **do**
6:         **if** $V_i \in \{1, 8, 27\}$ **then**
7:             $\mu \leftarrow \mu + 1$
8:         **else if** $(V_i) = 6$ **then**
9:             $\nu \leftarrow \nu + 1$
10:        **end if**
11:     **end for**
12:     $\mu \leftarrow \frac{2}{n(n-1)}\mu$
13:     $\nu \leftarrow \frac{2}{n(n-1)}\nu$
14:     **return** μ, ν
15: **end function**

Remark 1 Once again it is desirable to reiterate that the degree of "disagreement" in this case confers the meaning of the inherent discrepancy presented by the triple of criteria rather than measuring the opposing behavior as is the case of comparing a pair of criteria.

As a result 3D-IM [9] with index sets $K = \{C_1, C_2, \ldots, C_{m-2}\}, L = \{C_2, C_3, \ldots, C_{m-1}\}$ and $H = \{C_3, C_4, \ldots, C_m\}$ is produced by the Algorithm 2 over the Eq. (8). The following illustration is provided in the form of the first few and the last slice:

$$\begin{array}{c|c} C_3 & C_2 \\ \hline C_1 & a_{C_1,C_2,C_3} \end{array},$$

$$\begin{array}{c|cc} C_4 & C_2 & C_3 \\ \hline C_1 & a_{C_1,C_2,C_4} & a_{C_1,C_3,C_4} \\ C_2 & \ldots & a_{C_2,C_3,C_4} \end{array},$$

$$\begin{array}{c|ccc} C_5 & C_2 & C_3 & C_4 \\ \hline C_1 & a_{C_1,C_2,C_5} & a_{C_1,C_3,C_5} & a_{C_1,C_4,C_5} \\ C_2 & \ldots & a_{C_2,C_3,C_5} & a_{C_2,C_4,C_5} \\ C_3 & \ldots & \ldots & a_{C_3,C_4,C_5} \end{array},$$

$$\vdots$$

$$\begin{array}{c|ccccc} C_m & C_2 & \ldots & C_j & \ldots & C_{m-1} \\ \hline C_1 & a_{C_1,C_2,C_m} & \ldots & a_{C_1,C_j,C_m} & \ldots & a_{C_1,C_{m-1},C_m} \\ \vdots & \vdots & \ldots & \vdots & \ldots & \vdots \\ C_i & \ldots & \ldots & a_{C_i,C_j,C_m} & \ldots & a_{C_i,C_{m-1},C_m} \\ \vdots & \vdots & \ldots & \vdots & \ldots & \vdots \\ C_{m-2} & \ldots & \ldots & \ldots & \ldots & a_{C_{m-2},C_{m-1},C_m} \end{array},$$

where

$$a_{C_i,C_j,C_k} \stackrel{\text{def}}{=} \langle \mu_{C_i,C_j,C_k}, \nu_{C_i,C_j,C_k} \rangle.$$

4 Numerical Results and Discussion

In order to obtain reliable results for computation time, objective function value and model parameters estimations, thirty independent runs have been performed for each SGA. Twenty SGA with different *xovr* and *mutr* values are ran for the both examined here Case studies. The obtained results from 30 runs have been averaged and two IMs have been constructed for each Case study, involving values for *xovr* and *mutr*, respectively. Thus, four IMs are constructed altogether: IMs $A_{1(xovr)}$ (Eq. (14)) and $A_{1(mutr)}$ (Eq. (15)) for the Case study 1 and IMs $A_{2(xovr)}$ (Eq. (16)) and $A_{2(mutr)}$ (Eq. (17)) for the Case study 2.

Case study 1, IM $A_{1(xovr)}$:

$$A_{1(xovr)} = \begin{array}{r|cccc} & SGA_{1,1}^{xovr} & SGA_{1,2}^{xovr} & SGA_{1,3}^{xovr} & SGA_{1,4}^{xovr} \\ \hline J & 0.0222 & 0.0222 & 0.0222 & 0.0221 \\ T & 69.140600 & 70.212400 & 69.475000 & 71.359200 \\ xovr & 0.65 & 0.75 & 0.85 & 0.95 \\ \mu_{2S} & 0.962120 & 0.949840 & 0.974790 & 0.923920 \\ \mu_{2E} & 0.103840 & 0.107940 & 0.115320 & 0.129580 \\ k_S & 0.124640 & 0.119580 & 0.128700 & 0.119780 \\ k_E & 0.799020 & 0.798700 & 0.798860 & 0.798960 \\ Y_{S/X} & 0.417885 & 0.413705 & 0.413850 & 0.409500 \end{array} \quad (14)$$

IM $A_{1(xovr)}$ presents the average estimates of the ICrA criteria, namely the model parameters μ_{2S}, μ_{2E}, k_S, k_E, and $Y_{S/X}$, the resulting computation time T and objective function value J (according to Eq. (7)), as well as the respective *xovr* values—*xovr* = {0.65; 0.75; 0.85; 0.95}. The ICrA objects are denoted as $SGA_{1,1}^{xovr} \div SGA_{1,4}^{xovr}$, corresponding to each of the considered *xovr* values.

Case study 1, IM $A_{1(mutr)}$:

$$A_{1(mutr)} = \begin{array}{r|ccccc} & SGA_{1,1}^{mutr} & SGA_{1,2}^{mutr} & SGA_{1,3}^{mutr} & SGA_{1,4}^{mutr} & SGA_{1,5}^{mutr} \\ \hline J & 0.022200 & 0.022167 & 0.022133 & 0.022300 & 0.022100 \\ T & 71.677000 & 76.104333 & 90.479000 & 101.400667 & 98.161667 \\ mutr & 0.02 & 0.04 & 0.06 & 0.08 & 0.1 \\ \mu_{2S} & 0.963433 & 0.987333 & 0.943333 & 0.960033 & 0.914933 \\ \mu_{2E} & 0.113100 & 0.111900 & 0.129733 & 0.094967 & 0.146100 \\ k_S & 0.124000 & 0.123333 & 0.128167 & 0.117033 & 0.121300 \\ k_E & 0.799867 & 0.799500 & 0.799600 & 0.792433 & 0.797833 \\ Y_{S/X} & 0.410841 & 0.411348 & 0.407914 & 0.421965 & 0.398290 \end{array} \quad (15)$$

In the same way, IM $A_{1(mutr)}$ presents the results for the mentioned above ICrA criteria—model parameters, T, J and *mutr*, respectively for *mutr* = {0.02; 0.04; 0.06; 0.08; 0.1}. The ICrA objects are denoted as $SGA_{1,1}^{mutr} \div SGA_{1,5}^{mutr}$, corresponding to each of the considered *mutr* values.

By analogy with the Case study 1, IMs $A_{2(xovr)}$ and $A_{2(mutr)}$ have been created for the Case study 2.

Case study 2, IM $A_{2(xovr)}$:

$$A_{2(xovr)} = \begin{array}{r|cccccc} & \text{SGA}_{2,1}^{xovr} & \text{SGA}_{2,2}^{xovr} & \text{SGA}_{2,3}^{xovr} & \text{SGA}_{2,4}^{xovr} & \text{SGA}_{2,5}^{xovr} & \text{SGA}_{2,6}^{xovr} \\ \hline J & 0.010700 & 0.000310 & 0.000320 & 0.000170 & 0.000450 & 0.000310 \\ T & 143.156 & 77.782 & 218.234 & 104.719 & 158.078 & 86.953 \\ xovr & 0.5 & 0.6 & 0.7 & 0.8 & 0.9 & 1 \\ \mu_{max} & 0.553000 & 0.549000 & 0.550000 & 0.551000 & 0.549000 & 0.548000 \\ k_S & 0.011700 & 0.009800 & 0.010100 & 0.010000 & 0.009800 & 0.009900 \\ Y_{S/X} & 0.500275 & 0.499975 & 0.499950 & 0.500000 & 0.500250 & 0.500500 \end{array}$$

$$(16)$$

Case study 2, IM $A_{2(mutr)}$:

$$A_{2(mutr)} = \begin{array}{r|ccccc} & \text{SGA}_{2,1}^{mutr} & \text{SGA}_{2,2}^{mutr} & \text{SGA}_{2,3}^{mutr} & \text{SGA}_{2,4}^{mutr} & \text{SGA}_{2,5}^{mutr} \\ \hline J & 0.019000 & 0.000360 & 0.007300 & 4.130700 & 25.622800 \\ T & 53.250000 & 116.594000 & 193.641000 & 70.937000 & 39.234000 \\ mutr & 0.001 & 0.01 & 0.1 & 0.5 & 1 \\ \mu_{max} & 0.546000 & 0.550000 & 0.554000 & 0.599000 & 0.432000 \\ k_S & 0.007800 & 0.010200 & 0.011000 & 0.044400 & 0.002300 \\ Y_{S/X} & 0.499500 & 0.500250 & 0.500501 & 0.500325 & 0.518403 \end{array} \quad (17)$$

4.1 ICrA by Pairs

ICrA by pairs has been applied following the Algorithm 1. ICrA algorithm calculates the IFP $\langle \mu_{C,C'}, \nu_{C,C'} \rangle$ for every two pairs of considered criteria based on the constructed IMs $A_{1(xovr)}$, $A_{1(mutr)}$, $A_{2(xovr)}$ and $A_{2(mutr)}$. Values of $\pi_{C,C'}$ (Eq. (13)) are calculated too. The obtained results are collected in Table 1 for both Case studies, considering dependences between crossover and mutation rates, objective function value, computation time and model parameters themselves. The obtained results are visualized on Figs. 1 and 2 in the intuitionistic fuzzy interpretation triangle.

Here and further, the scheme for defining the consonance and dissonance between each pair of criteria (Table 2), presented in [10], is going to be used.

The considered here non-linear models for two Case studies (respectively Eqs. (1)–(6)) are a prerequisite some closer relations between observed criteria to be expected after ICrA application. On the other hand, some differences in the model parameters relations might appear caused by the different specific growth rates in *S. cerevisiae* and *E. coli* FP.

As it could be seen from Table 1, there is a strong relation between $T \leftrightarrow xovr/mutr$ for the Case study 1 (weak positive consonance and positive consonance), while in the Case study 2 a lack of relations is observed (dissonance and strong dissonance). The similar discrepancy is identified in the correlation between $Y_{S/X} \leftrightarrow xovr/mutr$: in the Case study 2 there is a strong relation for GA parameter *mutr* (positive consonance), while in the Case study 1 the criteria are in dissonance. These discrepancies might be explained by the stochastic nature of GA. Crossover

Table 1 Results from ICrA by pairs: influence of *xovr* and *mutr* to the parameter identification of *S. cerevisiae* and *E. coli* fed-batch FP models

Correlation	*S. cerevisiae* fed-batch fermentation process				*E. coli* fed-batch fermentation process			
	xovr		*mutr*		*xovr*		*mutr*	
	$\langle \mu, \nu \rangle$	π	$\langle \mu, \nu \rangle$	π	$\langle \mu, \nu \rangle$	π	$\langle \mu, \nu \rangle$	π
$T \leftrightarrow xovr/mutr$	0.8, 0.2	0	0.9, 0.1	0	0.5, 0.5	0	0.4, 0.6	0
$J \leftrightarrow xovr/mutr$	0, 0.5	0.5	0.3, 0.7	0	0.3, 0.6	0.1	0.8, 0.2	0
$\mu_{2S} \leftrightarrow xovr/mutr$	0.3, 0.7	0	0.2, 0.8	0				
$\mu_{2E} \leftrightarrow xovr/mutr$	1, 0	0	0.6, 0,4	0				
$\mu_{max} \leftrightarrow xovr/mutr$					0.2, 0.7	0.1	0.6, 0.4	0
$Y_{S/X} \leftrightarrow xovr/mutr$	0.2, 0.8	0	0.4, 0.6	0	0.7, 0.3	0	0.9, 0.1	0
$k_S \leftrightarrow xovr/mutr$	0.5, 0.5	0	0.3, 0.7	0	0.3, 0.7	0	0.6, 0.4	0
$k_E \leftrightarrow xovr/mutr$	0.5, 0.5	0	0.2, 0.8	0				
$T \leftrightarrow J$	0, 0.5	0.5	0.4, 0.6	0	0.6, 0.3	0.1	0.2, 0.8	0
$\mu_{2S} \leftrightarrow J$	0.5, 0	0.5	0.7, 0.3	0				
$\mu_{2E} \leftrightarrow J$	0, 0.5	0.5	0.1, 0.9	0				
$\mu_{max} \leftrightarrow J$					0.5, 0.3	0.2	0.4, 0.6	0
$Y_{S/X} \leftrightarrow J$	0.5, 0	0	0.9, 0.1	0	0.5, 0.4	0.1	0.7, 0.3	0
$k_S \leftrightarrow J$	0.3, 0.2	0.5	0.4, 0.6	0	0.5, 0.3	0.2	0.4, 0.6	0
$k_E \leftrightarrow J$	0.2, 0.3	0.5	0.5, 0.5	0				
$\mu_{2S} \leftrightarrow T$	0.2, 0.8	0	0.3, 0.7	0				
$\mu_{2E} \leftrightarrow T$	0.8, 0.2	0	0.5, 0.5	0				
$\mu_{max} \leftrightarrow T$					0.6, 0.3	0.1	0.8, 0.2	0
$Y_{S/X} \leftrightarrow T$	0, 1	0	0.5, 0.5	0	0.4, 0.6	0	0.5, 0.5	0
$k_S \leftrightarrow T$	0.3, 0.7	0	0.2, 0.8	0	0.7, 0.3	0	0.8, 0.2	0
$k_S \leftrightarrow T$	0.3, 0.7	0	0.1, 0.9	0				
$\mu_{2S} \leftrightarrow \mu_{2E}$	0.3, 0.7	0	0.2, 0.8	0				
$\mu_{2S} \leftrightarrow k_S$	0.8, 0.2	0	0.5, 0.5	0				
$\mu_{2E} \leftrightarrow k_S$	0.5, 0.5	0	0.7, 0.3	0				
$\mu_{max} \leftrightarrow k_S$					0.8, 0.2	0	1, 0	0
$\mu_{2S} \leftrightarrow k_E$	0.5, 0.5	0	0.6, 0.4	0				
$\mu_{2E} \leftrightarrow k_E$	0.5, 0.5	0	0.6, 0.4	0				
$k_S \leftrightarrow k_E$	0.7, 0.3	0	0.9, 0.1	0				
$Y_{S/X} \leftrightarrow \mu_{2S}$	0.8, 0.2	0	0.8, 0.2	0				
$Y_{S/X} \leftrightarrow \mu_{2E}$	0.2, 0.8	0	0, 1	0				
$Y_{S/X} \leftrightarrow \mu_{max}$					0.4, 0.5	0.1	0.5, 0.5	0
$Y_{S/X} \leftrightarrow k_S$	0.7, 0.3	0	0.3, 0.7	0	0.5, 0.5	0	0.5, 0.5	0
$Y_{S/X} \leftrightarrow k_E$	0.7, 0.3	0	0.4, 0.6	0				

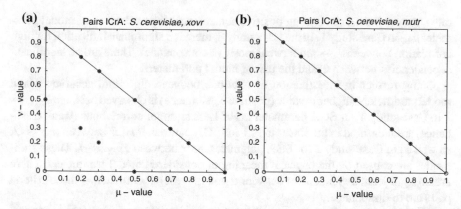

Fig. 1 Presentation of ICrA by pairs in the intuitionistic fuzzy interpretation triangle: Case study 1 (*S. cerevisiae*). **a** *xovr*. **b** *mutr*

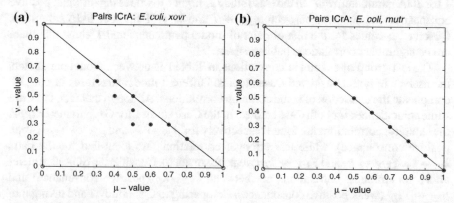

Fig. 2 Presentation of ICrA by pairs in the intuitionistic fuzzy interpretation triangle: Case study 2 (*E. coli*). **a** *xovr*. **b** *mutr*

Table 2 Consonance and dissonance scale	Interval of $\mu_{C,C'}$, %	Meaning
	[0–5]	Strong negative consonance
	(5–15]	Negative consonance
	(15–25]	Weak negative consonance
	(25–33]	Weak dissonance
	(33–43]	Dissonance
	(43–57]	Strong dissonance
	(57–67]	Dissonance
	(67–75]	Weak dissonance
	(75–85]	Weak positive consonance
	(85–95]	Positive consonance
	(95–100]	Strong positive consonance

rate strongly influences (strong positive consonance) the evaluation of model parameter μ_{2E} in Case study 1. In the Case study 2, there is a significant indication for high correlation between $J \leftrightarrow mutr$ (weak positive consonance). There are no significant dependencies between T and the rest of model parameters.

Going further in investigation of relations between algorithm accuracy J and model parameters, higher μ-value (positive consonance) is observed between $Y_{S/X} \leftrightarrow J$ in Case study 1 for SGA parameter $mutr$. Less stronger correlations (weak dissonance) are identified in the Case study 1 for SGA parameter $mutr$ between $\mu_{2S} \leftrightarrow J$, as well as in Case study 2 for SGA parameter $mutr$ between $Y_{S/X} \leftrightarrow J$. These similarities are caused by the physical meaning of considered model parameters. There are no significant dependencies between J and the rest of parameters—those criteria pairs are in dissonance.

When considering the influence of computation time T over the model parameters, higher μ (weak positive consonance) is observed in the pair $\mu_{2E} \leftrightarrow T$ in Case study 1 for SGA parameter $xovr$. In the Case study 2, higher μ-values (again weak positive consonance) are observed between $\mu_{max} \leftrightarrow T$ and $k_S \leftrightarrow T$ for $mutr$ SGA parameter. Observed μ-values for the rest of pairs of model parameters and T show that there are no significant correlations between them.

The last group of examined correlations in Table 1 is between model parameters themselves in both considered Case studies. Different model structures in both FP complicate the extraction of some common correlations. Although that fact, there are some coincidences for both Case studies. In the Case study 1 for GA parameter $xovr$, the strongest correlations are found respectively for $\mu_{2S} \leftrightarrow k_S$ and $\mu_{2S} \leftrightarrow Y_{S/X}$ (weak positive consonance), while less stronger correlations are identified for the pairs $k_S \leftrightarrow k_E$, $Y_{S/X} \leftrightarrow k_S$ and $Y_{S/X} \leftrightarrow k_E$ (weak dissonance). Considering SGA parameter $mutr$, the strongest correlations are between $k_S \leftrightarrow k_E$ (positive consonance) and $\mu_{2S} \leftrightarrow Y_{S/X}$ (weak positive consonance). Comparing to Case study 2 and taking into account the more simple specific growth rate model structure, a similar result (weak positive consonance and strong positive consonance, respectively for $xovr$ and $mutr$) for the pair $\mu_{max} \leftrightarrow k_S$ is observed. The highest correlation is observed for both GA parameters $xovr$ and $mutr$. These strong parameter dependencies are again caused by the physical meaning of FP models parameters. For the remaining correlations between model parameters themselves, the μ-values are low and, as such, there are no significant dependencies.

It is also interesting to be noted that during the investigation of $xovr$ influence, there are some pairs of considered criteria with reported degree of uncertainty π. For the Case study 1, all observed appearances of π are in pairs with the objective function value, while in Case study 2—in pairs of objective function value or specific growth rate. All these observations have an obvious explanation—as it can be seen from IM $A_{1(xovr)}$ for Case study 1, there are equal values for objective function value. In analogy, as seen from IM $A_{2(xovr)}$, there are equal values of objective function value and specific growth rate in Case study 2. It is logical, that equal values cause an uncertainty and makes difficult the process of decision making.

As a summary of the application of ICrA by pairs, the following main results might be outlined:

- Considered GA parameters *xovr* and *mutr* show a high correlation with T in Case study 1. In Case study 2, parameter *mutr* is in weak positive consonance with J and model parameter $Y_{S/X}$. The values of *xovr* and *mutr* reflect on T because of the more complex model used in Case study 1 [1, 21]. In opposite, the more simple model structure in Case study 2 allows the relations between *mutr* and J and one of the most sensitive model parameter $Y_{S/X}$ [24] to be outlined.
- When looking at T and J relations, positive consonance is observed for $J \leftrightarrow Y_{S/X}$, especially in Case study 1. Weak positive consonance is reported between specific growth rates (respectively μ_{2E} and μ_{max}) and T in both Case studies, as well as for $k_S \leftrightarrow T$ in Case study 2. The stochastic nature of GA is a preposition of a relatively small number of observed strong relations [16, 19].
- In the last group of examined correlations between model parameters themselves, weak positive consonance and strong positive consonance are obtained between specific growth rates μ_{2S} and μ_{max} and model parameter k_S in both Case studies, especially in Case study 2 for GA parameter *mutr*. Considering Case study 1, weak positive consonance is observed for $Y_{S/X} \leftrightarrow \mu_{2S}$. The ascertained results are caused by the physical meaning of FP model parameters, as well as by the strong non-linearity of FP model structures [13, 15, 18, 21].

4.2 ICrA by Triples

Going further in the desire for thorough analysis of the correlations between FP model parameters and SGA parameters and outcomes, one can consider the investigation of the mentioned above criteria by triples. In this case the application of ICrA uses the same input data, as in the case of ICrA by pairs, namely IMs $A_{1(xovr)}$ (Eq. (14)) and $A_{1(mutr)}$ (Eq. (15)) for the Case study 1, and IMs $A_{2(xovr)}$ (Eq. (16)) and $A_{2(mutr)}$ (Eq. (17)) for the Case study 2.

Case study 1

The application of ICrA by triples follows the Algorithm 2, presented in Sect. 3.2. The obtained results for the degrees of "agreement" and "disagreement" of the criteria triples are listed in Table 3 as original. As it can be seen, it is more than obvious that there are very large values of the degree of "uncertainty" π (Eq. (13)). The logical explanation of this fact might be found in the chosen way of calculation of μ-, ν- and π-values (see *Remark 1*). For getting over such a problem, the idea for application of a topological operator intuitively appears. In this investigation, the one of the most recently defined in [3] topological operator is applied.

For every IFS $A \neq U^*$, the recently defined in [3] topological operator has the form

Table 3 Results from ICrA by triples: influence of $xovr$ to the parameter identification of *S. cerevisiae* fed-batch FP model

Criteria triples	Original		Modified		Criteria triples	Original		Modified	
	μ	ν	μ	ν		μ	ν	μ	ν
$J \leftrightarrow T \leftrightarrow xovr$	0	0.20	0	0.22	$xovr \leftrightarrow \mu_{2E} \leftrightarrow k_E$	0	0	0	0
$J \leftrightarrow T \leftrightarrow \mu_{2S}$	0	0.30	0	0.33	$\mu_{2S} \leftrightarrow \mu_{2E} \leftrightarrow k_E$	0	0.40	0	0.44
$J \leftrightarrow xovr \leftrightarrow \mu_{2S}$	0.30	0.10	0.33	0.11	$J \leftrightarrow k_S \leftrightarrow k_E$	0.20	0.10	0.22	0.11
$T \leftrightarrow xovr \leftrightarrow \mu_{2S}$	0.10	0	0.11	0	$T \leftrightarrow k_S \leftrightarrow k_E$	0	0.40	0	0.44
$J \leftrightarrow T \leftrightarrow \mu_{2E}$	0	0.10	0	0.11	$xovr \leftrightarrow k_S \leftrightarrow k_E$	0.20	0.30	0.22	0.33
$J \leftrightarrow xovr \leftrightarrow \mu_{2E}$	0	0.10	0	0.11	$\mu_{2S} \leftrightarrow k_S \leftrightarrow k_E$	0.20	0	0.22	0
$T \leftrightarrow xovr \leftrightarrow \mu_{2E}$	0.50	0	0.56	0	$\mu_{2E} \leftrightarrow k_S \leftrightarrow k_E$	0	0.30	0	0.33
$J \leftrightarrow \mu_{2S} \leftrightarrow \mu_{2E}$	0	0.20	0	0.22	$J \leftrightarrow T \leftrightarrow Y_{S/X}$	0	0.30	0	0.33
$T \leftrightarrow \mu_{2S} \leftrightarrow \mu_{2E}$	0.10	0	0.11	0	$J \leftrightarrow xovr \leftrightarrow Y_{S/X}$	0.30	0.20	0.33	0.22
$xovr \leftrightarrow \mu_{2S} \leftrightarrow \mu_{2E}$	0.20	0	0.22	0	$T \leftrightarrow xovr \leftrightarrow Y_{S/X}$	0	0.10	0	0.11
$J \leftrightarrow T \leftrightarrow k_S$	0	0.30	0	0.33	$J \leftrightarrow \mu_{2S} \leftrightarrow Y_{S/X}$	0.60	0.10	0.67	0.11
$J \leftrightarrow xovr \leftrightarrow k_S$	0.20	0.30	0.22	0.33	$T \leftrightarrow \mu_{2S} \leftrightarrow Y_{S/X}$	0	0.10	0	0.11
$T \leftrightarrow xovr \leftrightarrow k_S$	0.10	0	0.11	0	$xovr \leftrightarrow \mu_{2S} \leftrightarrow Y_{S/X}$	0.40	0	0.44	0
$J \leftrightarrow \mu_{2S} \leftrightarrow k_S$	0.40	0.10	0.44	0.11	$J \leftrightarrow \mu_{2E} \leftrightarrow Y_{S/X}$	0	0.30	0	0.33
$T \leftrightarrow \mu_{2S} \leftrightarrow k_S$	0.10	0.10	0.11	0.11	$T \leftrightarrow \mu_{2E} \leftrightarrow Y_{S/X}$	0	0.10	0	0.11
$xovr \leftrightarrow \mu_{2S} \leftrightarrow k_S$	0.40	0.10	0.44	0.11	$xovr \leftrightarrow \mu_{2E} \leftrightarrow Y_{S/X}$	0	0	0	0
$J \leftrightarrow \mu_{2E} \leftrightarrow k_S$	0	0.20	0	0.22	$\mu_{2S} \leftrightarrow \mu_{2E} \leftrightarrow Y_{S/X}$	0	0	0	0
$T \leftrightarrow \mu_{2E} \leftrightarrow k_S$	0.30	0	0.33	0	$J \leftrightarrow k_S \leftrightarrow Y_{S/X}$	0.40	0.20	0.44	0.22
$xovr \leftrightarrow \mu_{2E} \leftrightarrow k_S$	0.20	0	0.22	0	$T \leftrightarrow k_S \leftrightarrow Y_{S/X}$	0	0.20	0	0.22
$\mu_{2S} \leftrightarrow \mu_{2E} \leftrightarrow k_S$	0.20	0.10	0.22	0.11	$xovr \leftrightarrow k_S \leftrightarrow Y_{S/X}$	0.20	0.10	0.22	0.11
$J \leftrightarrow T \leftrightarrow k_E$	0	0.40	0	0.44	$\mu_{2S} \leftrightarrow k_S \leftrightarrow Y_{S/X}$	0.50	0	0.56	0
$J \leftrightarrow xovr \leftrightarrow k_E$	0.30	0.30	0.33	0.33	$\mu_{2E} \leftrightarrow k_S \leftrightarrow Y_{S/X}$	0	0.10	0	0.11
$T \leftrightarrow xovr \leftrightarrow k_E$	0	0.10	0	0.11	$J \leftrightarrow k_E \leftrightarrow Y_{S/X}$	0.40	0	0.44	0
$J \leftrightarrow \mu_{2S} \leftrightarrow k_E$	0.30	0	0.33	0	$T \leftrightarrow k_E \leftrightarrow Y_{S/X}$	0	0.50	0	0.56
$T \leftrightarrow \mu_{2S} \leftrightarrow k_E$	0	0.50	0	0.56	$xovr \leftrightarrow k_E \leftrightarrow Y_{S/X}$	0.40	0.50	0.44	0.56
$xovr \leftrightarrow \mu_{2S} \leftrightarrow k_E$	0.40	0.40	0.44	0.44	$\mu_{2S} \leftrightarrow k_E \leftrightarrow Y_{S/X}$	0.40	0.10	0.44	0.11
$J \leftrightarrow \mu_{2E} \leftrightarrow k_E$	0	0.40	0	0.44	$\mu_{2E} \leftrightarrow k_E \leftrightarrow Y_{S/X}$	0	0.50	0	0.56
$T \leftrightarrow \mu_{2E} \leftrightarrow k_E$	0	0.10	0	0.11	$k_S \leftrightarrow k_E \leftrightarrow Y_{S/X}$	0.20	0.10	0.22	0.11

$$T(A) = \left\{ \left\langle x, \frac{\mu_A(x)}{\sup_{y \in E}(\mu_A(y) + \nu_A(y))}, \frac{\nu_A(x)}{\sup_{y \in E}(\mu_A(y) + \nu_A(y))} \right\rangle \middle| x \in E \right\}. \qquad (18)$$

The operator $T(A)$ transforms the element x with respect to its degrees $\mu_A(x)$ and $\nu_A(x)$, if $\mu_A(x) + \nu_A(x) = \sup_{y \in E}(\mu_A(y) + \nu_A(y))$, and the element z with $\mu_A(z) + \nu_A(z) < \sup_{y \in E}(\mu_A(y) + \nu_A(y))$, as it is shown in Fig. 3b [3]. Therefore, the topological

Fig. 3 Topological operator $T(A)$ [3]. **a** Original. **b** Modified

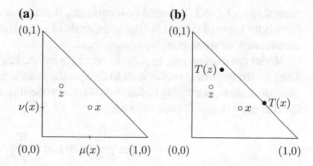

operator $T(A)$ decreases the degree of "uncertainty", while increases both the degrees of "agreement" and "disagreement".

The topological operator $T(A)$ (Eq. (18)) is applied for considered here Case study 1, and for GA parameter $xovr$. The results are presented in Table 3 as modified. The obtained results are also visualized in the intuitionistic fuzzy interpretation triangle (Fig. 4), for both before (original) and after (modified) the application of the topological operator $T(A)$.

The denotations original and modified is going to be used for the cases and GA parameters considered further.

Also, the authors assume the scheme for defining the consonance and dissonance between each pair of criteria (Table 2) [10] to be appropriate for the definition of consonance and dissonance when the criteria are considered by triples as well.

Investigation of the influence of SGA parameter $xovr$

In Case study 1, when considering the parameter $xovr$, the highest μ value is observed for the triple $J \leftrightarrow \mu_{2S} \leftrightarrow Y_{S/X}$. But even with the highest μ, this triple is in dissonance

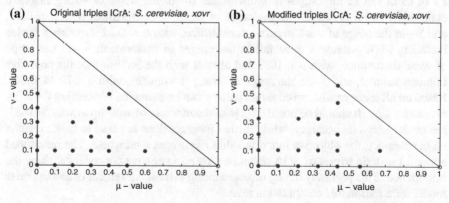

Fig. 4 Presentation of ICrA by triples in the intuitionistic fuzzy interpretation triangle: Case study 1 (*S. cerevisiae*), SGA parameter $xovr$. **a** Original. **b** Modified

according to Table 2. It should be mentioned, that for the considered parameter *xovr* there is no triple of criteria falling in the scale of strong positive consonance, positive consonance, or weak positive consonance.

Model parameters $\mu_{2S}, \mu_{2E}, k_S, k_E$ and $Y_{S/X}$ are the key growth kinetic parameters. They are the empirical coefficients in the specific rates in the so called *mixed oxidative functional state*, according to functional state modelling approach [21]. As such, the specific growth rate is presented as:

$$\mu = \mu_{2S}\frac{S}{S+k_S} + \mu_{2E}\frac{E}{E+k_E}, \tag{19}$$

while the substrate utilization rate (Monod-like) is described by:

$$q_S = -\frac{\mu_{2S}}{Y_{S/X}}\frac{S}{S+k_S}. \tag{20}$$

Among all triples between the model parameters themselves, the triple $\mu_{2S} \leftrightarrow k_S \leftrightarrow Y_{S/X}$ is in strong dissonance. This is quite logical, if one considers that these are the parameters in the Monod-like Eq. (20). Three of the other "model parameters" triples, namely $\mu_{2S} \leftrightarrow \mu_{2E} \leftrightarrow k_S, \mu_{2S} \leftrightarrow k_S \leftrightarrow k_E$ and $k_S \leftrightarrow k_E \leftrightarrow Y_{S/X}$, are in weak negative consonance, while the rest of the triples are in strong negative consonance. All these observations have a logical explanation as well, having in mind the very strong non-linear dependences between the model parameters in Eq. (19) [21].

The results show that the main parameters that define the process kinetics are in strong negative consonance with the resulting SGA computation time T and the obtained objective function value J. This result might be explained with a statement, that if one is looking for a model with higher degree of accuracy (which means lower objective function value J), it is mostly to the account of the computation time T.

Concerning the criteria triples that reflect the influence of SGA parameter *xovr*, 13 of them fall in the ranges of consonance: 6—in the range of strong negative consonance, with $\mu = 0$; 2—in the range of negative consonance, with $\mu = 0.11$, and 5—in the range of weak negative consonance, with $\mu = 0.22$. Another 8 triples including SGA parameter *xovr* fall in the ranges of dissonance: 3—in the range of weak dissonance, with $\mu = 0.33$ and always with the inclusion of the objective function value J, and 5—in the range of strong dissonance, with $\mu \in [0.44, 0.56]$. Based on all these results, no common pattern can be extracted concerning the SGA parameter *xovr*. It should be noted, that with the increase of *xovr* up to *xovr* $= 0.85$, the model keeps its accuracy, while further increase does not lead to better results; on the contrary, the objective function value J becomes a bit worse. The mentioned above 13 criteria triples of SGA parameter *xovr* in negative consonance, show the importance of the precisely tuning of *xovr* aiming at the achievement of more precise model for a reasonable computation time.

Considering the results for the computation time T, 19 out of 21 criteria triples fall in ranges of consonance, distributed as follows: 15—in the range of strong negative consonance, with $\mu = 0$, and 4—in the range of negative consonance, with $\mu = 0.11$. The remaining two triples are distributed, respectively, one in the range of weak dissonance, with $\mu = 0.33$, and one in the range of strong dissonance, with $\mu = 0.56$. Again no pattern of the computation time change can be outlined. This fact might be logically explained by the stochastic nature of the GA. The resulting computation time T greatly depends on the randomly chosen initial conditions. If the initial population leads to a solution close to the global minimum of the objective function, then the computation time T decreases drastically. And vice versa, if the initial conditions are far from the global minimum, T increases dramatically.

Investigation of the influence of SGA parameter *mutr*

The obtained in this case results are listed in Table 4, again without (original) and with the application of the topological operator $T(A)$ (modified). Figure 5 visualizes the results in the intuitionistic fuzzy interpretation triangle, before (original) and after (modified) the application of $T(A)$.

In Case study 1, when considering the influence of the other SGA parameter *mutr*, there are two criteria triples with the highest μ-value—$\mu = 1.00$, namely $\mu_{2E} \leftrightarrow k_S \leftrightarrow k_E$ and $J \leftrightarrow \mu_{2S} \leftrightarrow Y_{S/X}$. The second triple was with the highest value of μ when the SGA operator *xovr* was considered, but there it was in the range of dissonance, while now it is in the range of strong positive consonance. There are two other triples with a relatively high μ-values—$\mu_{2S} \leftrightarrow k_S \leftrightarrow k_E$ and $T \leftrightarrow mutr \leftrightarrow \mu_{2E}$, which fall in the range of weak positive consonance with $\mu = 0.83$. For the considered SGA parameter *mutr* there is no triple of criteria falling in the scale of positive consonance ($\mu \in (0.85, 0.95]$) and weak dissonance ($\mu \in (0.67, 0.75]$).

Among all triples between the model parameters themselves, two of the triples were mentioned in the previous paragraph as representatives of strong positive consonance ($\mu_{2E} \leftrightarrow k_S \leftrightarrow k_E$) and of weak positive consonance ($\mu_{2S} \leftrightarrow k_S \leftrightarrow k_E$). The other 8 triples are distributed as follows (Table 4): 1 is in dissonance, 2 are in strong dissonance, 2 are in weak dissonance, while 3 are in strong negative consonance. All these observations might be again explained with a very strong non-linear dependences between model parameters in Eq. (19) [21].

When considering the triples reflecting the resulting SGA computation time T, obtained objective function value J and main parameters that defined the process kinetics, the results are distributed as follows: respectively one triple in strong dissonance, one—in weak dissonance, and one—in dissonance, while the remaining three are in strong negative consonance, as it was with all of the triples in the case of *xovr*. It is again a consequence of what was mentioned above, that if one is looking for a model with higher degree of accuracy it is usually to the account of the computation time.

Concerning the criteria triples, that reflect the influence of SGA parameter *mutr*, 12 of them fall in the ranges of consonance: 9—in the range of strong negative consonance, with $\mu = 0$; 2—in the range of weak negative consonance, with $\mu = 0.17$, and 1 in the range of weak positive consonance, with $\mu = 0.83$. The remaining 9 triples including SGA parameter *mutr* fall in the ranges of dissonance: 3 are in the range of strong dissonance, with $\mu = 0.50$, 5—in the range of weak dissonance, with

Table 4 Results from ICrA by triples: influence of *mutr* to the parameter identification of *S. cerevisiae* fed-batch FP model

Criteria triples	Original		Modified		Criteria triples	Original		Modified	
	μ	ν	μ	ν		μ	ν	μ	ν
$J \leftrightarrow T \leftrightarrow mutr$	0.30	0	0.50	0	$mutr \leftrightarrow \mu_{2E} \leftrightarrow k_E$	0.20	0	0.33	0
$J \leftrightarrow T \leftrightarrow \mu_{2S}$	0.20	0.10	0.33	0.17	$\mu_{2S} \leftrightarrow \mu_{2E} \leftrightarrow k_E$	0.20	0	0.33	0
$J \leftrightarrow mutr \leftrightarrow \mu_{2S}$	0.10	0.10	0.17	0.17	$J \leftrightarrow k_S \leftrightarrow k_E$	0.20	0	0.33	0
$T \leftrightarrow mutr \leftrightarrow \mu_{2S}$	0.20	0	0.33	0	$T \leftrightarrow k_S \leftrightarrow k_E$	0.10	0	0.17	0
$J \leftrightarrow T \leftrightarrow \mu_{2E}$	0	0.10	0	0.17	$mutr \leftrightarrow k_S \leftrightarrow k_E$	0.20	0	0.33	0
$J \leftrightarrow mutr \leftrightarrow \mu_{2E}$	0	0.10	0	0.17	$\mu_{2S} \leftrightarrow k_S \leftrightarrow k_E$	0.50	0	0.83	0
$T \leftrightarrow mutr \leftrightarrow \mu_{2E}$	0.50	0	0.83	0	$\mu_{2E} \leftrightarrow k_S \leftrightarrow k_E$	0.60	0	1.00	0
$J \leftrightarrow \mu_{2S} \leftrightarrow \mu_{2E}$	0	0.20	0	0.33	$J \leftrightarrow T \leftrightarrow Y_{S/X}$	0.40	0.10	0.67	0.17
$T \leftrightarrow \mu_{2S} \leftrightarrow \mu_{2E}$	0	0	0	0	$J \leftrightarrow mutr \leftrightarrow Y_{S/X}$	0.30	0.10	0.50	0.17
$mutr \leftrightarrow \mu_{2S} \leftrightarrow \mu_{2E}$	0	0	0	0	$T \leftrightarrow mutr \leftrightarrow Y_{S/X}$	0.40	0	0.67	0
$J \leftrightarrow T \leftrightarrow k_S$	0	0.20	0	0.33	$J \leftrightarrow \mu_{2S} \leftrightarrow Y_{S/X}$	0.60	0	1.00	0
$J \leftrightarrow mutr \leftrightarrow k_S$	0	0.20	0	0.33	$T \leftrightarrow \mu_{2S} \leftrightarrow Y_{S/X}$	0.30	0	0.50	0
$T \leftrightarrow mutr \leftrightarrow k_S$	0.20	0	0.33	0	$mutr \leftrightarrow \mu_{2S} \leftrightarrow Y_{S/X}$	0.20	0	0.33	0
$J \leftrightarrow \mu_{2S} \leftrightarrow k_S$	0.20	0.10	0.33	0.17	$J \leftrightarrow \mu_{2E} \leftrightarrow Y_{S/X}$	0	0.20	0	0.33
$T \leftrightarrow \mu_{2S} \leftrightarrow k_S$	0	0	0	0	$T \leftrightarrow \mu_{2E} \leftrightarrow Y_{S/X}$	0	0	0	0
$mutr \leftrightarrow \mu_{2S} \leftrightarrow k_S$	0	0	0	0	$mutr \leftrightarrow \mu_{2E} \leftrightarrow Y_{S/X}$	0	0	0	0
$J \leftrightarrow \mu_{2E} \leftrightarrow k_S$	0	0.10	0	0.17	$\mu_{2S} \leftrightarrow \mu_{2E} \leftrightarrow Y_{S/X}$	0	0	0	0
$T \leftrightarrow \mu_{2E} \leftrightarrow k_S$	0.20	0	0.33	0	$J \leftrightarrow k_S \leftrightarrow Y_{S/X}$	0.20	0.10	0.33	0.17
$mutr \leftrightarrow \mu_{2E} \leftrightarrow k_S$	0.30	0	0.50	0	$T \leftrightarrow k_S \leftrightarrow Y_{S/X}$	0	0	0	0
$\mu_{2S} \leftrightarrow \mu_{2E} \leftrightarrow k_S$	0.20	0	0.33	0	$mutr \leftrightarrow k_S \leftrightarrow Y_{S/X}$	0	0	0	0
$J \leftrightarrow T \leftrightarrow k_E$	0	0.20	0	0.33	$\mu_{2S} \leftrightarrow k_S \leftrightarrow Y_{S/X}$	0.30	0	0.50	0
$J \leftrightarrow mutr \leftrightarrow k_E$	0	0.20	0	0.33	$\mu_{2E} \leftrightarrow k_S \leftrightarrow Y_{S/X}$	0	0	0	0
$T \leftrightarrow mutr \leftrightarrow k_E$	0.10	0	0.17	0	$J \leftrightarrow k_E \leftrightarrow Y_{S/X}$	0.30	0.10	0.50	0.17
$J \leftrightarrow \mu_{2S} \leftrightarrow k_E$	0.30	0.10	0.50	0.17	$T \leftrightarrow k_E \leftrightarrow Y_{S/X}$	0	0	0	0
$T \leftrightarrow \mu_{2S} \leftrightarrow k_E$	0	0	0	0	$mutr \leftrightarrow k_E \leftrightarrow Y_{S/X}$	0	0	0	0
$mutr \leftrightarrow \mu_{2S} \leftrightarrow k_E$	0	0	0	0	$\mu_{2S} \leftrightarrow k_E \leftrightarrow Y_{S/X}$	0.40	0	0.67	0
$J \leftrightarrow \mu_{2E} \leftrightarrow k_E$	0	0.10	0	0.17	$\mu_{2E} \leftrightarrow k_E \leftrightarrow Y_{S/X}$	0	0	0	0
$T \leftrightarrow \mu_{2E} \leftrightarrow k_E$	0.10	0	0.17	0	$k_S \leftrightarrow k_E \leftrightarrow Y_{S/X}$	0.30	0	0.50	0

$\mu = 0.33$ and 1—in dissonance. Analysing the obtained results, again no common pattern can be drawn concerning the SGA parameter *mutr*. It should be noted, that when *mutr* is increased up to *mutr* = 0.06, the model accuracy is slightly improved, but the computation time increases as well. The best results concerning the objective function value is obtained for *mutr* = 0.1, but to the account of computation time. Mentioned above 11 criteria triples of SGA parameter *mutr* in negative consonance show the importance of the precisely tuning of this SGA parameter as well, aiming at the achievement of more precise model for a reasonable computation time.

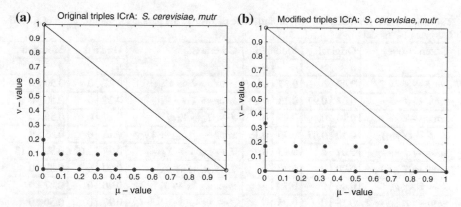

Fig. 5 Presentation of ICrA by triples in the intuitionistic fuzzy interpretation triangle: Case study 1 (*S. cerevisiae*), SGA parameter *mutr*. **a** Original. **b** Modified

Considering the results for the computation time T, 13 out of 21 criteria triples fall in the ranges of consonance, distributed as follows: 9 are in the range of strong negative consonance, with $\mu = 0$; 3—in the range of weak negative consonance, with $\mu = 0.17$, and 1—in the range of weak positive consonance, with $\mu = 0.83$. The remaining 8 triples are distributed as follows: 2 are in dissonance, with $\mu = 0.67$, 2—in the range of strong dissonance, with $\mu = 0.50$, and 4—in the range of weak dissonance, with $\mu = 0.33$. It is again the case, where no pattern of the computation time change can be outlined with the explanation as given in the case of SGA parameter *xovr*.

Case study 2

Investigation of the influence of SGA parameter *xovr*

ICrA by triples has been applied for Case study 2 as well, again following the Algorithm 2, presented in Sect. 3.2. The obtained results for the degrees of "agreement" and "disagreement" are presented in Table 5 as original. As it can be seen from the presented results, there is again a big values of the degree of "uncertainty" π (Eq. (13)) (see *Remark 1*). That implies the application of the topological operator $T(A)$ (Eq. (18)) in this case as well. As such, Table 5 presents both original (obtained by Algorithm 2) and modified (after the application of $T(A)$) results. Figure 6 visualizes the results in the intuitionistic fuzzy interpretation triangle, before (original) and after (modified) the application of $T(A)$.

In the Case study 2, when considering the influence of the other SGA parameter *xovr*, higher μ-value is observed for the triple $T \leftrightarrow \mu_{max} \leftrightarrow k_S$. This triple is in strong positive consonance, according to Table 2. The triple $J \leftrightarrow \mu_{max} \leftrightarrow k_S$ is in positive consonance, while the triples $J \leftrightarrow T \leftrightarrow \mu_{max}$ and $J \leftrightarrow T \leftrightarrow k_S$ are in weak positive consonance.

Table 5 Results from ICrA by triples: influence of $xovr$ to the parameter identification of *E. coli* fed-batch FP model

Criteria triples	Original		Modified		Criteria triples	Original		Modified	
	μ	ν	μ	ν		μ	ν	μ	ν
$xovr \leftrightarrow J \leftrightarrow T$	0.20	0	0.37	0	$xovr \leftrightarrow J \leftrightarrow Y_{S/X}$	0.27	0	0.50	0
$xovr \leftrightarrow J \leftrightarrow \mu_{max}$	0.07	0.07	0.13	0.13	$xovr \leftrightarrow T \leftrightarrow Y_{S/X}$	0.27	0	0.50	0
$xovr \leftrightarrow T \leftrightarrow \mu_{max}$	0.13	0	0.25	0	$J \leftrightarrow T \leftrightarrow Y_{S/X}$	0.27	0	0.50	0
$J \leftrightarrow T \leftrightarrow \mu_{max}$	0.40	0.07	0.75	0.13	$xovr \leftrightarrow \mu_{max} \leftrightarrow Y_{S/X}$	0.13	0	0.25	0
$xovr \leftrightarrow J \leftrightarrow k_S$	0.07	0	0.13	0	$J \leftrightarrow \mu_{max} \leftrightarrow Y_{S/X}$	0.27	0.07	0.50	0.13
$xovr \leftrightarrow T \leftrightarrow k_S$	0.20	0	0.37	0	$T \leftrightarrow \mu_{max} \leftrightarrow Y_{S/X}$	0.20	0	0.37	0
$J \leftrightarrow T \leftrightarrow k_S$	0.40	0	0.75	0	$xovr \leftrightarrow k_S \leftrightarrow Y_{S/X}$	0.20	0	0.37	0
$xovr \leftrightarrow \mu_{max} \leftrightarrow k_S$	0.13	0	0.25	0	$J \leftrightarrow k_S \leftrightarrow Y_{S/X}$	0.27	0	0.50	0
$J \leftrightarrow \mu_{max} \leftrightarrow k_S$	0.47	0.07	0.87	0.13	$T \leftrightarrow k_S \leftrightarrow Y_{S/X}$	0.27	0	0.50	0
$T \leftrightarrow \mu_{max} \leftrightarrow k_S$	0.53	0	1.00	0	$\mu_{max} \leftrightarrow k_S \leftrightarrow Y_{S/X}$	0.33	0	0.62	0

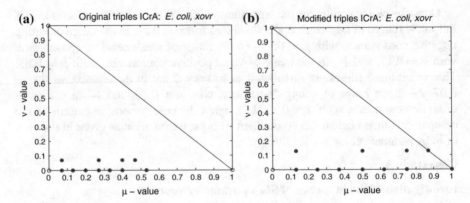

Fig. 6 Presentation of ICrA by triples in the intuitionistic fuzzy interpretation triangle: Case study 2 (*E. coli*), SGA parameter $xovr$. **a** Original. **b** Modified

Model parameters μ_{max} and k_S are the key growth kinetic parameters. In this case these parameters are empirical coefficients of the Monod equation (Eq. (21)) [13] used in the considered model (Eqs. (4)–(6)):

$$\mu = \mu_{max} \frac{S}{k_S + S}. \tag{21}$$

The results show that the main parameters that define the process kinetics are in positive consonance with the resulting SGA computation time T and obtained criteria value J. Therefore, the results from ICrA confirm that the applied SGA calculate adequate model parameters estimates, i.e. SGA work properly. Moreover, it can be seen that ICrA defines the parameters μ_{max} and k_S as more sensitive ones in the model, according to [24].

The results that show negative consonance include the SGA parameter $xovr$. The triples $xovr \leftrightarrow J \leftrightarrow \mu_{max}, xovr \leftrightarrow J \leftrightarrow k_S, xovr \leftrightarrow T \leftrightarrow \mu_{max}, xovr \leftrightarrow \mu_{max} \leftrightarrow k_S$ and $xovr \leftrightarrow \mu_{max} \leftrightarrow Y_{S/X}$ have $\mu \in [0.12, 0.25]$. Based on these results, some conclusion about the optimal value or about the influence of $xovr$ could be drawn. Firstly, with the increase of $xovr$ to the value of 0.8, there is some improvements of J. The best performance SGA shows for $xovr = 0.8$. Further increase of $xovr$ does not lead to better results; on contrary, the value of J becomes progressively worse. Considering the results for T there is absence of pattern of change in the computation time. Beyond the fact that large values of $xovr$ lead to some increase of the algorithm calculations, of significant importance to the execution time is the stochastic nature of the GA. The resulting time T depends on the randomly chosen initial population. If the objective value of the initial population is close to the global minimum, then SGA will find the solution for shorter T; and vice versa, if the initial conditions are far from the global minimum—the computation time T will be greater.

The negative consonance, especially for the triples $xovr \leftrightarrow J \leftrightarrow \mu_{max}, xovr \leftrightarrow J \leftrightarrow k_S$ and $xovr \leftrightarrow T \leftrightarrow \mu_{max}$, shows the importance of the precisely tuning of $xovr$ aiming at the achievement of higher accuracy of the SGA for a reasonable T.

The remaining triples are in dissonance, i.e. there are no any dependences between the criteria in the triples. In this instance no further conclusions about SGA performance and model parameters dependencies can be outlined.

Investigation of the influence of SGA parameter $mutr$

The obtained in this case results for $\mu_{C,C',C''}$ and $\nu_{C,C',C''}$ values are listed in Table 6 as original and modified by $T(A)$ topological operator. Figure 7 visualizes the results

Table 6 Results from ICrA by triples: influence of $mutr$ to the parameter identification of *E. coli* fed-batch FP model

Criteria triples	Original		Modified		Criteria triples	Original		Modified	
	μ	ν	μ	ν		μ	ν	μ	ν
$mutr \leftrightarrow J \leftrightarrow T$	0.20	0	0.25	0	$mutr \leftrightarrow J \leftrightarrow Y_{S/X}$	0.70	0	0.88	0
$mutr \leftrightarrow J \leftrightarrow \mu_{max}$	0.40	0	0.50	0	$mutr \leftrightarrow T \leftrightarrow Y_{S/X}$	0.40	0	0.50	0
$mutr \leftrightarrow T \leftrightarrow \mu_{max}$	0.40	0	0.50	0	$J \leftrightarrow T \leftrightarrow Y_{S/X}$	0.20	0	0.25	0
$J \leftrightarrow T \leftrightarrow \mu_{max}$	0.20	0	0.25	0	$mutr \leftrightarrow \mu_{max} \leftrightarrow Y_{S/X}$	0.50	0	0.63	0
$mutr \leftrightarrow J \leftrightarrow k_S$	0.40	0	0.50	0	$J \leftrightarrow \mu_{max} \leftrightarrow Y_{S/X}$	0.30	0	0.38	0
$mutr \leftrightarrow T \leftrightarrow k_S$	0.40	0	0.50	0	$T \leftrightarrow \mu_{max} \leftrightarrow Y_{S/X}$	0.40	0	0.50	0
$J \leftrightarrow T \leftrightarrow k_S$	0.20	0	0.25	0	$mutr \leftrightarrow k_S \leftrightarrow Y_{S/X}$	0.50	0	0.63	0
$mutr \leftrightarrow \mu_{max} \leftrightarrow k_S$	0.60	0	0.75	0	$J \leftrightarrow k_S \leftrightarrow Y_{S/X}$	0.30	0	0.38	0
$J \leftrightarrow \mu_{max} \leftrightarrow k_S$	0.40	0	0.50	0	$T \leftrightarrow k_S \leftrightarrow Y_{S/X}$	0.40	0	0.50	0
$T \leftrightarrow \mu_{max} \leftrightarrow k_S$	0.80	0	1.00	0	$\mu_{max} \leftrightarrow k_S \leftrightarrow Y_{S/X}$	0.50	0	0.63	0

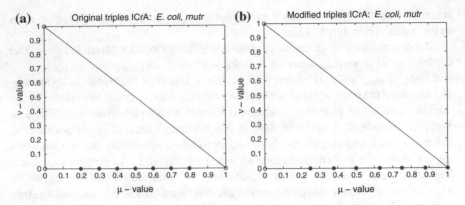

Fig. 7 Presentation of ICrA by triples in the intuitionistic fuzzy interpretation triangle: Case study 2 (*E. coli*), SGA parameter *mutr*. **a** Original. **b** Modified

in the intuitionistic fuzzy interpretation triangle, again before (original) and after (modified) the application of $T(A)$.

During the investigation of the SGA parameter *mutr*, again the highest correlation ($\mu = 1.00$) is observed for the triple $T \leftrightarrow \mu_{max} \leftrightarrow k_S$, which means that this triple is in strong positive consonance. The next two largest μ-values ($\mu = 0.88$ and $\mu = 0.75$) are observed for the triples *mutr* $\leftrightarrow J \leftrightarrow Y_{S/X}$ and *mutr* $\leftrightarrow \mu_{max} \leftrightarrow k_S$, which are in positive consonances. Results show that the important dependencies between model parameters (μ_{max} and k_S) in the Monod kinetic are again established. On the other hand, the estimates of these model parameters are determinative for the SGA total computation time. Moreover, here, the estimation of the model parameter $Y_{S/X}$ is in positive consonances with the criterion value J. This is the third model parameter that also has high sensitivity [24].

The next meaningful criteria triples are these that are in negative consonance. The triples *mutr* $\leftrightarrow J \leftrightarrow T$, $J \leftrightarrow T \leftrightarrow \mu_{max}$, $J \leftrightarrow T \leftrightarrow k_S$ and $J \leftrightarrow T \leftrightarrow Y_{S/X}$ confirm that in the considered optimization problem, i.e. parameter estimation of a model of *E. coli* fed-batch FP, the applied SGA give better estimates of model parameters (μ_{max}, k_S and $Y_{S/X}$) aiming at higher accuracy (lower J) for less computation time (T). Here, again the absence of pattern of change of computation time is observed.

It makes an impression the fact that as opposed to the results with variation of the SGA parameter *xovr*, here it is evident that SGA parameter *mutr* is not so strongly linked to the model parameters (μ_{max}, k_S and $Y_{S/X}$) and resulting T and J. As such, the parameter *xovr* is more decisive than *mutr*. Therefore, the tuning of the *xovr* is very important to achieve high accuracy for a reasonable computation time.

5 Conclusion

In this investigation two kinds of the InterCriteria Analysis approach are applied for the purposes of parameter identification of FP models. Two fed-batch FP—of yeast *S. cerevisiae* and bacteria *E. coli*—are considered. A series of parameter identification procedures using SGA are performed. Altogether twenty differently tuned SGA by means of various crossover and mutation rates are tested. For the purposes of ICrA, the following criteria are chosen: *S. cerevisiae* and *E. coli* FP models parameters, SGA parameters *xovr* and *mutr*, and SGA outcomes—computation time and objective function value. The ICrA implementation is expected to establish relations and dependencies between the considered criteria, both by pairs and by triples.

The application of ICrA by triples led to the first real problem implementation of one of the most recently defined topological operator $T(A)$. Applying the topological operator $T(A)$ in both considered Case studies and both examined SGA parameters *xovr* and *mutr* leads to the decrease of the degree of "uncertainty" and the increase of both degrees of "agreement" and "disagreement".

The obtained results from ICrA show some existing relations and dependencies that result from the physical meaning of the model parameters, on the one hand, and from stochastic nature of the considered meta-heuristic, on the other hand. Moreover, the derived additional knowledge for the ascertained correlations might be useful in further identification procedures of FP models and, in general, for more accurate SGA tuning. The obtained promising results of ICrA provoke the idea for further approach applications to other optimization techniques, as well as to different combinatorial problems.

Acknowledgments The work is supported by the Bulgarian National Scientific Fund under the grant DFNI-I-02-5 "InterCriteria Analysis—A New Approach to Decision Making".

References

1. Angelova, M.: Modified genetic algorithms and intuitionistic fuzzy logic for parameter identification of fed-batch cultivation Model. Ph.D. Thesis. Sofia (2014) (in Bulgarian)
2. Angelova, M., Tzonkov, St., Pencheva, T.: Modified multi-population genetic algorithm for yeast fed-batch cultivation parameter identification. Int. J. Bioautomation **13**(4), 163–172 (2009)
3. Atanassov, K.: A new topological operator over intuitionistic fuzzy sets. Notes Intuit. Fuzzy Sets **21**(3), 90–92 (2015)
4. Atanassov, K.: Generalized index matrices. Compt. rend. Acad. Bulg. Sci. **40**(11), 15–18 (1987)
5. Atanassov, K.: Index Matrices: Towards an Augmented Matrix Calculus. Springer International Publishing, Switzerland (2014)
6. Atanassov, K.: On Intuitionistic Fuzzy Sets Theory. Springer, Berlin (2012)
7. Atanassov, K.: On index matrices, part 1: standard cases. Adv. Stud. Contemp. Math. **20**(2), 291–302 (2010)
8. Atanassov, K.: On index matrices, part 2: intuitionistic fuzzy case. Proc. Jangjeon Math. Soc. **13**(2), 121–126 (2010)

9. Atanassov, K.: On index matrices. part 5: 3-dimensional index matrices. Adv. Stud. Contemp. Math. **24**(4), 423–432 (2014)
10. Atanassov, K., Atanassova, V., Gluhchev, G.: InterCriteria analysis: ideas and problems. Notes Intuit. Fuzzy Sets **21**(1), 81–88 (2015)
11. Atanassov, K., Mavrov, D., Atanassova, V.: Intercriteria decision making: a new approach for multicriteria decision making, based on index matrices and intuitionistic fuzzy sets. Issues Intuit. Fuzzy Sets Gen. Nets **11**, 1–8 (2014)
12. Atanassov, K., Szmidt, E., Kacprzyk, J.: On intuitionistic fuzzy pairs. Notes Intuit. Fuzzy Sets **19**(3), 1–13 (2013)
13. Bastin, G., Dochain, D.: On-line Estimation and Adaptive Control of Bioreactors. Elsevier Scientific Publications, Amsterdam (1991)
14. Boussaid, I., Lepagnot, J., Siarry, P.: A survey on optimization metaheuristics. Inf. Sci. **237**, 82–117 (2013)
15. Dickinson, R.J., Schweizer, M.: Metabolism and Molecular Physiology of *Saccharomyces cerevisiae*, 2nd edn. CRC Press, Florida (2004)
16. Goldberg, D.E.: Genetic algorithms in search, optimization and machine learning. Addison Wesley Longman, London (2006)
17. Lin, W., Lee, W., Hong, T.: Adapting crossover and mutation rates in genetic algorithms. J. Inf. Sci. Eng. **19**, 889–903 (2003)
18. Matsuoka, Y., Shimizu, K.: Importance of understanding the main metabolic regulation in response to the specific pathway mutation for metabolic engineering of Escherichia coli. Comput. Struct. Biotechnol. J. **3**(4), e201210018 (2012)
19. Obitko, M.: Genetic Algorithms, http://www.obitko.com/tutorials/genetic-algorithms/
20. Pencheva, T., Angelova, M., Atanassova, V., Roeva, O.: InterCriteria analysis of genetic algorithm parameters in parameter identification. Notes Intuit. Fuzzy Sets **21**(2), 99–110 (2015)
21. Pencheva, T., Hristozov, I., Huell, D., Hitzmann, B., Tzonkov, St.: Modelling of functional states during *Saccharomyces cerevisiae* fed-batch cultivation. Int. J. Bioautomation **2**, 8–16 (2005)
22. Roeva, O., Pencheva, T., Hitzmann, B., Tzonkov, St.: A genetic algorithms based approach for identification of *Escherichia coli* fed-batch fermentation. Int. J. Bioautomation **1**, 30–41 (2004)
23. Roeva, O., Fidanova, S.: Chapter 13. Application of genetic algorithms and ant colony optimization for modeling of *E. coli* cultivation process, In: Real-World Application of Genetic Algorithms. In Tech (2012)
24. Roeva, O.: Sensitivity analysis of E. coli fed-batch cultivation local models. Mathematica Balkanica. New Series **25**(4), 395–411 (2011)
25. Vassilev, P., Todorova, L., Andonov, V.: An auxiliary technique for InterCriteria Analysis via a three dimensional index matrix. Notes Intuit. Fuzzy Sets **21**(2), 71–76 (2015)

Genetic Algorithms for Constrained Tree Problems

Riham Moharam and Ehab Morsy

Abstract Given an undirected weighted connected graph $G = (V, E)$ with vertex set V, edge set E and a designated vertex $r \in V$, this chapter studies the following constrained tree problems in G. The first problem, called *Constrained Minimum Spanning Tree Problem* (CMST), asks for a rooted tree T in G that minimizes the total weight of T such that the distance between the r and any vertex v in T is at most a given constant C times the shortest distance between the two vertices in G. The second problem, *Constrained Shortest Path Tree Problem* (CSPT) requires a rooted tree T in G that minimizes the maximum distance between r and all vertices in V such that the total weight of T is at most a given constant C times the minimum tree weight in G. It is easy to conclude from the literatures that the above problems are NP-hard. This chapter presents efficient genetic algorithms that return (as shown by our experimental results) high quality solutions for those two problems.

Keywords Minimum spanning tree · Shortest paths tree · Balanced spanning tree · Genetic algorithms · Graph algorithms

1 Introduction

Let $G = (V, E)$ be an edge-weighted connected graph with vertex set V and edge set E. Let n and m denote the cardinalities of V and E respectively (i.e., $|V| = n$ and $|E| = m$). A spanning tree T_m in G is a Minimum Spanning Tree (MST) if there is no other spanning tree in G that attains a total weight less than that of T_m. Kruskal's and Prim's algorithms are well known polynomial time algorithms for computing a minimum spanning tree in weighted graphs. For any vertex r in G, a spanning tree T_s rooted at r is a shortest path tree if, for every vertex $v \in V$, the

This work is partially supported by Alexander von Humboldt foundation.

R. Moharam (✉) · E. Morsy
Department of Mathematics, Suez Canal University, Ismailia 41522, Egypt
e-mail: Riham.Sci@gmail.com

E. Morsy
e-mail: ehabmorsy@gmail.com

© Springer International Publishing Switzerland 2016
S. Fidanova (ed.), *Recent Advances in Computational Optimization*,
Studies in Computational Intelligence 655, DOI 10.1007/978-3-319-40132-4_13

distance between r and v in T_s equals the shortest distance between the two vertices in G. Dijkstra's algorithm is one of the well known polynomial time algorithms for computing shortest path tree in weighted graphs [31].

The shortest paths tree and minimum spanning tree are widely used in network routing. In particular, the shortest path tree minimizes the delay from the source to every destination through a routing tree, and the minimum spanning tree minimizes the total routing cost along a tree. See [6, 17, 20, 30] and the references therein. Thus, balanced spanning tree is an appropriate routing tree for networks that aim to balance the above two objectives.

Note that, there exist weighted graphs in which the total weight of a shortest path tree may be much more than that of a minimum spanning tree, and vertices that are close to the designated root can be far away from the root in a minimum spanning tree (see [19] for an illustrative example). A rooted tree in a given graph balances a minimum spanning tree and a shortest path tree if its total weight is at most a constant times the minimum tree weight, and the distance between the root and any vertex in the tree is at most a constant times the shortest distance between the two vertices in the graph. Formally, for any $\alpha, \beta \geq 1$, a rooted spanning tree T in G is called (α, β)-balanced spanning tree if, (i) for every vertex v, the distance between the root and v in T is at most α times the shortest distance between the two vertices in G, and (ii) the total weight of T is at most β times the minimum tree weight in G.

Given a constant $C \geq 1$, we first consider the following two variants of the (α, β)-balanced spanning tree problem. The Constrained Minimum Spanning Tree Problem (CMST) asks for a rooted tree T in G that minimizes the total weight of T (i.e., minimizes β) such that the distance between the root and any vertex v in T is at most C times the shortest distance between the two vertices in G. A formal definition for the CMST can be described as follow. For any pair of vertices u and v in G, $d_T(u, v)$ (respectively, $d_G(u, v)$) denotes the shortest distance between u and v in T (respectively, G). For a subgraph G' of G, let $w(G')$ denote the total weight of G' (i.e., the total weights of all edges in G').

Constrained Minimum Spanning Tree Problem (CMST):
Input: An edge-weighted graph $G = (V, E)$, a root $r \in V$, and a constant $C \geq 1$.
Feasible solution: A spanning tree T in G such that $d_T(r, v) \leq C \cdot d_G(r, v)$ for all $v \in G$.
Goal: Find a feasible solution that minimizes $w(T)/w(T_m)$.

Similarly, the Constrained Shortest Path Tree Problem (CSPT) requires a tree rooted T in G that minimizes the maximum distance between the root and all vertices in G (i.e., minimizes α) such that the total weight of T is at most C times the minimum tree weight in G. This problem can be defined formally as follows.

Constrained Shortest Path Tree Problem (CSPT):
Input: An edge-weighted graph $G = (V, E)$, a root $r \in V$, and a constant $C \geq 1$.
Feasible solution: A spanning tree T in G such that $w(T) \leq C \cdot w(T_m)$.
Goal: Find a feasible solution that minimizes $\max_{v \in V(G)}(d_T(r, v)/d_G(r, v))$.

It is well known that, for any $\alpha, \beta \geq 1$, the problem of deciding whether G contains an (α, β)-balanced spanning tree is NP-complete [19]. Consequently, the CMST and CSPT problems are NP-hard problems.

The rest of this chapter is organized as follows. Section 2 reviews some results on related problems. Section 3 presents the proposed genetic algorithm. Section 4 evaluates our algorithm by applying it to randomly generated instances of the two problems. Section 5 makes some concluding remarks.

2 Related Work

In this section, we present results on related problems. Let $G = (V, E)$ be a given edge-weighted graph. The eccentricity of a vertex $v \in V$ is the greatest distance between v and any other vertex in V. The radius of G is the minimum eccentricity of any vertex. The diameter of G is maximum eccentricity of any vertex in the graph.

Bharath-Kumar and Jaffe [4] studied the problem of finding a rooted tree in G such that the total distances from the root to all vertices is at most a constant times the minimum total distances from the root to all vertices.

A tree in G is called shallow-light tree if its diameter is at most a constant (greater than or equal 1) times the diameter of G and with total weight at most a constant times the minimum tree weight. Awerbuch et al. [2] proved that each graph has a shallow-light tree.

Cong et al. [8] proposed a model of timing-driven global routing for cell-based design to improve the construction of a shallow-light tree based on the idea of finding minimum spanning tree with bounded radius. They designed an algorithm to find, for any constant $\epsilon > 0$, a spanning tree with radius $(1 + \epsilon) \cdot R$ (using an analog of the classical Prim's minimum spanning tree structure), where R is the minimum possible tree radius. That is, they found a smooth trade-off between the radius and the cost of the tree. Afterwards, they proposed a new method [9] to improve their previous algorithm based on a provably good algorithm that simultaneously minimizes both the total weight and the longest interconnection path length of the tree. More specifically, their algorithm produced a tree with radius at most $(1 + \epsilon) \cdot R$ and of total weight at most $(1 + 2/\epsilon)$ times the minimum tree weight.

Given any $\alpha \geq 1$, Awerbuch et al. [3] proposed an algorithm that approximates a minimum spanning tree and a shortest paths tree in G. Namely, they modified the algorithm described in [2, 8] to compute an $(\alpha, 1 + \frac{4}{\alpha-1})$-balanced spanning tree in $O(m + n \log n)$ time. Afterwards, Khullar et al. [19] improved the above result and presented a constructive linear time algorithm that outputs an $(\alpha, 1 + \frac{2}{\alpha-1})$-balanced spanning tree in G. In other words, for any $\gamma > 0$, the algorithm of Khullar et al. [19] outputs an $(1 + \sqrt{2}\gamma, 1 + \frac{\sqrt{2}}{\gamma})$-balanced spanning tree in linear time.

3 Genetic Algorithm

In this section, we propose genetic algorithms to CMST and CSPT problems defined in Sect. 1. We first briefly describe the main steps of typical genetic algorithms in the next subsection. For a subgraph G' of a given graph G, the sets $V(G')$ and $E(G')$ denote the sets of vertices and edges of G', respectively.

3.1 Algorithm Overview

The Genetic Algorithm (GA) is an iterative optimization approach based on the principles of genetics and natural selection [11]. The first step of genetic algorithms is to determine a suitable data structure to represent individual solutions (chromosomes), and then construct an initial population (first generation) of prescribed cardinality *pop-size*. A typical intermediate iteration of genetic algorithms can be outlined as follows. Starting with the current generation, we use a predefined selection technique to repeatedly choose a pair of individuals (parents) in the current generation to reproduce, with probability p_c, a new set of individuals (offsprings) by applying crossover operation to the selected pair. To keep an appropriate diversity among different generations, we apply mutation operation with specific probability p_m to genes of individuals of the current generation to get more offsprings. A new generation is then selected from both the offspring and the current generation based on their fitness values (the more suitable solutions have more chances to reproduce). The algorithm terminates when it meets prescribed stopping criteria.

A formal description of the proposed genetic algorithm is given in Algorithm 1.

Algorithm 1 Genetic Algorithm for the Balanced Spanning Tree

1. Compute an initial population I_0 (cf. Sect. 3.3).
2. $gen \leftarrow 1$.
3. **While** $(gen \leq maxgen)$ **do**
4.   **For** $i = 1$ to *pop-size* **do**
5.   Select a pair of chromosomes from I_{gen-1} (cf. Sect. 3.5).
6.   Apply crossover operator with probability p_c to the selected pair of chromosomes to get two offsprings (cf. Sect. 3.6).
7.   **Endfor**
8.   For each chromosome in I_{gen-1}, apply mutation operator with probability p_m to get an offspring (cf. Sect. 3.7).
9.   Extend I_{gen-1} with feasible offsprings output from lines 6 and 8.
10.   Find a chromosome T_{gen-1} in I_{gen-1} that has the minimum fitness value (cf. Sect. 3.4).
11.   If $gen \geq 2$ and $w(T_{gen-2}) = w(T_{gen-1}) = w(T_{gen})$, then **break**.
12.   Select *pop-size* chromosomes from I_{gen-1} to form I_{gen} (cf. Sect. 3.5).
13. $gen \leftarrow gen + 1$.
14. **Endwhile**
15. Output T_{gen}.

In genetic algorithms, determining representation method, population size, selection technique, crossover and mutation probabilities, and stopping criteria are crucial since they mainly affect the convergence of the algorithm (see [1, 16, 21, 26, 29]).

The rest of this section is devoted to describe steps of Algorithm 1 in details.

3.2 Representation

Choosing a proper representation to encode an individual solution (chromosome) may affect the performance of the algorithm. There are three common encoding methods that can be used to encode spanning trees in genetic algorithms; characteristic vectors-based encoding [10, 24], node-based encoding [22], and edge-based encoding [28]. Recently, many studies have showed that edge-based encoding is the most appropriate method for representing spanning trees for the problems affected by edge weights [14, 18, 25, 28]. This motivates us to use this representation in the proposed algorithms.

Let $G = (V, E)$ be a given undirected graph such that $V = \{1, 2, \ldots, n\}$. Note that, each edge $e \in E$ with end points i and j can be defined by an unordered pair $\{i, j\}$, and consequently, any subgraph G' of G is uniquely defined by the set of unordered pairs of all its edges. In particular, any spanning tree T in G is induced by a set of exactly $n - 1$ pairs corresponding to its edges since T is a subgraph of G that spans all vertices in V and has no cycles. Therefore, each chromosome (spanning tree) can be represented as a sequence of unordered pairs of integers each of which represent a gene (edge) in the chromosome as illustrated in Fig. 1.

3.3 Initial Population

Constructing an initial generation is the first step in typical genetic algorithms. We first have to decide the population size *pop-size*, one of the decisions that affect

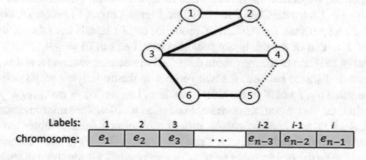

Fig. 1 The representation of a chromosome in Algorithm 1

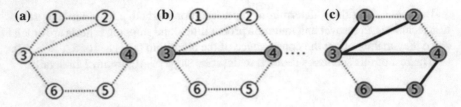

Fig. 2 Constructing a chromosome in the initial population: in **a** a random vertex 4 is chosen to form the current tree T. A random neighbor 3 of 4 is chosen and the edge $\{3, 4\}$ is then added to the current tree in **b**. This procedure will be repeated $n - 1$ times until the set of all vertices in the underlying graph is visited as in **c**

the convergence of the genetic algorithm. It is expected that small population size may lead to weak solutions, while, large population size increases the space and time complexity of the algorithm. Many literatures studied the influence of the population size to the performance of genetic algorithms (see [26] and the references therein). In this paper, we discuss the effect of the population size on the convergence time of the algorithm (cf. Sect. 4).

On of the most common methods is to apply random initialization to get an initial population. Namely, we compute each chromosome in the initial population by repeatedly applying the following simple procedure as long as the cardinality of the set of visited vertices is less than n (or as long as the cardinality of the set of traversed edges is less than $n - 1$). Let T denote the tree constructed so far by the procedure (initially, T consists of a random vertex from $V(G)$). We first select a random vertex $v \notin V(T)$ from the set of the neighbors of all vertices in T, and then add the edge $e = (u, v)$ to T, where u is the neighbor of v in T. It is easy to verify that the above procedure returns a tree after exactly $n - 1$ iterations as illustrated in Fig. 2. The generated tree T is added to the initial population.

The above algorithm is repeated as long as the number of constructed population is less than *pop-size*.

3.4 Fitness Function

In this section, we define the fitness value of an individual (chromosome) tree T, denoted by $fv(T)$, for the CMST and CSPT problems. Let $ov(T)$ denote the objective value of T. Recall that, for the CMST problem, $ov(T)$ equals the ratio of the total weight of T to that of the minimum spanning tree, i.e., $ov(T) = \frac{w(T)}{w(T_m)}$.

Similarly, the ratio of the maximum distance between the root and a vertex in T to the minimum distance between the two vertices in the underlying graph defines the objective value $ov(T)$ of T in the CSPT problem, i.e., $ov(T) = \max_{v \in V(G)} \frac{d_T(r,v)}{d_G(r,v)}$.

Note that, the above combinatorial optimization problems are minimization problems (cf. Sect. 1), and hence chromosomes of minimum objective values have good chances to be survive and reproduce. Thus, the fitness value $fv(T)$ of a chromosome T is defined to be the reciprocal of its objective value $ov(T)$ throughout the execution of the proposed genetic algorithms, i.e., $fv(T) = 1/ov(T)$.

3.5 Selection Process

In this paper, we present three common selection techniques: roulette wheel selection, stochastic universal sampling selection, and tournament selection. All these techniques are called fitness-proportionate selection techniques since they are based on a predefined fitness function used to evaluate the quality of individual chromosomes. Throughout the execution of the proposed algorithm, the reverse of this ratio is used as the fitness function of the corresponding chromosome. We assume that the same selection technique is used throughout the whole algorithm. The rest of this section is devoted to briefly describe these selection techniques.

Roulette Wheel Selection (RWS): [7, 11] Here, the probability of selecting a chromosome is based on its fitness value. More precisely, each chromosome is selected with the probability that equals to its normalized fitness value, i.e., the ratio of its fitness value to the total fitness values of all chromosomes in the set from which it will be selected.

Stochastic Universal Sampling Selection (SUS): [5, 7] Instead of a single selection pointer used in roulette wheel approach, SUS uses h equally spaced pointers, where h is the number of chromosomes to be selected from the underlying population. All chromosomes are represented in number line randomly and a single pointer $ptr \in (0, \frac{1}{h}]$ is generated to indicate the first chromosome to be selected. The remaining $h - 1$ individuals whose fitness spans the positions of the pointers $ptr + i/h$, $i = 1, 2, \ldots, h - 1$ are then chosen.

Tournament Selection (TRWS): [5, 11] This is a two stages selection technique. We first select a set of $k < pop\text{-}size$ chromosomes randomly from the current population. From the selected set, we choose the more fit chromosome by applying the roulette wheel selection approach. Tournament selection is performed according to the required number of chromosomes.

3.6 Crossover Process

In each iteration of the algorithm we repeatedly select a pair of chromosomes (parents) from the current generation and then apply crossover operator with probability p_c to the selected chromosomes to get new chromosomes (offsprings). Simulations and experimental results of the literatures show that a typical crossover probability lies between 0.75 and 0.95. There are two common crossover techniques: single-point crossover and multi-point crossover. Many researchers studied the influence of crossover approach and crossover probability to the efficiency of the whole genetic algorithm, see for example [20, 29] and the references therein. In this paper, we use a multi-point crossover approach by exchanging a randomly selected set of edges between the two parents. In particular, we repeat the following procedures $pop\text{-}size$ times. We select a pair of chromosomes T_1 and T_2 from the current generation, we

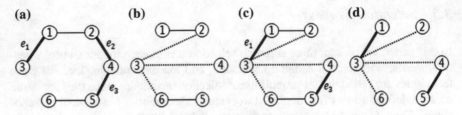

Fig. 3 An example of crossover process applying to parents T_1 and T_2 in **a** and **b**, respectively. $E_1 = E(T_1) - E(T_2) = \{e_1, e_2, e_3\}$ (*solid edges* in **a**). **c** $k = 2$ edges are selected from E_1 (*solid edges* in **c**) and added to T_2. **d** Fixing the cycles to get a new offspring

generate a random number $s \in (0, 1]$, and apply the following crossover operation to T_1 and T_2 if $s < p_c$ holds.

Define the two sets $E_1 = E(T_1) - E(T_2)$ and $E_2 = E(T_2) - E(T_1)$ ($|E_1| = |E_2|$ holds). Let $t = |E_1| = |E_2|$, and generate a random number k from $[1, t]$. We first choose a random subset E_1' of cardinality k from E_1, and then add E_1' to T_2 to get a subgraph T' (i.e., $T' = T_2 \cup E_1'$). Clearly, T' contains k cycles each of which contains a distinct edge from E_1'. For every edge $e = (u, v)$ in E_1', we apply the following procedure to fix a cycle containing e. Let $\widetilde{T}$ be the current subgraph (initially, $\widetilde{T} = T'$). We first find a path $P_{\widetilde{T}}(u, v)$ between u and v in the set of edges $E(\widetilde{T}) - \{e\}$. We then choose an edge $\widetilde{e}$ in $P_{\widetilde{T}}(u, v)$ randomly and delete it from the subgraph $\widetilde{T}$. See Fig. 3 for an illustrative example of the crossover operation.

Similarly, we apply the above crossover technique by interchanging the roles of T_1 and T_2 to get one more offspring. Note that, we only consider the feasible offsprings as candidate chromosomes for the next generation.

3.7 Mutation Process

To maintain the diversity among different generations of the population (and hence avoid local minimum), we apply a genetic (mutation) operator to chromosomes of the current generation with predefined (usually small) probability p_m. Namely, for each chromosome T, we generate a random number $s \in (0, 1]$, and then mutate T if $s < p_m$ holds by replacing a random edge (gene) in T with a random edge from $E(G) - E(T)$. Many results analyzed the role of mutation operator in genetic algorithms [1, 16, 20].

Formally, a chromosome T is mutated as follows. We first select a random edge $e = (u, v)$ in the graph G but not in the chromosome T, i.e., e is randomly chosen from the set $E(G) - E(T)$ of edges, It is easy to see that the subgraph $E(T) \cup \{e\}$ contains exactly one cycle including e. We then select a random edge e' in the path $P_T(u, v)$ between u and v in T. Let T' denote the offspring obtained from T by exchanging the two edges e and e', i.e., $E(T') = (E(T) - \{e\}) \cup \{e'\}$. See Fig. 4 for

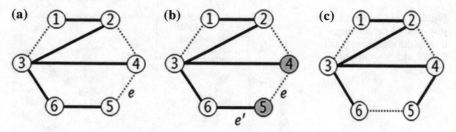

Fig. 4 An example of mutation process. **a** a Random edge e is selected from the set $E(G) - E(T)$. **b** The cycle is fixed by selecting a random edge e' in $P_T(4, 5)$. **c** The resulting offspring

an illustrative example of the mutation process. The resulting spanning tree T' is considered as a candidate chromosome for the next iteration if it is feasible solution of the corresponding problem.

4 Experimental Results

In this section, we evaluate the proposed genetic algorithms by applying them to several random instances of the studied problems. These instances are generated by applying Erdos and Renyi [12] approach in which an edge is independently included between each pair of nodes with randomly chosen probability $p \in [0, 1]$. Here, we generate random graphs with sizes 20, 50, and 100, and the edge weights are randomly assigned from the range [1, 100].

All results presented in this section were performed in MATLAB R2014b on a computer powered by a core i7 processor and 16 GB RAM.

For the sake of determining the appropriate population size *pop-size*, we apply the proposed algorithms with population sizes $n/3, 2n/3, n, 4n/3, 5n/3, 2n, 7n/3$, and $8n/3$, where n is the graph size. The obtained results show that the graph size n is the most appropriate value for the population size (see Figs. 5, 6, 7 and 8). We apply the proposed algorithms with population size *pop-size* = n, maximum number of iterations $maxgen = 300$, crossover probability $p_c = 0.9$, and mutation probability $p_m = 0.2$. The algorithms terminates if either the number of iterations exceeds $maxgen$ or the solution does not change for three consecutive iterations. All obtained solutions (for graphs of sizes 20, 50 and 100) are compared with the corresponding optimal solutions computed by considering all possibilities of feasible spanning trees in the underlying graphs. It is seen from the results presented in this section that the obtained trees are optimal solutions for almost all instances the algorithms apply to.

Tables 1, 2 and 3 show the results of the algorithm for CMST instances of sizes 20, 50 and 100, respectively.

The results of the algorithm for CSPT instances of sizes 20, 50 and 100 are shown in Tables 4, 5 and 6, respectively.

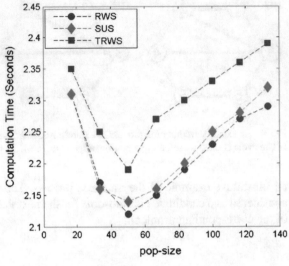

Fig. 5 The influence of *pop-size* on the running time of the algorithm for the CMST with $n = 50$ and $C = 1.3$

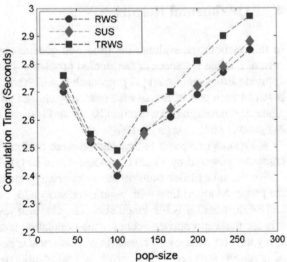

Fig. 6 The influence of *pop-size* on the running time of the algorithm for the CMST with $n = 100$ and $C = 1.3$

It is difficult to compute the optimal solutions for the proposed problems especially for graphs of large sizes. We observe that the results obtained by running the algorithms multiple times for the same CMST and CSPT instances of size 1000 lie in a very small range; the recorded results in Table 7 are the minimum obtained values.

Fig. 7 The influence of *pop-size* on the running time of the algorithm for the CSPT with $n = 50$ and $C = 1.3$

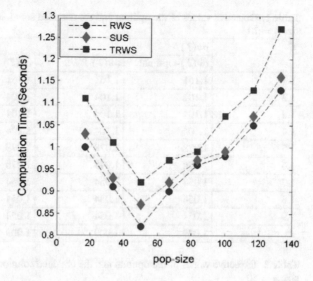

Fig. 8 The influence of *pop-size* on the running time of the algorithm for the CSPT with $n = 100$ and $C = 1.3$

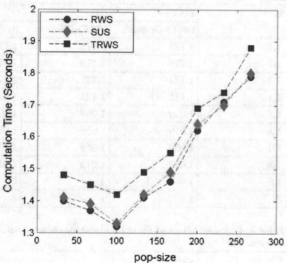

Table 1 Objective values of the optimal and the obtained solutions of the CMST in a graph with size $n = 20$

C	$ov(T)$			
	$ov(T)$-Optimal	$ov(T)$-RWS	$ov(T)$-SUS	$ov(T)$-TRWS
1.1	1.161	1.161	1.161	1.161
1.2	1.104	1.104	1.104	1.104
1.3	1.104	1.104	1.104	1.104
1.4	1.104	1.104	1.104	1.104
1.5	1.078	1.078	1.078	1.078
1.6	1.078	1.078	1.078	1.078
1.7	1.054	1.054	1.054	1.054
1.8	1.054	1.054	1.054	1.054
1.9	1.054	1.054	1.054	1.054
2	1.009	1.009	1.009	1.009

Table 2 Objective values of the optimal and the obtained solutions of the CMST in a graph with size $n = 50$

C	$ov(T)$			
	$ov(T)$-Optimal	$ov(T)$-RWS	$ov(T)$-SUS	$ov(T)$-TRWS
1.1	1.176	1.176	1.176	1.176
1.2	1.176	1.176	1.176	1.176
1.3	1.121	1.121	1.176	1.121
1.4	1.121	1.121	1.121	1.121
1.5	1.099	1.099	1.099	1.099
1.6	1.059	1.059	1.099	1.059
1.7	1.059	1.059	1.059	1.059
1.8	1.012	1.012	1.012	1.012
1.9	1	1	1	1
2	1	1	1	1

Table 3 Objective values of the optimal and the obtained solutions of the CMST in a graph with size $n = 100$

C	$ov(T)$			
	$ov(T)$-Optimal	$ov(T)$-RWS	$ov(T)$-SUS	$ov(T)$-TRWS
1.1	1.128	1.128	1.128	1.128
1.2	1.128	1.128	1.128	1.128
1.3	1.108	1.108	1.177	1.108
1.4	1.108	1.108	1.108	1.108
1.5	1.072	1.083	1.083	1.072
1.6	1.072	1.072	1.072	1.072
1.7	1.049	1.049	1.072	1.049
1.8	1.049	1.049	1.049	1.049
1.9	1	1	1	1
2	1	1	1	1

Table 4 Objective values of the optimal and the obtained solutions of the CSPT in a graph with size $n = 20$

C	$ov(T)$			
	$ov(T)$-Optimal	$ov(T)$-RWS	$ov(T)$-SUS	$ov(T)$-TRWS
1.1	1.444	1.444	1.444	1.444
1.2	1.112	1.112	1.112	1.112
1.3	1.112	1.112	1.112	1.112
1.4	1.048	1.048	1.048	1.048
1.5	1.048	1.048	1.048	1.048
1.6	1.048	1.048	1.048	1.048
1.7	1	1	1	1

Table 5 Objective values of the optimal and the obtained solutions of the CSPT in a graph with size $n = 50$

C	$ov(T)$			
	$ov(T)$-Optimal	$ov(T)$-RWS	$ov(T)$-SUS	$ov(T)$-TRWS
1.1	1.320	1.320	1.416	1.320
1.2	1.320	1.320	1.320	1.320
1.3	1.187	1.187	1.320	1.187
1.4	1.187	1.187	1.187	1.187
1.5	1.115	1.115	1.115	1.115
1.6	1.101	1.101	1.101	1.101
1.7	1.075	1.075	1.075	1.075
1.8	1.075	1.075	1.075	1.075
1.9	1	1	1	1

Table 6 Objective values of the optimal and the obtained solutions of the CSPT in a graph with size $n = 100$

C	$ov(T)$			
	$ov(T)$-Optimal	$ov(T)$-RWS	$ov(T)$-SUS	$ov(T)$-TRWS
1.1	1.113	1.113	1.113	1.113
1.2	1.113	1.113	1.113	1.113
1.3	1.096	1.096	1.096	1.096
1.4	1.096	1.096	1.096	1.096
1.5	1.064	1.064	1.064	1.064
1.6	1.064	1.064	1.064	1.064
1.7	1.033	1.033	1.033	1.033
1.8	1	1	1.033	1
1.9	1	1	1	1

Table 7 Objective values of the obtained solutions of CMST and CSPT instances of size $n = 1000$

$ov(T)$	C									
	1.1	1.2	1.3	1.4	1.5	1.6	1.7	1.8	1.9	2
$ov(T)$-CMST	1.437	1.437	1.211	1.197	1.197	1.141	1.141	1.010	1	1
$ov(T)$-CSPT	1.211	1.211	1.178	1.178	1.178	1.065	1.065	1	1	1

5 Conclusion

In this chapter, we have studied two constrained tree problems in a given edge-weighted graph G, the Constrained Minimum Spanning Tree (CMST) and the Constrained Shortest Path Tree (CSPT). In the CMST (respectively, CSPT), we are given an upper bound on how far the required tree is from the shortest path tree (respectively, the minimum spanning tree), and the objective is to find the closest tree to the minimum spanning tree (respectively, the shortest path tree) under this bound. It is easy to conclude from the literatures that the above problems are NP-hard. We have designed genetic algorithms for these problems evaluated by applying them to random graph instances. The proposed algorithms output hight quality spanning trees for all instances they have been applied to. It will be interesting to relax our model to balanced subgraphs instead of balanced trees, that is, the problem of finding a minimum weight subgraph such that the distance between any two vertices u and v in the subgraph is at most a given constant times the shortest distance between the two vertices in the underlying graph (this problem is known as t-spanner problem in the literatures). See [13, 15, 23, 27] and the references therein.

References

1. Abdoun, O., Abouchabaka, J., Tajani, C.: Analyzing the performance of mutation operators to solve the travelling salesman problem. IJES, Int. J. Emerg. Sci. **2**(1), 61–77 (2012)
2. Awerbuch, B., Baratz, A., Peleg, D.: Cost-sensetive analysis of communication protocols. In: Proceedings on Principles of Distributed Computing, pp. 177–187 (1990)
3. Awerbuch, B., Baratz, A. Peleg, D.: Efficient broadcast and light-weight spanners. Manuscript (1991)
4. Bharath-Kumar, K., Jaffe, J.M.: Routing to multiple destinations in computer networks. IEEE Trans. Commun. **31**(3), 343–351 (1983)
5. Blickle, T., Thiele, L.: A comparison of selection schemes used in genetic algorithms (Technical Report No. 11), Swiss Federal Institute of Technology (ETH) Zurich, Computer Engineering and Communications Networks Lab (TIK) (1995)
6. Campos, R., Ricardo, M.: A fast algorithm for computing minimum routing cost spanning trees. Comput. Netw. **52**(17), 3229–3247 (2008)
7. Chipperfield, A., Fleming, P., Pohlheim, H., Fonseca, C.: The matlab genetic algorithm user's guide, UK SERC (1994)

8. Cong, J., Kahng, A., Robins, G., Sarrafzadeh, M., Wong, C.K.: Performance-driven global routing for cell based IC's. In: Proceedings of the IEEE International Conference on Computer Design, pp. 170-173 (1991)
9. Cong, J., Kahng, A., Robins, G., Sarrafzadeh, M., Wong C.K.: Provably good performance-driven global routing. IEEE Transaction on CAD, pp. 739–752 (1992)
10. Davis, L., Orvosh, D., Cox, A., Qiu, Y.: A genetic algorithm for survivable network design. In: Proceedings 5th International Conference on Genetic Algorithms, pp. 408–415 (1993)
11. Engelbrecht, A.P.: Computational Intelligence: An Introduction. Wiley, New York (2007)
12. Erdos, P., Renyi, A.: On random graphs. Publ. Math **6**, 290–297 (1959)
13. Farley, A.M., Zappala D., Proskurowski, A., Windisch, K.: Spanners and message distribution in networks, Dicret. Appl. Math. **137**, 159–171 (2004)
14. Gottlieb, J., Julstrom, B.A., Rothlauf, F., Raidl, G.R.: Prüfer numbers: a poor representation of spanning trees for evolutionary search. In: Proceedings of the 2001 Genetic and Evolutionary Computation Conference, pp. 343350. Morgan Kaufmann (2000)
15. Gudmundsson, J., Levcopoulos, C., Narasimhan, G.: Fast greedy algorithms for constructing sparse geometric spanners. SIAM J. Comput. **31**, 1479–1500 (2002)
16. Hesser, J., Mnner, R.: Towards an optimal mutation probability for genetic algorithms. In: Proceedings of the 1st Workshop in Parallel Problem Solving from Nature, pp. 23-32 (1991)
17. Huang, G., Li, X., He, J.: Dynamic minimal spanning tree routing protocol for large wireless sensor networks. In: Proceedings of the 1st IEEE Conference on Industrial Electronics and Applications, Singapore, pp. 1-5 (2006)
18. Julstrom, B. A.: Encoding rectilinear Steiner trees as lists of edges. In: Proceedings of the 16th ACM Symposium on Applied Computing, pp. 356–360. ACM Press (2001)
19. Khullar, S., Raghavachari, B., Young, N.: Balancing minimum spanning trees and shortest-path trees. Algorithmica **14**, 305–322 (1995)
20. Li, C., Zhang, H., Hao, B., Li, J.: A survey on routing protocols for large-scale wireless sensor networks. Sensors **11**, 3498–3526 (2011)
21. Lin, W.-Y. Lee, W.-Y., T.-P. Hong (2001) Adapting Crossover and Mutation Rates in Genetic Algorithms, the Sixth Conference on Artificial Intelligence and Applications, Kaohsiung, Taiwan (2001)
22. Mathur, R., Khan, I., Choudhary, V.: Genetic algorithm for dynamic capacitated minimum spanning tree. Int.l J. Comput. Tech. Appl. **4**, 404–413 (2013)
23. Navarro, G., Paredes, R., Chavez, E.: t-spanners as a data structure for metric space searching. In: International Symposium on String Processing and Information Retrieval, SPIRE. LNCS, vol. 2476, 298–309 (2002)
24. Piggott, P., Suraweera, F.: Encoding graphs for genetic algorithms: an investigation using the minimum spanning tree problem. In: Yao X. (ed.) Progress in Evol. Comput. LNAI, vol. 956, pp. 305–314. Springer, New York (1995)
25. Raidl, G.R., Julstrom, B.A.: Edge-sets: an effective evolutionary coding of spanning trees, IEEE Trans. Evol. Comput. **7**, 225–239 (2003)
26. Roeva, O., Fidanova, S., Paprzycki, M.: Influence of the population size on the genetic algorithm performance in case of cultivation process modelling. In: Proceedings of the Federated Conference on Computer Science and Information Systems, pp. 371-376 (2013)
27. Sigurd, M., Zachariasen, M.: Construction of Minimum-Weight Spanners. Springer, Berlin (2004)
28. Tan, Q.P.: A genetic approach for solving minimum routing cost spanning tree problem. Int. J. Mach. Learn. Comput. **2**(4), 410–414 (2012)
29. Vekaria, K., Clack, C.: Selective Crossover in Genetic Algorithms: An Empirical Study. Lecture Notes in Computer Science, vol. 1498, pp. 438–447. Springer, Berlin (1998)
30. Xiao, B., ZhuGe, Q., Sha, E.H.-M.: Minimum dynamic update for shortest path tree construction, global telecommunications conference, San Antonio, TX, pp. 126-130 (2001)
31. Ye We, B., Chao, K.: Spanning Trees and Optimization Problems. Chapman & Hall, Boca Raton (2004)

InterCriteria Analysis of Genetic Algorithms Performance

Olympia Roeva, Peter Vassilev, Stefka Fidanova
and Marcin Paprzycki

Abstract In this paper we apply InterCriteria Analysis (ICrA) approach based on the apparatus of Index Matrices and Intuitionistic Fuzzy Sets. The main idea is to use ICrA to establish the existing relations and dependencies of defined parameters in a non-linear model of an *E. coli* fed-batch cultivation process. We perform a series of model identification procedures applying Genetic Algorithms (GAs). We proposed a schema of ICrA of ICrA results to examine the obtained model identification results. The discussion about existing relations and dependencies is performed according to criteria defined in terms of ICrA. We consider as ICrA criteria model parameters and GAs outcomes on the one hand, and 14 differently tuned GAs on the other. Based on the results, we observe the mutual relations between model parameters and GAs outcomes, such as computation time and objective function value. Moreover, some conclusions about the preferred tuned GAs for the considered model parameter identification in terms of achieved accuracy for given computation time are presented.

Keywords InterCriteria analysis · Index matrices · Intuitionistic fuzzy sets · Genetic algorithm · Parameter identification · *E. coli* · Cultivation process

O. Roeva · P. Vassilev
Institute of Biophysics and Biomedical Engineering, Bulgarian Academy of Sciences,
Sofia, Bulgaria
e-mail: olympia@biomed.bas.bg

P. Vassilev
e-mail: peter.vassilev@gmail.com

S. Fidanova (✉)
Institute of Information and Communication Technology,
Bulgarian Academy of Sciences, Sofia, Bulgaria
e-mail: stefka@parallel.bas.bg

M. Paprzycki
System Research Institute Polish Academy of Sciences,
Warsaw and Management Academy, Warsaw, Poland
e-mail: marcin.paprzycki@ibspan.waw.pl

© Springer International Publishing Switzerland 2016
S. Fidanova (ed.), *Recent Advances in Computational Optimization*,
Studies in Computational Intelligence 655, DOI 10.1007/978-3-319-40132-4_14

1 Introduction

InterCriteria Analysis (ICrA) is developed with the aim to gain additional insight into the nature of the criteria involved and to discover on this basis existing relations between the criteria themselves [6]. It is based on the apparatus of Index Matrices (IM) [2–4] and Intuitionistic Fuzzy Sets [5] and can be applied for decision making in different areas of science and practice.

Up to now, the ICrA approach has been discussed in several papers considering parameter estimation problems of cultivation processes. In [13], ICrA implementation allowed to establish relations and dependencies between two of the main genetic algorithms (GAs) parameters [11]—numbers of individuals and number of generations, on the one hand, and computation time, model accuracy and model parameters, on the other hand. In [12], ICrA has been applied to establish fundamental correlation between the kinetic variables of fed-batch processes for *E. coli* cultivation. Further, ICrA has been applied to explore the existing relations and dependencies of defined model parameters and GA outcomes—computation time and objective function value—in case of *S. cerevisiae* cultivation [1]. Moreover, ICrA is applied for establishing the relations and dependencies between GAs parameter *generation gap* as well as the computation time, model accuracy and model parameters in case of *E. coli* cultivation [18] and *S. cerevisiae* cultivation [14]. Finally, ICrA has been applied to establish the relations and dependencies of the considered parameters based on different criteria for metaheuristic hybrid schemes using GAs and Ant Colony Optimization (ACO) [22].

Encouraged by the results of these first applications of ICrA, we continue to investigate further the relations between the model parameters of an *E. coli* fed-batch cultivation process and the outcomes of the applied GAs.

Cultivation processes are characterized with complex, non-linear dynamic and their modelling is a hard combinatorial problem [21]. To support and bring some lights to this optimization problem, ICrA could be applied. On the one hand, the parameter identification is of key importance for process modelling, and ICrA based additional knowledge about the model parameters relations will be extremely useful to improve the model accuracy. On the other hand, ICrA could be used to improve the performance of the used optimization algorithms if, for instance, some algorithm outcomes are added to the defined ICrA criteria. Thus, the relations between model parameters and optimization algorithm performance will be established.

Metaheuristic techniques such as GAs have received more and more attention as optimization algorithms. These methods offer good solutions, even global optima, within reasonable computation time [9].

The performance of GAs largely depends on the proper selection of their parameters values, including population size, crossover mechanism, probability of crossover and mutation, etc. The GAs performance has been investigated using different crossover techniques [15], different parent selection strategies [16], etc. In [10], the effects of tuning parameters on the performance of GAs have been evaluated based on Design of Experiments approach and regression modeling.

To establish the correlation between the applied GAs, we propose a schema of ICrA of ICrA results. ICrA enables us to compare different GAs and to find similar GAs performances. Moreover, ICrA of ICrA results gives us the possibility to choose better GA performance in terms of obtained model accuracy (objective function value) and total computation time for an *E. coli* cultivation model parameter identification problem.

The paper is organized as follows. The background of InterCriteria Analysis is given in Sect. 2. The problem formulation is described in Sect. 3. The numerical results and a discussion are presented in Sect. 4. Conclusion remarks are made in Sect. 5.

2 InterCriteria Analysis

Following [5, 6], we will obtain an Intuitionistic Fuzzy Pair (IFP) as the degrees of "agreement" and "disagreement" between two criteria applied on different objects. We remind briefly that an IFP is an ordered pair of real non-negative numbers $\langle a, b \rangle$ such that:

$$a + b \leq 1.$$

For clarity, let us be given an IM [3] whose index sets consist of the names of the criteria (for rows) and objects (for columns). The elements of this IM are further supposed to be real numbers (in the general case, this is not required). We will obtain an IM with index sets consisting of the names of the criteria (for rows and for columns) with elements IFPs corresponding to the "agreement" and "disagreement" of the respective criteria.

Two things are further supposed (which are not always guaranteed in practice and, when not fulfilled, present an interesting direction for new research by themselves):

1. All criteria provide an evaluation for all objects (i.e. there are no inapplicable criteria for a given object) and all these evaluations are available (no missing evaluations).
2. All the evaluations of a given criteria can be compared amongst themselves.

Further, by O we denote the set of all objects $O_1, O_2, \ldots, O_n$ being evaluated, and by $C(O)$ the set of values assigned by a given criteria C to the objects, i.e.

$$O \stackrel{\text{def}}{=} \{O_1, O_2, \ldots, O_n\},$$
$$C(O) \stackrel{\text{def}}{=} \{C(O_1), C(O_2), \ldots, C(O_n)\}.$$

Let $x_i = C(O_i)$. Then the following set can be defined:

$$C^*(O) \overset{\text{def}}{=} \{\langle x_i, x_j \rangle | i \neq j \,\&\, \langle x_i, x_j \rangle \in C(O) \times C(O)\}.$$

In order to compare two criteria, we must construct the vector of all internal comparisons of each criteria which fulfill exactly one of three relations R, $\overline{R}$ and $\tilde{R}$. In other words, we require that for a fixed criterion C and any ordered pair $\langle x, y \rangle \in C^*(O)$ it is true:

$$\langle x, y \rangle \in R \Leftrightarrow \langle y, x \rangle \in \overline{R}, \tag{1}$$

$$\langle x, y \rangle \in \tilde{R} \Leftrightarrow \langle x, y \rangle \notin (R \cup \overline{R}), \tag{2}$$

$$R \cup \overline{R} \cup \tilde{R} = C^*(O). \tag{3}$$

From the above it is seen that we only need to consider a subset of $C(O) \times C(O)$ for the effective calculation of the vector of internal comparisons (denoted further by $V(C)$) since from Eqs. (1)–(3) it follows that if we know what is the relation between x and y, we also know what is the relation between y and x. Thus, we will only consider lexicographically ordered pairs $\langle x, y \rangle$.

Let, for brevity:
$$C_{i,j} = \langle C(O_i), C(O_j) \rangle.$$

Then, for a fixed criterion C we construct the vector:

$$V(C) = \{C_{1,2}, C_{1,3}, \ldots, C_{1,n}, C_{2,3}, C_{2,4}, \ldots, C_{2,n}, C_{3,4}, \ldots, C_{3,n}, \ldots, C_{n-1,n}\}.$$

It can be easily seen that it has exactly $\frac{n(n-1)}{2}$ elements.

Further, to simplify our considerations, we replace the vector $V(C)$ with $\hat{V}(C)$, where for each $1 \leq k \leq \frac{n(n-1)}{2}$ for the kth component it is true:

$$\hat{V}_k(C) = \begin{cases} 1 & \text{iff } V_k(C) \in R, \\ -1 & \text{iff } V_k(C) \in \overline{R}, \\ 0 & \text{otherwise.} \end{cases}$$

Then, when comparing two criteria, we determine the degree of "agreement" between the two as the number of matching components (divided by the length of the vector for normalization purposes). This can be done in several ways, e.g. by counting the matches or by taking the complement of the Hamming distance. The degree of "disagreement" is the number of components of opposing signs in the two vectors (again normalized by the length). This may also be done in various ways. A pseudocode of the algorithm (Algorithm 1) [18], used in this study for calculating the degrees of "agreement" and "disagreement" between two criteria C and C', is presented below.

Algorithm 1 Calculating "agreement" and "disagreement" between two criteria

Require: Vectors $\hat{V}(C)$ and $\hat{V}(C')$

1: **function** DEGREE OF AGREEMENT $(\hat{V}(C), \hat{V}(C'))$
2:     $V \leftarrow \hat{V}(C) - \hat{V}(C')$
3:     $\mu_{C,C'} \leftarrow 0$
4:     **for** $i \leftarrow 1$ to $\frac{n(n-1)}{2}$ **do**
5:         **if** $V_i = 0$ **then**
6:             $\mu_{C,C'} \leftarrow \mu_{C,C'} + 1$
7:         **end if**
8:     **end for**
9:     $\mu_{C,C'} \leftarrow \frac{2}{n(n-1)}\mu_{C,C'}$
10:     **return** $\mu_{C,C'}$
11: **end function**

12: **function** DEGREE OF DISAGREEMENT $(\hat{V}(C), \hat{V}(C'))$
13:     $V \leftarrow \hat{V}(C) - \hat{V}(C')$
14:     $\nu_{C,C'} \leftarrow 0$
15:     **for** $i \leftarrow 1$ to $\frac{n(n-1)}{2}$ **do**
16:         **if** $abs(V_i) = 2$ **then**   ▷ abs: absolute value
17:             $\nu_{C,C'} \leftarrow \nu_{C,C'} + 1$
18:         **end if**
19:     **end for**
20:     $\nu_{C,C'} \leftarrow \frac{2}{n(n-1)}\nu_{C,C'}$
21:     **return** $\nu_{C,C'}$
22: **end function**

It is obvious (from the way of calculation) that for $\mu_{C,C'}$, $\nu_{C,C'}$, we have:

$$\mu_{C,C'} = \mu_{C',C}, \nu_{C,C'} = \nu_{C',C}.$$

Also, $\langle \mu_{C,C'}, \nu_{C,C'} \rangle$ is an IFP.

In the most of the obtained pairs $\langle \mu_{C,C'}, \nu_{C,C'} \rangle$, the sum $\mu_{C,C'} + \nu_{C,C'}$ is equal to 1. However, there may be some pairs for which this sum is less than 1. The difference

$$\pi_{C,C'} = 1 - \mu_{C,C'} - \nu_{C,C'} \tag{4}$$

is considered as a degree of "uncertainty".

3 Problem Formulation

Let us use the following non-linear differential equation system to describe the *E. coli* fed-batch cultivation process [8]:

$$\frac{dX}{dt} = \mu X - \frac{F_{in}}{V}X, \tag{5}$$

$$\frac{dS}{dt} = -q_S X + \frac{F_{in}}{V}(S_{in} - S),\tag{6}$$

$$\frac{dV}{dt} = F_{in},\tag{7}$$

where

$$\mu = \mu_{max}\frac{S}{k_S + S}, \quad q_S = \frac{1}{Y_{S/X}}\mu\tag{8}$$

and X is the biomass concentration, [g/l]; S is the substrate concentration, [g/l]; F_{in} is the feeding rate, [l/h]; V is the bioreactor volume, [l]; S_{in} is the substrate concentration in the feeding solution, [g/l]; μ and q_S are the specific rate functions, [1/h]; μ_{max} is the maximum value of the μ, [1/h]; k_S is the saturation constant, [g/l]; $Y_{S/X}$ is the yield coefficient, [−].

For the model (Eqs. (5)–(8)) the parameters that will be estimated are: μ_{max}, k_S and $Y_{S/X}$.

Let $Z_{\text{mod}} \stackrel{\text{def}}{=} [X_{\text{mod}} \ S_{\text{mod}}]$ (model predictions for biomass and substrate) and $Z_{\text{exp}} \stackrel{\text{def}}{=} [X_{\text{exp}} \ S_{\text{exp}}]$ (known experimental data for biomass and substrate). Then, putting $Z = Z_{\text{mod}} - Z_{\text{exp}}$, we define the objective function as:

$$J = \|Z\|^2 \to \min,\tag{9}$$

where $\|\ \|$ denotes the ℓ^2-vector norm.

For the model parameters identification we use experimental data for biomass and glucose concentration of an *E. coli* MC4110 fed-batch cultivation process. The detailed description of the process condition and experimental data are presented in [19].

To estimate the model parameters we applied consistently 14 differently tuned GAs. In each GA we use various population sizes—from 5 to 200 chromosomes in the population, i.e. a GA with $5, 10, 20, \ldots, 100, 110, 150$ and 200 chromosomes. For all 14 GAs the number of generations is fixed to 200. The main GA operators and parameters are summarized in Table 1. The detailed description of identification procedure is given in [20].

Due to the stochastic nature of the applied GAs, we perform series of 30 runs for each population size. Thus, we obtain the average, best and worst estimates of the model parameters μ_{max}, k_S and $Y_{S/X}$ (Eqs. (5)–(8)), as well as of the values of objective function J (Eq. (9)) and algorithms computation times T.

3.1   ICrA of Model Parameter Identification Results

In order to establish the relations and dependencies between the parameters in a non-linear model of an *E. coli* fed-batch cultivation process and between the model

Table 1 Main GA operators
and parameters

Operator	Type
Fitness function	Linear ranking
Selection function	Roulette wheel selection
Crossover function	Simple crossover
Mutation function	Binary mutation
Reinsertion	Fitness-based
Parameter	Value
Generation gap	0.97
Crossover probability	0.75
Mutation probability	0.01
Number of generations	200

parameters and GAs objective function J values and computation times T, ICrA is
performed. Based on the obtained data from the all model parameters identification
procedures, the following IMs are constructed:

- $IM_{average}$ (Eq. (10)) with the calculated average results for model parameters
  μ_{max}, k_S and $Y_{S/X}$, and GAs outcomes J and T;
- IM_{best} (Eq. (11)) with the best obtained results;
- IM_{worst} (Eq. (12)) with the worst obtained results.

The presented data in the (Eqs. (10)–(12)) are rounded to the third decimal, but for
the further calculations the full number format is used.

In terms of ICrA, the five considered criteria are as follows: C_1 is parameter μ_{max},
C_2 is parameter k_S, C_3 is parameter $Y_{S/X}$, C_4 is objective function value J and C_5 is
resulting computation time T.

$$
IM_{average} = \begin{array}{c|ccccc}
 & C_1 & C_2 & C_3 & C_4 & C_5 \\
\hline
O_1 & 0.552 & 0.022 & 2.032 & 6.271 & 4.649 \\
O_2 & 0.525 & 0.021 & 2.025 & 5.838 & 6.053 \\
O_3 & 0.514 & 0.018 & 2.019 & 4.760 & 7.472 \\
O_4 & 0.491 & 0.013 & 2.023 & 4.561 & 11.248 \\
O_5 & 0.486 & 0.010 & 2.023 & 4.646 & 12.917 \\
O_6 & 0.508 & 0.016 & 2.020 & 4.607 & 14.649 \\
O_7 & 0.497 & 0.014 & 2.022 & 4.580 & 16.973 \\
O_8 & 0.498 & 0.013 & 2.022 & 4.568 & 19.719 \\
O_9 & 0.498 & 0.014 & 2.022 & 4.578 & 21.793 \\
O_{10} & 0.496 & 0.014 & 2.021 & 4.570 & 24.196 \\
O_{11} & 0.494 & 0.012 & 2.023 & 4.553 & 26.848 \\
O_{12} & 0.500 & 0.015 & 2.021 & 4.547 & 29.515 \\
O_{13} & 0.492 & 0.013 & 2.021 & 4.560 & 39.406 \\
O_{14} & 0.489 & 0.011 & 2.023 & 4.545 & 51.917 \\
\end{array}
\qquad (10)
$$

$$
\text{IM}_{best} = \begin{array}{c|ccccc}
 & C_1 & C_2 & C_3 & C_4 & C_5 \\
\hline
O_1 & 0.491 & 0.013 & 2.023 & 4.833 & 4.867 \\
O_2 & 0.480 & 0.011 & 2.024 & 4.855 & 5.912 \\
O_3 & 0.494 & 0.013 & 2.018 & 4.475 & 7.675 \\
O_4 & 0.491 & 0.012 & 2.023 & 4.482 & 11.295 \\
O_5 & 0.488 & 0.012 & 2.020 & 4.444 & 13.229 \\
O_6 & 0.492 & 0.012 & 2.023 & 4.449 & 15.007 \\
O_7 & 0.488 & 0.012 & 2.020 & 4.463 & 17.316 \\
O_8 & 0.490 & 0.013 & 2.019 & 4.438 & 20.062 \\
O_9 & 0.484 & 0.011 & 2.019 & 4.447 & 22.667 \\
O_{10} & 0.491 & 0.013 & 2.020 & 4.450 & 24.757 \\
O_{11} & 0.488 & 0.012 & 2.019 & 4.425 & 26.926 \\
O_{12} & 0.486 & 0.012 & 2.021 & 4.433 & 30.015 \\
O_{13} & 0.488 & 0.012 & 2.019 & 4.458 & 39.780 \\
O_{14} & 0.488 & 0.012 & 2.018 & 4.436 & 52.323 \\
\end{array}
\tag{11}
$$

$$
\text{IM}_{worst} = \begin{array}{c|ccccc}
 & C_1 & C_2 & C_3 & C_4 & C_5 \\
\hline
O_1 & 0.577 & 0.015 & 2.037 & 9.296 & 5.600 \\
O_2 & 0.538 & 0.026 & 1.995 & 9.618 & 5.632 \\
O_3 & 0.544 & 0.024 & 2.019 & 5.363 & 7.301 \\
O_4 & 0.518 & 0.018 & 2.021 & 5.009 & 10.827 \\
O_5 & 0.521 & 0.019 & 2.021 & 4.967 & 12.496 \\
O_6 & 0.517 & 0.018 & 2.021 & 4.864 & 14.399 \\
O_7 & 0.515 & 0.018 & 2.020 & 4.808 & 16.801 \\
O_8 & 0.510 & 0.016 & 2.022 & 4.736 & 19.500 \\
O_9 & 0.510 & 0.017 & 2.021 & 4.746 & 21.715 \\
O_{10} & 0.505 & 0.015 & 2.022 & 4.721 & 23.915 \\
O_{11} & 0.489 & 0.012 & 2.022 & 4.702 & 27.051 \\
O_{12} & 0.510 & 0.016 & 2.023 & 4.732 & 29.188 \\
O_{13} & 0.504 & 0.015 & 2.022 & 4.672 & 39.921 \\
O_{14} & 0.511 & 0.016 & 2.022 & 4.721 & 51.309 \\
\end{array}
\tag{12}
$$

The objects in the ICrA, $O_1, O_2, \ldots, O_{14}$ are as follows: O_1 is the GA with a population of 5 chromosomes (GA_5), O_2—GA with a population of 10 chromosomes (GA_{10}), O_3—GA with a population of 20 chromosomes (GA_{20}), ..., O_{11}—GA with a population of 100 chromosomes (GA_{100}), O_{12}—GA with a population of 110 chromosomes (GA_{110}), O_{13}—GA with a population of 150 chromosomes (GA_{150}), and O_{14}—GA with a population of 200 chromosomes (GA_{200}).

3.2 ICrA of Genetic Algorithms Performance

Another important point is to identify the correlation between the performance of the considered 14 differently tuned GAs, namely GA_5, GA_{10}, GA_{20}, ..., GA_{100},

GA_{110}, GA_{150}, and GA_{200}. The GAs performance is evaluated in terms of obtained model accuracy, value J and computation time T.

Based on the IMs $\text{IM}_{average}$ (Eq. (10)), IM_{best} (Eq. (11)) and IM_{worst} (Eq. (12)), we cannot do any analysis about the relation between the 14 GAs presented as objects in ICrA. To investigate the performances correlation of considered GAs, we will apply the proposed here scheme of ICrA of ICrA results. For this purpose, 14 IMs are constructed as follows:

- In the case of GA with 5 chromosomes, we construct IM_{GA_5} (Eq. 13) with the following objects: $O_1, O_2, \ldots, O_{30}$ (30 algorithm runs) and the following criteria: C_1 is the obtained all 30 results for the parameter μ_{max}, C_2—30 results for the parameter k_S, C_3—30 results for the parameter $Y_{S/X}$, C_4—30 results for the J and C_5—30 results for the T;
- In the case of GA with 10 chromosomes, we construct $\text{IM}_{GA_{10}}$ (Eq. 14) with the same objects and criteria;
- In the case of GA with 20 chromosomes, we construct $\text{IM}_{GA_{20}}$ (Eq. 15) with the same objects and criteria;

$\vdots$ (Eqs. (16)–(22));

- In the case of GA with 100 chromosomes, we construct $\text{IM}_{GA_{100}}$ (Eq. 23) with the same objects and criteria;
- In the case of GA with 110 chromosomes, we construct $\text{IM}_{GA_{110}}$ (Eq. 24) with the same objects and criteria;
- In the case of GA with 150 chromosomes, we construct $\text{IM}_{GA_{150}}$ (Eq. 25) with the same objects and criteria;
- In the case of GA with 200 chromosomes, we construct $\text{IM}_{GA_{200}}$ (Eq. 26) with the same objects and criteria.

$$
\text{IM}_{GA_5} =
\begin{array}{c|ccccccc}
 & O_1 & O_2 & O_3 & O_{15} & O_{28} & O_{29} & O_{30} \\
\hline
C_1 & 0.556 & 0.528 & 0.550 & \ldots\ 0.569\ \ldots & 0.514 & 0.482 & 0.520 \\
C_2 & 0.026 & 0.016 & 0.026 & \ldots\ 0.031\ \ldots & 0.017 & 0.010 & 0.019 \\
C_3 & 2.023 & 2.022 & 2.022 & \ldots\ 2.022\ \ldots & 2.023 & 2.032 & 2.023 \\
C_4 & 5.948 & 5.605 & 5.814 & \ldots\ 6.478\ \ldots & 5.243 & 5.235 & 4.960 \\
C_5 & 4.742 & 4.867 & 5.288 & \ldots\ 4.727\ \ldots & 4.836 & 4.836 & 4.711 \\
\end{array}
\tag{13}
$$

$$
\text{IM}_{GA_{10}} =
\begin{array}{c|ccccccc}
 & O_1 & O_2 & O_3 & \ldots\ O_{15}\ \ldots & O_{28} & O_{29} & O_{30} \\
\hline
C_1 & 0.520 & 0.504 & 0.545 & \ldots\ 0.533\ \ldots & 0.509 & 0.481 & 0.517 \\
C_2 & 0.019 & 0.015 & 0.025 & \ldots\ 0.022\ \ldots & 0.016 & 0.011 & 0.019 \\
C_3 & 2.022 & 2.024 & 2.021 & \ldots\ 2.022\ \ldots & 2.024 & 2.024 & 2.022 \\
C_4 & 5.292 & 5.024 & 6.001 & \ldots\ 5.796\ \ldots & 5.151 & 4.916 & 5.409 \\
C_5 & 6.006 & 6.256 & 6.022 & \ldots\ 6.084\ \ldots & 6.224 & 5.834 & 6.068 \\
\end{array}
\tag{14}
$$

$$
\mathrm{IM}_{GA_{20}} =
\begin{array}{c|cccccccc}
 & O_1 & O_2 & O_3 & \ldots & O_{15} & \ldots & O_{28} & O_{29} & O_{30} \\
\hline
C_1 & 0.498 & 0.489 & 0.525 & \ldots & 0.493 & \ldots & 0.487 & 0.532 & 0.487 \\
C_2 & 0.014 & 0.012 & 0.020 & \ldots & 0.013 & \ldots & 0.012 & 0.019 & 0.012 \\
C_3 & 2.022 & 2.023 & 2.021 & \ldots & 2.022 & \ldots & 2.024 & 2.020 & 2.022 \\
C_4 & 4.570 & 4.538 & 5.002 & \ldots & 4.573 & \ldots & 4.665 & 4.987 & 4.517 \\
C_5 & 7.504 & 7.722 & 7.660 & \ldots & 7.738 & \ldots & 7.816 & 7.301 & 7.784
\end{array}
\tag{15}
$$

$$
\mathrm{IM}_{GA_{30}} =
\begin{array}{c|cccccccc}
 & O_1 & O_2 & O_3 & \ldots & O_{15} & \ldots & O_{28} & O_{29} & O_{30} \\
\hline
C_1 & 0.514 & 0.513 & 0.493 & \ldots & 0.498 & \ldots & 0.495 & 0.514 & 0.509 \\
C_2 & 0.017 & 0.016 & 0.013 & \ldots & 0.014 & \ldots & 0.014 & 0.017 & 0.016 \\
C_3 & 2.021 & 2.022 & 2.021 & \ldots & 2.022 & \ldots & 2.020 & 2.021 & 2.022 \\
C_4 & 4.753 & 4.660 & 4.585 & \ldots & 4.555 & \ldots & 4.552 & 4.708 & 4.682 \\
C_5 & 11.201 & 10.967 & 11.513 & \ldots & 11.029 & \ldots & 11.045 & 11.092 & 11.217
\end{array}
\tag{16}
$$

$$
\mathrm{IM}_{GA_{40}} =
\begin{array}{c|cccccccc}
 & O_1 & O_2 & O_3 & \ldots & O_{15} & \ldots & O_{28} & O_{29} & O_{30} \\
\hline
C_1 & 0.498 & 0.497 & 0.510 & \ldots & 0.502 & \ldots & 0.507 & 0.487 & 0.510 \\
C_2 & 0.014 & 0.014 & 0.016 & \ldots & 0.014 & \ldots & 0.016 & 0.012 & 0.017 \\
C_3 & 2.022 & 2.023 & 2.019 & \ldots & 2.023 & \ldots & 2.022 & 2.022 & 2.021 \\
C_4 & 4.547 & 4.582 & 4.679 & \ldots & 4.655 & \ldots & 4.700 & 4.546 & 4.755 \\
C_5 & 13.026 & 13.104 & 12.948 & \ldots & 13.026 & \ldots & 12.901 & 13.057 & 12.823
\end{array}
\tag{17}
$$

$$
\mathrm{IM}_{GA_{50}} =
\begin{array}{c|cccccccc}
 & O_1 & O_2 & O_3 & \ldots & O_{15} & \ldots & O_{28} & O_{29} & O_{30} \\
\hline
C_1 & 0.498 & 0.487 & 0.490 & \ldots & 0.481 & \ldots & 0.507 & 0.508 & 0.492 \\
C_2 & 0.014 & 0.012 & 0.013 & \ldots & 0.010 & \ldots & 0.015 & 0.016 & 0.013 \\
C_3 & 2.022 & 2.022 & 2.022 & \ldots & 2.023 & \ldots & 2.023 & 2.020 & 2.020 \\
C_4 & 4.588 & 4.506 & 4.630 & \ldots & 4.504 & \ldots & 4.739 & 4.607 & 4.452 \\
C_5 & 14.961 & 14.461 & 14.929 & \ldots & 15.117 & \ldots & 14.992 & 14.649 & 15.007
\end{array}
\tag{18}
$$

$$
\mathrm{IM}_{GA_{60}} =
\begin{array}{c|cccccccc}
 & O_1 & O_2 & O_3 & \ldots & O_{15} & \ldots & O_{28} & O_{29} & O_{30} \\
\hline
C_1 & 0.492 & 0.496 & 0.492 & \ldots & 0.500 & \ldots & 0.494 & 0.493 & 0.498 \\
C_2 & 0.013 & 0.013 & 0.012 & \ldots & 0.014 & \ldots & 0.013 & 0.012 & 0.014 \\
C_3 & 2.022 & 2.021 & 2.022 & \ldots & 2.023 & \ldots & 2.021 & 2.023 & 2.019 \\
C_4 & 4.497 & 4.484 & 4.496 & \ldots & 4.561 & \ldots & 4.527 & 4.557 & 4.483 \\
C_5 & 20.311 & 17.472 & 17.909 & \ldots & 17.254 & \ldots & 17.223 & 17.145 & 16.973
\end{array}
\tag{19}
$$

$$
\mathrm{IM}_{GA_{70}} =
\begin{array}{c|cccccccc}
 & O_1 & O_2 & O_3 & \ldots & O_{15} & \ldots & O_{28} & O_{29} & O_{30} \\
\hline
C_1 & 0.490 & 0.486 & 0.498 & \ldots & 0.500 & \ldots & 0.493 & 0.495 & 0.496 \\
C_2 & 0.013 & 0.012 & 0.013 & \ldots & 0.014 & \ldots & 0.013 & 0.013 & 0.014 \\
C_3 & 2.019 & 2.021 & 2.022 & \ldots & 2.021 & \ldots & 2.021 & 2.022 & 2.020 \\
C_4 & 4.438 & 4.497 & 4.568 & \ldots & 4.708 & \ldots & 4.533 & 4.511 & 4.541 \\
C_5 & 20.062 & 19.999 & 19.719 & \ldots & 19.719 & \ldots & 19.984 & 19.765 & 19.547
\end{array}
\tag{20}
$$

$$
\text{IM}_{GA_{80}} =
\begin{array}{c|cccccccc}
 & O_1 & O_2 & O_3 & \cdots & O_{15} & \cdots & O_{28} & O_{29} & O_{30} \\
\hline
C_1 & 0.499 & 0.490 & 0.498 & \cdots & 0.495 & \cdots & 0.498 & 0.496 & 0.504 \\
C_2 & 0.014 & 0.013 & 0.014 & \cdots & 0.013 & \cdots & 0.014 & 0.014 & 0.015 \\
C_3 & 2.023 & 2.019 & 2.018 & \cdots & 2.023 & \cdots & 2.023 & 2.023 & 2.020 \\
C_4 & 4.630 & 4.457 & 4.581 & \cdots & 4.584 & \cdots & 4.649 & 4.611 & 4.636 \\
C_5 & 22.433 & 22.449 & 21.996 & \cdots & 22.246 & \cdots & 22.105 & 22.074 & 22.012
\end{array}
\tag{21}
$$

$$
\text{IM}_{GA_{90}} =
\begin{array}{c|cccccccc}
 & O_1 & O_2 & O_3 & \cdots & O_{15} & \cdots & O_{28} & O_{29} & O_{30} \\
\hline
C_1 & 0.488 & 0.483 & 0.498 & \cdots & 0.495 & \cdots & 0.505 & 0.491 & 0.491 \\
C_2 & 0.012 & 0.011 & 0.014 & \cdots & 0.013 & \cdots & 0.015 & 0.013 & 0.012 \\
C_3 & 2.021 & 2.020 & 2.020 & \cdots & 2.023 & \cdots & 2.021 & 2.020 & 2.022 \\
C_4 & 4.459 & 4.497 & 4.598 & \cdots & 4.577 & \cdots & 4.666 & 4.450 & 4.528 \\
C_5 & 24.555 & 24.352 & 24.321 & \cdots & 24.305 & \cdots & 24.118 & 24.757 & 24.539
\end{array}
\tag{22}
$$

$$
\text{IM}_{GA_{100}} =
\begin{array}{c|cccccccc}
 & O_1 & O_2 & O_3 & \cdots & O_{15} & \cdots & O_{28} & O_{29} & O_{30} \\
\hline
C_1 & 0.493 & 0.485 & 0.492 & \cdots & 0.487 & \cdots & 0.490 & 0.489 & 0.505 \\
C_2 & 0.013 & 0.011 & 0.013 & \cdots & 0.011 & \cdots & 0.012 & 0.012 & 0.015 \\
C_3 & 2.022 & 2.017 & 2.022 & \cdots & 2.023 & \cdots & 2.020 & 2.021 & 2.022 \\
C_4 & 4.605 & 4.497 & 4.562 & \cdots & 4.502 & \cdots & 4.459 & 4.520 & 4.669 \\
C_5 & 26.832 & 27.004 & 27.175 & \cdots & 27.066 & \cdots & 27.051 & 26.848 & 26.551
\end{array}
\tag{23}
$$

$$
\text{IM}_{GA_{110}} =
\begin{array}{c|cccccccc}
 & O_1 & O_2 & O_3 & \cdots & O_{15} & \cdots & O_{28} & O_{29} & O_{30} \\
\hline
C_1 & 0.490 & 0.509 & 0.495 & \cdots & 0.489 & \cdots & 0.497 & 0.495 & 0.486 \\
C_2 & 0.013 & 0.015 & 0.013 & \cdots & 0.012 & \cdots & 0.014 & 0.013 & 0.012 \\
C_3 & 2.019 & 2.022 & 2.021 & \cdots & 2.019 & \cdots & 2.021 & 2.022 & 2.022 \\
C_4 & 4.510 & 4.694 & 4.539 & \cdots & 4.452 & \cdots & 4.514 & 4.568 & 4.469 \\
C_5 & 30.171 & 29.562 & 29.968 & \cdots & 29.484 & \cdots & 29.266 & 29.703 & 30.077
\end{array}
\tag{24}
$$

$$
\text{IM}_{GA_{150}} =
\begin{array}{c|cccccccc}
 & O_1 & O_2 & O_3 & \cdots & O_{15} & \cdots & O_{28} & O_{29} & O_{30} \\
\hline
C_1 & 0.492 & 0.492 & 0.491 & \cdots & 0.492 & \cdots & 0.496 & 0.487 & 0.489 \\
C_2 & 0.013 & 0.013 & 0.013 & \cdots & 0.013 & \cdots & 0.014 & 0.012 & 0.012 \\
C_3 & 2.022 & 2.021 & 2.022 & \cdots & 2.023 & \cdots & 2.021 & 2.019 & 2.024 \\
C_4 & 4.479 & 4.560 & 4.502 & \cdots & 4.631 & \cdots & 4.499 & 4.465 & 4.623 \\
C_5 & 39.812 & 39.406 & 40.030 & \cdots & 39.344 & \cdots & 40.108 & 39.843 & 40.108
\end{array}
\tag{25}
$$

$$
\text{IM}_{GA_{200}} =
\begin{array}{c|cccccccc}
 & O_1 & O_2 & O_3 & \cdots & O_{15} & \cdots & O_{28} & O_{29} & O_{30} \\
\hline
C_1 & 0.487 & 0.486 & 0.490 & \cdots & 0.511 & \cdots & 0.488 & 0.487 & 0.492 \\
C_2 & 0.012 & 0.012 & 0.012 & \cdots & 0.016 & \cdots & 0.012 & 0.012 & 0.013 \\
C_3 & 2.022 & 2.023 & 2.021 & \cdots & 2.022 & \cdots & 2.021 & 2.022 & 2.022 \\
C_4 & 4.483 & 4.679 & 4.486 & \cdots & 4.721 & \cdots & 4.443 & 4.676 & 4.472 \\
C_5 & 55.895 & 53.649 & 52.947 & \cdots & 51.309 & \cdots & 52.744 & 53.181 & 51.917
\end{array}
\tag{26}
$$

The presented data in the Eqs. (13)–(26) are rounded to the third decimal, but for the further calculations the full number format is used. The complete set of data in IMs IM_{GA_i} is available at http://intercriteria.net/studies/gengap/e-coli/.

4 Numerical Results and Discussion

Computer specification to run all identification procedures are Intel Core i5-2329 3.0 GHz, 8 GB Memory, Windows 7 (64bit) operating system.

Based on the presented Algorithm 1, the ICrA is implemented in the Matlab 7.5 environment. The constructed IMs $IM_{average}$, IM_{best}, IM_{worst} and IM_{GA_5}, $IM_{GA_{10}}$, $IM_{GA_{20}}, \ldots, IM_{GA_{100}}$, $IM_{GA_{110}}$, $IM_{GA_{150}}$, $IM_{GA_{200}}$ are used. We obtain IMs that determine the degrees of "agreement" ($\mu_{C,C'}$) and degrees of "disagreement" ($\nu_{C,C'}$) between criteria for the two considered problems.

4.1 Results of ICrA—Model Parameter Identification Results

Since in the IMs for ICrA the criteria are placed in rows and objects are places in columns, in this case we used the IMs $(IM_{average})^T$, $(IM_{best})^T$ and $(IM_{worst})^T$. The obtained results are presented below.

Case of average results
When applying ICrA on $(IM_{average})^T$, the resulting two IMs—IM_1^μ for the degrees of "agreement" ($\mu_{C,C'}$) and IM_2^ν for the degrees of "disagreement" ($\nu_{C,C'}$), are as follows:

$$
IM_1^\mu = \begin{array}{c|ccccc}
 & C_1 & C_2 & C_3 & C_4 & C_5 \\
\hline
C_1 & 1 & 0.91 & 0.41 & 0.74 & 0.26 \\
C_2 & 0.91 & 1 & 0.36 & 0.78 & 0.27 \\
C_3 & 0.41 & 0.36 & 1 & 0.55 & 0.38 \\
C_4 & 0.74 & 0.78 & 0.55 & 1 & 0.11 \\
C_5 & 0.26 & 0.27 & 0.38 & 0.11 & 1
\end{array},
\quad
IM_2^\nu = \begin{array}{c|ccccc}
 & C_1 & C_2 & C_3 & C_4 & C_5 \\
\hline
C_1 & 0 & 0.08 & 0.58 & 0.26 & 0.74 \\
C_2 & 0.08 & 0 & 0.62 & 0.21 & 0.71 \\
C_3 & 0.58 & 0.62 & 0 & 0.44 & 0.60 \\
C_4 & 0.26 & 0.21 & 0.44 & 0 & 0.89 \\
C_5 & 0.74 & 0.71 & 0.60 & 0.89 & 0
\end{array}
$$

Case of worst results
When applying ICrA on $(IM_{worst})^T$, the resulting two IMs—IM_3^μ ($\mu_{C,C'}$) and IM_4^ν ($\nu_{C,C'}$), are as follows:

$$
IM_3^\mu = \begin{array}{c|ccccc}
 & C_1 & C_2 & C_3 & C_4 & C_5 \\
\hline
C_1 & 1 & 0.79 & 0.34 & 0.88 & 0.14 \\
C_2 & 0.79 & 1 & 0.18 & 0.84 & 0.22 \\
C_3 & 0.34 & 0.18 & 1 & 0.33 & 0.64 \\
C_4 & 0.88 & 0.84 & 0.33 & 1 & 0.07 \\
C_5 & 0.14 & 0.22 & 0.64 & 0.07 & 1
\end{array},
\quad
IM_4^\nu = \begin{array}{c|ccccc}
 & C_1 & C_2 & C_3 & C_4 & C_5 \\
\hline
C_1 & 0 & 0.20 & 0.65 & 0.12 & 0.86 \\
C_2 & 0.20 & 0 & 0.80 & 0.15 & 0.77 \\
C_3 & 0.65 & 0.80 & 0 & 0.66 & 0.35 \\
C_4 & 0.12 & 0.15 & 0.66 & 0 & 0.93 \\
C_5 & 0.86 & 0.77 & 0.35 & 0.93 & 0
\end{array}
$$

Table 2 Observed $\langle \mu_{C,C'}, \nu_{C,C'} \rangle$ values of criteria pairs: sorted by $\mu_{C,C'}$ values of average results

Criteria pairs	Obtained $\langle \mu_{C,C'}, \nu_{C,C'} \rangle$ values in case of		
	Average results	Worst results	Best results
$C_1 \leftrightarrow C_2$	$\langle 0.91, 0.08 \rangle$	$\langle 0.79, 0.20 \rangle$	$\langle 0.74, 0.19 \rangle$
$C_2 \leftrightarrow C_4$	$\langle 0.78, 0.21 \rangle$	$\langle 0.84, 0.15 \rangle$	$\langle 0.53, 0.44 \rangle$
$C_1 \leftrightarrow C_4$	$\langle 0.74, 0.26 \rangle$	$\langle 0.88, 0.12 \rangle$	$\langle 0.63, 0.33 \rangle$
$C_3 \leftrightarrow C_4$	$\langle 0.55, 0.44 \rangle$	$\langle 0.33, 0.66 \rangle$	$\langle 0.71, 0.26 \rangle$
$C_1 \leftrightarrow C_3$	$\langle 0.41, 0.58 \rangle$	$\langle 0.34, 0.65 \rangle$	$\langle 0.49, 0.44 \rangle$
$C_3 \leftrightarrow C_5$	$\langle 0.38, 0.60 \rangle$	$\langle 0.64, 0.35 \rangle$	$\langle 0.30, 0.68 \rangle$
$C_2 \leftrightarrow C_3$	$\langle 0.36, 0.62 \rangle$	$\langle 0.18, 0.80 \rangle$	$\langle 0.35, 0.59 \rangle$
$C_2 \leftrightarrow C_5$	$\langle 0.27, 0.71 \rangle$	$\langle 0.22, 0.77 \rangle$	$\langle 0.51, 0.46 \rangle$
$C_1 \leftrightarrow C_5$	$\langle 0.26, 0.74 \rangle$	$\langle 0.14, 0.86 \rangle$	$\langle 0.36, 0.59 \rangle$
$C_4 \leftrightarrow C_5$	$\langle 0.11, 0.89 \rangle$	$\langle 0.07, 0.93 \rangle$	$\langle 0.25, 0.75 \rangle$

Case of best results

When applying ICrA on $(\mathrm{IM}_{best})^{\mathrm{T}}$, the resulting two IMs—$\mathrm{IM}_5^{\mu}$ ($\mu_{C,C'}$) and IM_6^{ν} ($\nu_{C,C'}$), are as follows:

$$\mathrm{IM}_5^{\mu} = \begin{array}{c|ccccc} & C_1 & C_2 & C_3 & C_4 & C_5 \\ \hline C_1 & 1 & 0.74 & 0.49 & 0.63 & 0.36 \\ C_2 & 0.74 & 1 & 0.35 & 0.53 & 0.51 \\ C_3 & 0.49 & 0.35 & 1 & 0.71 & 0.30 \\ C_4 & 0.63 & 0.53 & 0.71 & 1 & 0.25 \\ C_5 & 0.36 & 0.51 & 0.30 & 0.25 & 1 \end{array}, \qquad \mathrm{IM}_6^{\nu} = \begin{array}{c|ccccc} & C_1 & C_2 & C_3 & C_4 & C_5 \\ \hline C_1 & 0 & 0.19 & 0.44 & 0.33 & 0.59 \\ C_2 & 0.19 & 0 & 0.59 & 0.44 & 0.46 \\ C_3 & 0.44 & 0.59 & 0 & 0.26 & 0.68 \\ C_4 & 0.33 & 0.44 & 0.26 & 0 & 0.75 \\ C_5 & 0.59 & 0.46 & 0.68 & 0.75 & 0 \end{array}$$

Criteria relation, sorted by $\mu_{C,C'}$ values in the case of average results, are presented in Table 2. The resulting ICrA relations are compared with the obtained values of $\mu_{C,C'}, \nu_{C,C'}$ in the cases of best and worst results. The results show that there are some differences between ICrA evaluation for the degrees of "agreement" $\mu_{C,C'}$ and the degrees of "disagreement" $\nu_{C,C'}$ in the three cases—average, best and worst results. The same results are graphically presented in Figs. 1 and 2. In the figures the observed differences in the obtained $\mu_{C,C'}$ and $\nu_{C,C'}$ values can be seen more clearly. For example, Fig. 1 shows that for the criteria pairs $C_4 \leftrightarrow C_5$, $C_1 \leftrightarrow C_5$, $C_1 \leftrightarrow C_3$ and $C_3 \leftrightarrow C_4$ the obtained $\mu_{C,C'}$ values are higher for the best results compared to the average and worst ones. The largest values of $\mu_{C,C'}$ for the pairs $C_2 \leftrightarrow C_4$, $C_1 \leftrightarrow C_4$ and $C_3 \leftrightarrow C_5$ make an impression, too. Figure 2 is a mirror image of Fig. 1 with the exceptions for the pairs where there is a degree of "uncertainty" $\pi_{C,C'}$ (Eq. (4)). The observed "uncertainty" is presented in Table 3. In this case, the obtained $\pi_{C,C'}$ values are very small and we can assume that the results are adequate and there is no substantial uncertainty in them.

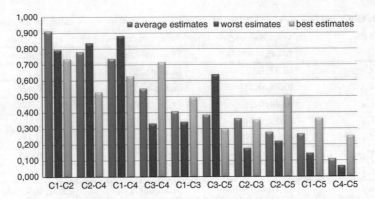

Fig. 1 Degrees of "agreement" ($\mu_{C,C'}$ values) for all cases

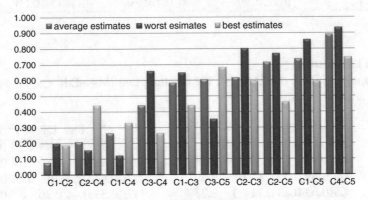

Fig. 2 Degrees of "disagreement" ($\nu_{C,C'}$ values) for all cases

Taking into account the stochastic nature of GAs, we will consider the ICrA results in the case of average estimates as those with the highest significance. For the analysis of the results, we use the scheme proposed in [7] for defining the consonance and dissonance between each pair of criteria. Further discussion is made following the scheme presented in Table 4.

In the case of the average values of the examined criteria, we found the following pair dependencies:

- There is no observed strong positive consonance or strong negative consonance between any of the ten criteria pairs. Since the observed values depend on the number of objects if we can expand their number, it is possible to obtain values in these intervals.
- For the pair $C_4 \leftrightarrow C_5$ (i.e. $T \leftrightarrow J$) a negative consonance is identified. Such dependence is logical—for a large number of algorithm iterations (i.e. greater computation time T) it is more likely to find a more accurate solution, i.e. smaller value of J.

Table 3 Observed $\pi_{C,C'}$ values of criteria pairs

Criteria pairs	Obtained $\pi_{C,C'}$ values in case of		
	Average results	Worst results	Best results
$C_1 \leftrightarrow C_2$	0.011	0.011	0.077
$C_1 \leftrightarrow C_3$	0.011	0.011	0.066
$C_1 \leftrightarrow C_4$	0	0	0.044
$C_1 \leftrightarrow C_5$	0	0	0.044
$C_2 \leftrightarrow C_3$	0.022	0.022	0.055
$C_2 \leftrightarrow C_4$	0.011	0.011	0.033
$C_2 \leftrightarrow C_5$	0.011	0.011	0.033
$C_3 \leftrightarrow C_4$	0.011	0.011	0.022
$C_3 \leftrightarrow C_5$	0.011	0.011	0.022
$C_4 \leftrightarrow C_5$	0	0	0

Table 4 Consonance and dissonance scale

Interval of $\mu_{C,C'}$, %	Meaning
[0–5]	Strong negative consonance
(5–15]	Negative consonance
(15–25]	Weak negative consonance
(25–33]	Weak dissonance
(33–43]	Dissonance
(43–57]	Strong dissonance
(57–67]	Dissonance
(67–75]	Weak dissonance
(75–85]	Weak positive consonance
(85–95]	Positive consonance
(95–100]	Strong positive consonance

- The pairs $C_2 \leftrightarrow C_5$ (i.e. $k_S \leftrightarrow T$) show identical results—these criteria are in weak dissonance. The third model parameter $Y_{S/X}$ and T are in dissonance. The conclusion is that the total computation time is not dependent solely on one of the model parameters. Logically, the triple of these parameters should be in consonance with T.
- For the pairs $C_1 \leftrightarrow C_3$ (i.e. $\mu_{max} \leftrightarrow Y_{S/X}$) and $C_2 \leftrightarrow C_3$ (i.e. $k_S \leftrightarrow Y_{S/X}$) a dissonance is observed. Considering the physical meaning of the model parameters [8], it is clear that there is no dependence between these criteria. A strong correlation is expected between criteria $C_1 \leftrightarrow C_2$ (i.e. $\mu_{max} \leftrightarrow k_S$) [8]. The results confirmed these expectation—these criteria are in a positive consonance.
- The observed low value of μ_{C_3,C_4}, i.e. strong dissonance between $Y_{S/X} \leftrightarrow J$ show the low sensitivity of this model parameter. According to [17], the parameter $Y_{S/X}$ has lower sensitivity compared to parameter μ_{max}.

- Due to the established strong correlation between criteria $C_1 \leftrightarrow C_2$ (i.e. $\mu_{max} \leftrightarrow k_S$), we observe that $C_1 \leftrightarrow C_4$ (i.e. $\mu_{max} \leftrightarrow J$) and $C_2 \leftrightarrow C_4$ (i.e. $k_S \leftrightarrow J$) are in respectively weak dissonance and weak positive consonance. Similarly to the relations with T, the conclusion is that the accuracy of the criterion is not dependent solely on one of the model parameters. Logically, the triple of these parameters should be in consonance (or strong consonance) with the criterion value. Moreover, taking into account the parameters sensitivity, it is clear that the more sensitive parameter will be more linked to the value of J.

Due to the stochastic nature of GAs considered here, we observed some different criteria dependences in the other two cases—of the worst and of the best results:

- In the case of the worst results, we found a weaker relation between $C_1 \leftrightarrow C_2$, $C_3 \leftrightarrow C_4$, $C_2 \leftrightarrow C_3$, $C_1 \leftrightarrow C_5$, $C_2 \leftrightarrow C_5$ and $C_4 \leftrightarrow C_5$. For the pairs $C_1 \leftrightarrow C_4$, $C_2 \leftrightarrow C_4$ and $C_3 \leftrightarrow C_5$, we observed a higher value of $\mu_{C,C'}$. Compared to the case of the average results, there are no large, strongly manifested discrepancies. In case of discrepancy, the considered criteria pair appears in an adjacent scale according to Table 4. For example, in the case of average results the pair $C_1 \leftrightarrow C_2$ is in positive consonance, while in the case of the worst results it is in weak positive consonance.
- In the case of the best results we identify the same results—in case of discrepancy, the considered criteria pairs appear in an adjacent scale. However, in this case we observed some larger discrepancies. Taking into account the nature of the GA, we consider that the results in case of average criteria values have the highest significance. Therefore, the best and the worst results are of a lower rank or importance.

4.2 Results of ICrA—Genetic Algorithms Performance

Based on the presented above results of ICrA, we can analyze only the relations of the considered criteria ($C_1, C_2, \ldots, C_5$), i.e. the relation between the model parameters (μ_{max}, k_S and $Y_{S/X}$) and GA outcomes (J and T). We cannot find any connections and dependencies between the 14 differently tuned GAs. Such a relation will be very useful information about the choice of an optimal tuned GA for the considered model parameter identification problem.

Here, we propose a schema for ICrA of GAs performance. Based on available data (model parameters estimates, and objective function value and computation time for 30 runs of all 14 GAs), the ICrA of GAs performance is performed following the schema:

Step 1. Perform ICrA based on the IMs IM_{GA_5}, $IM_{GA_{10}}$, $IM_{GA_{20}}$, $\ldots$, $IM_{GA_{100}}$, $IM_{GA_{110}}$, $IM_{GA_{150}}$, $IM_{GA_{200}}$.

Step 2. Perform ICrA of ICrA results from Step 1.

In Step 1 we obtain 14 IMs for $\mu_{C,C'}$ and 14 IMs for $\nu_{C,C'}$ of the criteria pairs. Therefore, for every criteria pair C, C' we have 14 different $\mu_{C,C'}$ estimates depending on the applied GA. Using these results, we construct new IM with the following objects: the 10 criteria pairs ($C_1 \leftrightarrow C_2$, $C_1 \leftrightarrow C_3$, $C_1 \leftrightarrow C_4$, $C_1 \leftrightarrow C_5$, $C_2 \leftrightarrow C_3$, $C_2 \leftrightarrow C_4$, $C_2 \leftrightarrow C_5$, $C_3 \leftrightarrow C_4$, $C_3 \leftrightarrow C_5$, $C_4 \leftrightarrow C_5$) and with the following criteria: the 14 differently tuned GAs (GA_5, GA_{10}, GA_{20}, ..., GA_{100}, GA_{110}, GA_{150}, GA_{200}). The resulting IM contains the $\mu_{C,C'}$ values obtained in Step 1. In Step 2, we perform ICrA of the ICrA results from Step 1. Now we can analyze the correlations between all the considered GAs applied for the model parameter identification of *E. coli* fed-batch process (Eqs. (5)–(8)).

ICrA of ICrA: Step 1

The ICrA is performed for the following 14 IMs:
IM_{GA_5}, $IM_{GA_{10}}$, $IM_{GA_{20}}$, ..., $IM_{GA_{100}}$, $IM_{GA_{110}}$, $IM_{GA_{150}}$, $IM_{GA_{200}}$.

The resulting IMs that determine the degrees of "agreement" ($\mu_{C,C'}$) and degrees of "disagreement" ($\nu_{C,C'}$) between criteria are as follows:

$$IM^{\mu}_{GA_5} = \begin{array}{c|ccccc} & C_1 & C_2 & C_3 & C_4 & C_5 \\ \hline C_1 & 1 & 0.88 & 0.40 & 0.84 & 0.33 \\ C_2 & 0.88 & 1 & 0.31 & 0.78 & 0.31 \\ C_3 & 0.40 & 0.31 & 1 & 0.47 & 0.58 \\ C_4 & 0.84 & 0.78 & 0.47 & 1 & 0.40 \\ C_5 & 0.33 & 0.31 & 0.58 & 0.40 & 1 \end{array}$$

$$IM^{\nu}_{GA_5} = \begin{array}{c|ccccc} & C_1 & C_2 & C_3 & C_4 & C_5 \\ \hline C_1 & 0 & 0.12 & 0.60 & 0.16 & 0.65 \\ C_2 & 0.12 & 0 & 0.68 & 0.22 & 0.67 \\ C_3 & 0.60 & 0.68 & 0 & 0.53 & 0.40 \\ C_4 & 0.16 & 0.22 & 0.53 & 0 & 0.58 \\ C_5 & 0.65 & 0.67 & 0.40 & 0.58 & 0 \end{array}$$

$$IM^{\mu}_{GA_{10}} = \begin{array}{c|ccccc} & C_1 & C_2 & C_3 & C_4 & C_5 \\ \hline C_1 & 1 & 0.94 & 0.26 & 0.91 & 0.30 \\ C_2 & 0.94 & 1 & 0.24 & 0.90 & 0.29 \\ C_3 & 0.26 & 0.24 & 1 & 0.27 & 0.61 \\ C_4 & 0.91 & 0.90 & 0.27 & 1 & 0.28 \\ C_5 & 0.30 & 0.29 & 0.61 & 0.28 & 1 \end{array}$$

$$IM^{\nu}_{GA_{10}} = \begin{array}{c|ccccc} & C_1 & C_2 & C_3 & C_4 & C_5 \\ \hline C_1 & 0 & 0.05 & 0.73 & 0.09 & 0.67 \\ C_2 & 0.05 & 0 & 0.75 & 0.10 & 0.68 \\ C_3 & 0.73 & 0.75 & 0 & 0.72 & 0.36 \\ C_4 & 0.09 & 0.10 & 0.72 & 0 & 0.70 \\ C_5 & 0.67 & 0.68 & 0.36 & 0.70 & 0 \end{array}$$

$$IM^{\mu}_{GA_{20}} = \begin{array}{c|ccccc} & C_1 & C_2 & C_3 & C_4 & C_5 \\ \hline C_1 & 1 & 0.97 & 0.23 & 0.90 & 0.22 \\ C_2 & 0.97 & 1 & 0.24 & 0.91 & 0.22 \\ C_3 & 0.23 & 0.24 & 1 & 0.32 & 0.70 \\ C_4 & 0.90 & 0.91 & 0.32 & 1 & 0.28 \\ C_5 & 0.22 & 0.22 & 0.70 & 0.28 & 1 \end{array}$$

$$IM^{\nu}_{GA_{20}} = \begin{array}{c|ccccc} & C_1 & C_2 & C_3 & C_4 & C_5 \\ \hline C_1 & 0 & 0.03 & 0.74 & 0.10 & 0.76 \\ C_2 & 0.03 & 0 & 0.73 & 0.08 & 0.75 \\ C_3 & 0.74 & 0.73 & 0 & 0.66 & 0.25 \\ C_4 & 0.10 & 0.08 & 0.66 & 0 & 0.70 \\ C_5 & 0.76 & 0.75 & 0.25 & 0.70 & 0 \end{array}$$

$$\text{IM}^{\mu}_{GA_{30}} = \begin{array}{c|ccccc} & C_1 & C_2 & C_3 & C_4 & C_5 \\ \hline C_1 & 1 & 0.94 & 0.37 & 0.80 & 0.33 \\ C_2 & 0.94 & 1 & 0.34 & 0.80 & 0.30 \\ C_3 & 0.37 & 0.34 & 1 & 0.48 & 0.53 \\ C_4 & 0.80 & 0.80 & 0.48 & 1 & 0.29 \\ C_5 & 0.33 & 0.30 & 0.53 & 0.29 & 1 \end{array},$$

$$\text{IM}^{\nu}_{GA_{30}} = \begin{array}{c|ccccc} & C_1 & C_2 & C_3 & C_4 & C_5 \\ \hline C_1 & 0 & 0.04 & 0.61 & 0.20 & 0.64 \\ C_2 & 0.04 & 0 & 0.62 & 0.18 & 0.66 \\ C_3 & 0.61 & 0.62 & 0 & 0.49 & 0.42 \\ C_4 & 0.20 & 0.18 & 0.49 & 0 & 0.67 \\ C_5 & 0.64 & 0.66 & 0.42 & 0.67 & 0 \end{array}$$

$$\text{IM}^{\mu}_{GA_{40}} = \begin{array}{c|ccccc} & C_1 & C_2 & C_3 & C_4 & C_5 \\ \hline C_1 & 1 & 0.94 & 0.30 & 0.82 & 0.23 \\ C_2 & 0.94 & 1 & 0.27 & 0.80 & 0.24 \\ C_3 & 0.30 & 0.27 & 1 & 0.38 & 0.62 \\ C_4 & 0.82 & 0.80 & 0.38 & 1 & 0.21 \\ C_5 & 0.23 & 0.24 & 0.62 & 0.21 & 1 \end{array},$$

$$\text{IM}^{\nu}_{GA_{40}} = \begin{array}{c|ccccc} & C_1 & C_2 & C_3 & C_4 & C_5 \\ \hline C_1 & 0 & 0.04 & 0.67 & 0.17 & 0.73 \\ C_2 & 0.04 & 0 & 0.69 & 0.18 & 0.72 \\ C_3 & 0.67 & 0.69 & 0 & 0.60 & 0.32 \\ C_4 & 0.17 & 0.18 & 0.60 & 0 & 0.75 \\ C_5 & 0.73 & 0.72 & 0.32 & 0.75 & 0 \end{array}$$

$$\text{IM}^{\mu}_{GA_{50}} = \begin{array}{c|ccccc} & C_1 & C_2 & C_3 & C_4 & C_5 \\ \hline C_1 & 1 & 0.93 & 0.48 & 0.80 & 0.35 \\ C_2 & 0.93 & 1 & 0.45 & 0.80 & 0.32 \\ C_3 & 0.48 & 0.45 & 1 & 0.57 & 0.54 \\ C_4 & 0.80 & 0.80 & 0.57 & 1 & 0.35 \\ C_5 & 0.35 & 0.32 & 0.54 & 0.35 & 1 \end{array},$$

$$\text{IM}^{\nu}_{GA_{50}} = \begin{array}{c|ccccc} & C_1 & C_2 & C_3 & C_4 & C_5 \\ \hline C_1 & 0 & 0.05 & 0.50 & 0.19 & 0.62 \\ C_2 & 0.05 & 0 & 0.51 & 0.18 & 0.64 \\ C_3 & 0.50 & 0.51 & 0 & 0.42 & 0.42 \\ C_4 & 0.19 & 0.18 & 0.42 & 0 & 0.63 \\ C_5 & 0.62 & 0.64 & 0.42 & 0.63 & 0 \end{array}$$

$$\text{IM}^{\mu}_{GA_{60}} = \begin{array}{c|ccccc} & C_1 & C_2 & C_3 & C_4 & C_5 \\ \hline C_1 & 1 & 0.93 & 0.48 & 0.77 & 0.26 \\ C_2 & 0.93 & 1 & 0.42 & 0.75 & 0.26 \\ C_3 & 0.48 & 0.42 & 1 & 0.57 & 0.52 \\ C_4 & 0.77 & 0.75 & 0.57 & 1 & 0.32 \\ C_5 & 0.26 & 0.26 & 0.52 & 0.32 & 1 \end{array},$$

$$\text{IM}^{\nu}_{GA_{60}} = \begin{array}{c|ccccc} & C_1 & C_2 & C_3 & C_4 & C_5 \\ \hline C_1 & 0 & 0.05 & 0.50 & 0.23 & 0.73 \\ C_2 & 0.05 & 0 & 0.54 & 0.23 & 0.71 \\ C_3 & 0.50 & 0.54 & 0 & 0.41 & 0.45 \\ C_4 & 0.23 & 0.23 & 0.41 & 0 & 0.66 \\ C_5 & 0.73 & 0.71 & 0.45 & 0.66 & 0 \end{array}$$

$$\text{IM}^{\mu}_{GA_{70}} = \begin{array}{c|ccccc} & C_1 & C_2 & C_3 & C_4 & C_5 \\ \hline C_1 & 1 & 0.95 & 0.43 & 0.75 & 0.28 \\ C_2 & 0.95 & 1 & 0.40 & 0.72 & 0.26 \\ C_3 & 0.43 & 0.40 & 1 & 0.54 & 0.54 \\ C_4 & 0.75 & 0.72 & 0.54 & 1 & 0.27 \\ C_5 & 0.28 & 0.26 & 0.54 & 0.27 & 1 \end{array},$$

$$\text{IM}^{\nu}_{GA_{70}} = \begin{array}{c|ccccc} & C_1 & C_2 & C_3 & C_4 & C_5 \\ \hline C_1 & 0 & 0.02 & 0.54 & 0.25 & 0.70 \\ C_2 & 0.02 & 0 & 0.54 & 0.25 & 0.70 \\ C_3 & 0.54 & 0.54 & 0 & 0.43 & 0.41 \\ C_4 & 0.25 & 0.25 & 0.43 & 0 & 0.71 \\ C_5 & 0.70 & 0.70 & 0.41 & 0.71 & 0 \end{array}$$

$$
\text{IM}^{\mu}_{GA_{80}} =
\begin{array}{c|ccccc}
 & C_1 & C_2 & C_3 & C_4 & C_5 \\
\hline
C_1 & 1 & 0.91 & 0.43 & 0.81 & 0.28 \\
C_2 & 0.91 & 1 & 0.40 & 0.81 & 0.26 \\
C_3 & 0.43 & 0.40 & 1 & 0.53 & 0.50 \\
C_4 & 0.81 & 0.81 & 0.53 & 1 & 0.32 \\
C_5 & 0.28 & 0.26 & 0.50 & 0.32 & 1
\end{array}, \quad
\text{IM}^{\nu}_{GA_{80}} =
\begin{array}{c|ccccc}
 & C_1 & C_2 & C_3 & C_4 & C_5 \\
\hline
C_1 & 0 & 0.07 & 0.54 & 0.19 & 0.71 \\
C_2 & 0.07 & 0 & 0.57 & 0.17 & 0.71 \\
C_3 & 0.54 & 0.57 & 0 & 0.45 & 0.47 \\
C_4 & 0.19 & 0.17 & 0.45 & 0 & 0.66 \\
C_5 & 0.71 & 0.71 & 0.47 & 0.66 & 0
\end{array}
$$

$$
\text{IM}^{\mu}_{GA_{90}} =
\begin{array}{c|ccccc}
 & C_1 & C_2 & C_3 & C_4 & C_5 \\
\hline
C_1 & 1 & 0.90 & 0.51 & 0.81 & 0.32 \\
C_2 & 0.90 & 1 & 0.45 & 0.77 & 0.29 \\
C_3 & 0.51 & 0.45 & 1 & 0.62 & 0.43 \\
C_4 & 0.81 & 0.77 & 0.62 & 1 & 0.33 \\
C_5 & 0.32 & 0.29 & 0.43 & 0.33 & 1
\end{array}, \quad
\text{IM}^{\nu}_{GA_{90}} =
\begin{array}{c|ccccc}
 & C_1 & C_2 & C_3 & C_4 & C_5 \\
\hline
C_1 & 0 & 0.08 & 0.46 & 0.18 & 0.66 \\
C_2 & 0.08 & 0 & 0.51 & 0.20 & 0.67 \\
C_3 & 0.46 & 0.51 & 0 & 0.35 & 0.52 \\
C_4 & 0.18 & 0.20 & 0.35 & 0 & 0.65 \\
C_5 & 0.66 & 0.67 & 0.52 & 0.65 & 0
\end{array}
$$

$$
\text{IM}^{\mu}_{GA_{100}} =
\begin{array}{c|ccccc}
 & C_1 & C_2 & C_3 & C_4 & C_5 \\
\hline
C_1 & 1 & 0.90 & 0.54 & 0.59 & 0.33 \\
C_2 & 0.90 & 1 & 0.47 & 0.57 & 0.30 \\
C_3 & 0.54 & 0.47 & 1 & 0.66 & 0.46 \\
C_4 & 0.59 & 0.57 & 0.66 & 1 & 0.39 \\
C_5 & 0.33 & 0.30 & 0.46 & 0.39 & 1
\end{array}, \quad
\text{IM}^{\nu}_{GA_{100}} =
\begin{array}{c|ccccc}
 & C_1 & C_2 & C_3 & C_4 & C_5 \\
\hline
C_1 & 0 & 0.07 & 0.44 & 0.41 & 0.65 \\
C_2 & 0.07 & 0 & 0.49 & 0.39 & 0.66 \\
C_3 & 0.44 & 0.49 & 0 & 0.33 & 0.51 \\
C_4 & 0.41 & 0.39 & 0.33 & 0 & 0.60 \\
C_5 & 0.65 & 0.66 & 0.51 & 0.60 & 0
\end{array}
$$

$$
\text{IM}^{\mu}_{GA_{110}} =
\begin{array}{c|ccccc}
 & C_1 & C_2 & C_3 & C_4 & C_5 \\
\hline
C_1 & 1 & 0.91 & 0.47 & 0.73 & 0.33 \\
C_2 & 0.91 & 1 & 0.40 & 0.72 & 0.30 \\
C_3 & 0.47 & 0.40 & 1 & 0.57 & 0.49 \\
C_4 & 0.73 & 0.72 & 0.57 & 1 & 0.35 \\
C_5 & 0.33 & 0.30 & 0.49 & 0.35 & 1
\end{array}, \quad
\text{IM}^{\nu}_{GA_{110}} =
\begin{array}{c|ccccc}
 & C_1 & C_2 & C_3 & C_4 & C_5 \\
\hline
C_1 & 0 & 0.06 & 0.51 & 0.27 & 0.66 \\
C_2 & 0.06 & 0 & 0.56 & 0.26 & 0.67 \\
C_3 & 0.51 & 0.56 & 0 & 0.41 & 0.49 \\
C_4 & 0.27 & 0.26 & 0.41 & 0 & 0.64 \\
C_5 & 0.66 & 0.67 & 0.49 & 0.64 & 0
\end{array}
$$

$$
\text{IM}^{\mu}_{GA_{150}} =
\begin{array}{c|ccccc}
 & C_1 & C_2 & C_3 & C_4 & C_5 \\
\hline
C_1 & 1 & 0.91 & 0.47 & 0.75 & 0.36 \\
C_2 & 0.91 & 1 & 0.42 & 0.72 & 0.32 \\
C_3 & 0.47 & 0.42 & 1 & 0.58 & 0.60 \\
C_4 & 0.75 & 0.72 & 0.58 & 1 & 0.37 \\
C_5 & 0.36 & 0.32 & 0.60 & 0.37 & 1
\end{array}, \quad
\text{IM}^{\nu}_{GA_{150}} =
\begin{array}{c|ccccc}
 & C_1 & C_2 & C_3 & C_4 & C_5 \\
\hline
C_1 & 0 & 0.05 & 0.51 & 0.24 & 0.61 \\
C_2 & 0.05 & 0 & 0.54 & 0.25 & 0.63 \\
C_3 & 0.51 & 0.54 & 0 & 0.40 & 0.36 \\
C_4 & 0.24 & 0.25 & 0.40 & 0 & 0.60 \\
C_5 & 0.61 & 0.63 & 0.36 & 0.60 & 0
\end{array}
$$

$$
\text{IM}^{\mu}_{GA_{200}} =
\begin{array}{c|ccccc}
 & C_1 & C_2 & C_3 & C_4 & C_5 \\
\hline
C_1 & 1 & 0.89 & 0.40 & 0.53 & 0.23 \\
C_2 & 0.89 & 1 & 0.37 & 0.53 & 0.25 \\
C_3 & 0.40 & 0.37 & 1 & 0.71 & 0.49 \\
C_4 & 0.53 & 0.53 & 0.71 & 1 & 0.39 \\
C_5 & 0.23 & 0.25 & 0.49 & 0.39 & 1
\end{array},
\qquad
\text{IM}^{\nu}_{GA_{200}} =
\begin{array}{c|ccccc}
 & C_1 & C_2 & C_3 & C_4 & C_5 \\
\hline
C_1 & 0 & 0.07 & 0.57 & 0.46 & 0.76 \\
C_2 & 0.07 & 0 & 0.57 & 0.44 & 0.71 \\
C_3 & 0.57 & 0.57 & 0 & 0.28 & 0.49 \\
C_4 & 0.46 & 0.44 & 0.28 & 0 & 0.61 \\
C_5 & 0.76 & 0.71 & 0.49 & 0.61 & 0
\end{array}
$$

According to the proposed schema, the obtained 14 different results for the degrees of "agreement" $\mu_{C,C'}$ ($\text{IM}^{\mu}_{GA_i}$) of the considered 10 criteria pairs are used to construct the following IM IM_{GA} (Eq. (27)):

$$\text{IM}_{GA} =$$

	$C_1 \leftrightarrow C_2$	$C_1 \leftrightarrow C_3$	$C_1 \leftrightarrow C_4$	$C_1 \leftrightarrow C_5$	$C_2 \leftrightarrow C_3$	$C_2 \leftrightarrow C_4$	$C_2 \leftrightarrow C_5$	$C_3 \leftrightarrow C_4$	$C_3 \leftrightarrow C_5$	$C_4 \leftrightarrow C_5$
GA_5	0.88	0.40	0.84	0.33	0.31	0.78	0.31	0.47	0.58	0.40
GA_{10}	0.94	0.26	0.91	0.30	0.24	0.90	0.29	0.27	0.61	0.28
GA_{20}	0.97	0.23	0.90	0.22	0.24	0.91	0.22	0.32	0.70	0.28
GA_{30}	0.94	0.37	0.80	0.33	0.34	0.80	0.30	0.48	0.53	0.29
GA_{40}	0.94	0.30	0.82	0.23	0.27	0.80	0.24	0.38	0.62	0.21
GA_{50}	0.93	0.48	0.80	0.35	0.45	0.80	0.32	0.57	0.54	0.35
GA_{60}	0.93	0.48	0.77	0.26	0.42	0.75	0.26	0.57	0.52	0.32
GA_{70}	0.95	0.43	0.75	0.28	0.40	0.72	0.26	0.54	0.54	0.27
GA_{80}	0.91	0.43	0.81	0.28	0.40	0.81	0.26	0.53	0.50	0.32
GA_{90}	0.90	0.51	0.81	0.32	0.45	0.77	0.29	0.62	0.43	0.33
GA_{100}	0.90	0.54	0.59	0.33	0.47	0.57	0.30	0.66	0.46	0.39
GA_{110}	0.91	0.47	0.73	0.33	0.40	0.72	0.30	0.57	0.49	0.35
GA_{150}	0.91	0.47	0.75	0.36	0.42	0.72	0.32	0.58	0.60	0.37
GA_{200}	0.89	0.40	0.53	0.23	0.37	0.53	0.25	0.71	0.49	0.39

$$\tag{27}$$

It will be interesting to compare the obtained $\mu_{C,C'}$ values for each criteria pair with those obtained in the first case—ICrA of the model parameter identification results. In the next figures we give the comparison for some of the criteria pairs. We compare the 14 $\mu_{C,C'}$ values with the $\mu_{C,C'}$ values for the average, the worst and the best results from the model parameters identification procedures.

In Fig. 3, $\mu_{C,C'}$ values for the criteria pair $C_1 \leftrightarrow C_2$ is compared. As can be seen, the $\mu_{C,C'}$ values are similar in all the cases. The average $\mu_{C,C'}$ value of 14 GAs results is very close to the $\mu_{C,C'}$ value for the average identification results. In the case of comparison of the $\mu_{C,C'}$ values criteria pair $C_1 \leftrightarrow C_3$, there are some fluctuations, but the average $\mu_{C,C'}$ value of 14 GAs results is equal to the $\mu_{C,C'}$ value for the average identification results—$\mu_{C,C'} = 0.41$ (see Fig. 4). Almost the same are the results about the criteria pair $C_2 \leftrightarrow C_3$ presented in Fig. 5—average $\mu_{C,C'} = 0.36$. As the criteria pairs $C_1 \leftrightarrow C_2$ and $C_1 \leftrightarrow C_3$ represent the relations of the three model parameters (μ_{max}, k_S and $Y_{S/X}$), it is logical to obtain closer results in all cases.

On the other hand, the largest differences in the $\mu_{C,C'}$ values are observed for the pair $C_4 \leftrightarrow C_5$ (Fig. 6). Due to the stochastic nature of the GAs, it is obvious that the relations between objective value and computation time are unpredictable. The GA could solve the problem with a high accuracy for a very short time if the algorithm initial solutions are close to the real ones. Conversely, the GA could reach

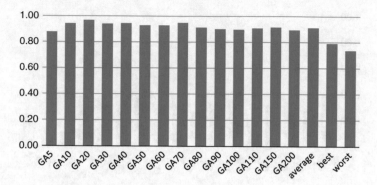

Fig. 3 μ_{C_1,C_2}-values for all cases

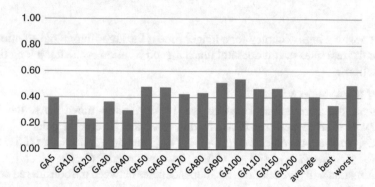

Fig. 4 μ_{C_1,C_3}-values for all cases

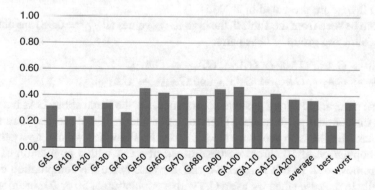

Fig. 5 μ_{C_2,C_3}-values for all cases

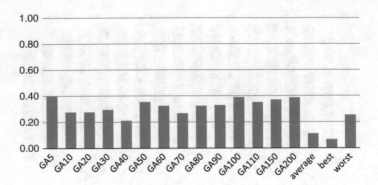

Fig. 6 μ_{C_4,C_5}-values for all cases

a solution with a high accuracy for a longer time if the algorithm initial solutions are far from the real ones or if it does not reach a good solution even after a long time of computations.

ICrA of ICrA: Step 2

To analyse the correlations between the 14 differently tuned GAs, the IM_{GA} (Eq. (27)) is used to perform ICrA. In this case, the 10 criteria pairs ($C_1 \leftrightarrow C_2$, $C_1 \leftrightarrow C_3$, etc.) become objects ($O_1, O_2, \ldots, O_{10}$) and the applied GAs (GA_5, GA_{10}, $GA_{20}, \ldots, GA_{100}, GA_{110}, GA_{150}, GA_{200}$) become criteria ($C_1, C_2, \ldots, C_{14}$). Thus, the ICrA results will give relations and dependencies between the considered criteria, i.e. the 14 GAs performances.

The obtained results for $\langle \mu_{C,C'}, \nu_{C,C'} \rangle$, sorted by μ-values, after application of ICrA to IM_{GA}, are presented in Table 5.

As can be seen from the Table 5, the high $\mu_{C,C'}$ values ($\mu_{C,C'} = 0.98$) are obtained for the following group of GAs pairs:

- $GA_{50} \leftrightarrow GA_{80}$, $GA_{50} \leftrightarrow GA_{110}$, $GA_{60} \leftrightarrow GA_{110}$,
  $GA_{70} \leftrightarrow GA_{150}$, $GA_{80} \leftrightarrow GA_{110}$, and $GA_{110} \leftrightarrow GA_{150}$.

These pairs are in strong positive consonance, i.e. the listed above GAs have very similar or identical performance. Therefore, considering the discussed here model parameter identification problem, the ICrA results show that we can apply the GA with a population of 50 chromosomes instead of a GA with a population of 80 or 110 chromosomes. In this manner we can decrease twice the computation efforts, especially in the case when we use a GA with a population of 50 or 70 chromosomes instead of a GA with a population of 110 or 150 chromosomes. Moreover, in the same time we reserve the solution accuracy.

The next group of criteria pairs that are still in strong positive consonance ($\mu_{C,C'} = 0.96$) are as follows:

- $GA_{30} \leftrightarrow GA_{40}$, $GA_{50} \leftrightarrow GA_{60}$, $GA_{90} \leftrightarrow GA_{100}$,
  $GA_{60} \leftrightarrow GA_{80}$, $GA_{90} \leftrightarrow GA_{110}$,

Table 5 Genetic algorithms relations sorted by $\mu_{C,C'}$, $\nu_{C,C'}$ values

GA pairs	$\langle \mu, \nu \rangle$	GA pairs	$\langle \mu, \nu \rangle$	GA pairs	$\langle \mu, \nu \rangle$
$GA_{50} \leftrightarrow GA_{80}$	$\langle 0.98, 0.00 \rangle$	$GA_{40} \leftrightarrow GA_{110}$	$\langle 0.91, 0.09 \rangle$	$GA_{30} \leftrightarrow GA_{90}$	$\langle 0.87, 0.13 \rangle$
$GA_{50} \leftrightarrow GA_{110}$	$\langle 0.98, 0.00 \rangle$	$GA_{60} \leftrightarrow GA_{200}$	$\langle 0.91, 0.07 \rangle$	$GA_{40} \leftrightarrow GA_{90}$	$\langle 0.87, 0.13 \rangle$
$GA_{60} \leftrightarrow GA_{110}$	$\langle 0.98, 0.00 \rangle$	$GA_{70} \leftrightarrow GA_{90}$	$\langle 0.91, 0.09 \rangle$	$GA_{40} \leftrightarrow GA_{200}$	$\langle 0.87, 0.13 \rangle$
$GA_{70} \leftrightarrow GA_{150}$	$\langle 0.98, 0.02 \rangle$	$GA_{100} \leftrightarrow GA_{110}$	$\langle 0.91, 0.09 \rangle$	$GA_{70} \leftrightarrow GA_{100}$	$\langle 0.87, 0.13 \rangle$
$GA_{80} \leftrightarrow GA_{110}$	$\langle 0.98, 0.02 \rangle$	$GA_{100} \leftrightarrow GA_{200}$	$\langle 0.91, 0.09 \rangle$	$GA_{70} \leftrightarrow GA_{200}$	$\langle 0.87, 0.13 \rangle$
$GA_{110} \leftrightarrow GA_{150}$	$\langle 0.98, 0.02 \rangle$	$GA_{110} \leftrightarrow GA_{200}$	$\langle 0.91, 0.09 \rangle$	$GA_{90} \leftrightarrow GA_{200}$	$\langle 0.87, 0.13 \rangle$
$GA_{30} \leftrightarrow GA_{40}$	$\langle 0.96, 0.04 \rangle$	$GA_{5} \leftrightarrow GA_{20}$	$\langle 0.89, 0.09 \rangle$	$GA_{5} \leftrightarrow GA_{30}$	$\langle 0.84, 0.13 \rangle$
$GA_{30} \leftrightarrow GA_{70}$	$\langle 0.96, 0.04 \rangle$	$GA_{5} \leftrightarrow GA_{70}$	$\langle 0.89, 0.09 \rangle$	$GA_{5} \leftrightarrow GA_{40}$	$\langle 0.84, 0.13 \rangle$
$GA_{40} \leftrightarrow GA_{70}$	$\langle 0.96, 0.04 \rangle$	$GA_{5} \leftrightarrow GA_{110}$	$\langle 0.89, 0.09 \rangle$	$GA_{5} \leftrightarrow GA_{90}$	$\langle 0.84, 0.13 \rangle$
$GA_{50} \leftrightarrow GA_{60}$	$\langle 0.96, 0.00 \rangle$	$GA_{20} \leftrightarrow GA_{80}$	$\langle 0.89, 0.11 \rangle$	$GA_{5} \leftrightarrow GA_{200}$	$\langle 0.84, 0.13 \rangle$
$GA_{50} \leftrightarrow GA_{150}$	$\langle 0.96, 0.02 \rangle$	$GA_{20} \leftrightarrow GA_{150}$	$\langle 0.89, 0.11 \rangle$	$GA_{5} \leftrightarrow GA_{10}$	$\langle 0.82, 0.16 \rangle$
$GA_{60} \leftrightarrow GA_{80}$	$\langle 0.96, 0.02 \rangle$	$GA_{30} \leftrightarrow GA_{60}$	$\langle 0.89, 0.09 \rangle$	$GA_{20} \leftrightarrow GA_{90}$	$\langle 0.82, 0.18 \rangle$
$GA_{60} \leftrightarrow GA_{150}$	$\langle 0.96, 0.02 \rangle$	$GA_{40} \leftrightarrow GA_{50}$	$\langle 0.89, 0.09 \rangle$	$GA_{30} \leftrightarrow GA_{100}$	$\langle 0.82, 0.18 \rangle$
$GA_{70} \leftrightarrow GA_{110}$	$\langle 0.96, 0.04 \rangle$	$GA_{40} \leftrightarrow GA_{80}$	$\langle 0.89, 0.11 \rangle$	$GA_{30} \leftrightarrow GA_{200}$	$\langle 0.82, 0.18 \rangle$
$GA_{80} \leftrightarrow GA_{150}$	$\langle 0.96, 0.04 \rangle$	$GA_{50} \leftrightarrow GA_{100}$	$\langle 0.89, 0.09 \rangle$	$GA_{40} \leftrightarrow GA_{100}$	$\langle 0.82, 0.18 \rangle$
$GA_{90} \leftrightarrow GA_{100}$	$\langle 0.96, 0.04 \rangle$	$GA_{50} \leftrightarrow GA_{200}$	$\langle 0.89, 0.09 \rangle$	$GA_{5} \leftrightarrow GA_{100}$	$\langle 0.80, 0.18 \rangle$
$GA_{90} \leftrightarrow GA_{110}$	$\langle 0.96, 0.04 \rangle$	$GA_{60} \leftrightarrow GA_{100}$	$\langle 0.89, 0.09 \rangle$	$GA_{20} \leftrightarrow GA_{100}$	$\langle 0.78, 0.22 \rangle$
$GA_{30} \leftrightarrow GA_{80}$	$\langle 0.93, 0.07 \rangle$	$GA_{80} \leftrightarrow GA_{100}$	$\langle 0.89, 0.11 \rangle$	$GA_{10} \leftrightarrow GA_{40}$	$\langle 0.78, 0.22 \rangle$
$GA_{30} \leftrightarrow GA_{150}$	$\langle 0.93, 0.07 \rangle$	$GA_{80} \leftrightarrow GA_{200}$	$\langle 0.89, 0.11 \rangle$	$GA_{10} \leftrightarrow GA_{70}$	$\langle 0.78, 0.22 \rangle$
$GA_{40} \leftrightarrow GA_{150}$	$\langle 0.93, 0.07 \rangle$	$GA_{100} \leftrightarrow GA_{150}$	$\langle 0.89, 0.11 \rangle$	$GA_{10} \leftrightarrow GA_{30}$	$\langle 0.78, 0.22 \rangle$
$GA_{50} \leftrightarrow GA_{70}$	$\langle 0.93, 0.04 \rangle$	$GA_{150} \leftrightarrow GA_{200}$	$\langle 0.89, 0.11 \rangle$	$GA_{10} \leftrightarrow GA_{150}$	$\langle 0.76, 0.24 \rangle$
$GA_{50} \leftrightarrow GA_{90}$	$\langle 0.93, 0.04 \rangle$	$GA_{5} \leftrightarrow GA_{50}$	$\langle 0.87, 0.09 \rangle$	$GA_{10} \leftrightarrow GA_{20}$	$\langle 0.73, 0.27 \rangle$
$GA_{60} \leftrightarrow GA_{70}$	$\langle 0.93, 0.04 \rangle$	$GA_{5} \leftrightarrow GA_{60}$	$\langle 0.87, 0.09 \rangle$	$GA_{10} \leftrightarrow GA_{110}$	$\langle 0.73, 0.27 \rangle$
$GA_{60} \leftrightarrow GA_{90}$	$\langle 0.93, 0.04 \rangle$	$GA_{5} \leftrightarrow GA_{80}$	$\langle 0.87, 0.11 \rangle$	$GA_{10} \leftrightarrow GA_{50}$	$\langle 0.71, 0.27 \rangle$
$GA_{70} \leftrightarrow GA_{80}$	$\langle 0.93, 0.07 \rangle$	$GA_{20} \leftrightarrow GA_{30}$	$\langle 0.87, 0.13 \rangle$	$GA_{10} \leftrightarrow GA_{60}$	$\langle 0.71, 0.27 \rangle$
$GA_{80} \leftrightarrow GA_{90}$	$\langle 0.93, 0.07 \rangle$	$GA_{20} \leftrightarrow GA_{40}$	$\langle 0.87, 0.13 \rangle$	$GA_{10} \leftrightarrow GA_{80}$	$\langle 0.71, 0.29 \rangle$
$GA_{90} \leftrightarrow GA_{150}$	$\langle 0.93, 0.07 \rangle$	$GA_{20} \leftrightarrow GA_{50}$	$\langle 0.87, 0.11 \rangle$	$GA_{10} \leftrightarrow GA_{90}$	$\langle 0.69, 0.31 \rangle$
$GA_{5} \leftrightarrow GA_{150}$	$\langle 0.91, 0.07 \rangle$	$GA_{20} \leftrightarrow GA_{60}$	$\langle 0.87, 0.11 \rangle$	$GA_{10} \leftrightarrow GA_{200}$	$\langle 0.69, 0.31 \rangle$
$GA_{30} \leftrightarrow GA_{50}$	$\langle 0.91, 0.07 \rangle$	$GA_{20} \leftrightarrow GA_{70}$	$\langle 0.87, 0.13 \rangle$	$GA_{10} \leftrightarrow GA_{100}$	$\langle 0.64, 0.36 \rangle$
$GA_{30} \leftrightarrow GA_{110}$	$\langle 0.91, 0.09 \rangle$	$GA_{20} \leftrightarrow GA_{110}$	$\langle 0.87, 0.13 \rangle$		
$GA_{40} \leftrightarrow GA_{60}$	$\langle 0.91, 0.07 \rangle$	$GA_{20} \leftrightarrow GA_{200}$	$\langle 0.87, 0.13 \rangle$		

$GA_{40} \leftrightarrow GA_{70}$, $GA_{70} \leftrightarrow GA_{110}$, $GA_{30} \leftrightarrow GA_{70}$, $GA_{50} \leftrightarrow GA_{150}$, $GA_{60} \leftrightarrow GA_{150}$, and $GA_{80} \leftrightarrow GA_{150}$.

For the pairs of the first row it is clear that these GAs will show similar performance. These GAs have closer population sizes—30–40, 50–60, etc. The similar performance of the GAs in the second and third rows could be explained similarly.

The interesting results from ICrA are in the last row. The GAs with populations of 50, 60 or 80 chromosomes have very similar performance compared to the GA with a population of 150 chromosomes. In other words, if we use a GA with a population of 50 instead of a GA with a population of 150 chromosomes, the algorithm will work over a three times smaller population, i.e. about three times less memory, preserving the solution accuracy.

The next group of criteria pairs, whose relations are in positive consonance ($\mu_{C,C'} = 0.93$), includes mainly pairs of GAs with a similar size of the population (60–70, 70–80, etc.) and such with a close size of the population (50–70, 60–90, etc.). There are some pairs for which ICrA shows similar performance, although the population sizes are very different. For example, such pairs in positive consonance are:

- $GA_{30} \leftrightarrow GA_{150}$, $GA_{40} \leftrightarrow GA_{150}$, $GA_{30} \leftrightarrow GA_{80}$ and $GA_{90} \leftrightarrow GA_{150}$.

Based on the results above, it can be concluded that it is not appropriate to use a GA with a population of 150 chromosomes—the resulting computation time is very long and the solution accuracy is commensurable with the accuracy of the GA with a population of 30 or 40 chromosomes, even more—the solution accuracy is equal to the accuracy of the GA with a population of 50 or 80 chromosomes.

The pairs in the next three groups (according to Table 5) with $\mu_{C,C'} = 0.91$, $\mu_{C,C'} = 0.89$ and $\mu_{C,C'} = 0.87$, respectively, are still in positive consonance (see Table 4). Since these values of $\mu_{C,C'}$ are close to the values for which we will have weak positive consonance, we do not establish so well-connected relations such as in the case of strong positive consonance—the first discussed group.

The last 12 criteria pairs in Table 5 are between GA_{10} and the rest considered GAs. This is an interesting result. ICrA shows that the GA with a population of 10 chromosomes is in weak positive consonance or in weak dissonance and dissonance in all pairs. The GA with a population of 10 chromosomes shows different performance compared to the rest of the applied GAs. We can explain the ICrA results with the fact that this GA gives worse results (Eq. (14)) compared to all other GAs, even with the GA with a population of 5 chromosomes (Eq. (13)).

5   Conclusion

In this paper, based on the apparatus of the Index Matrices and the Intuitionistic Fuzzy Sets, ICrA of a model parameters identification using GAs is performed. A non-linear model of an *E. coli* fed-batch cultivation process is considered. A series of model identification procedures using GAs are done. ICrA is applied to explore the existing relations and dependencies of the defined model parameters and GAs outcomes—computation time and objective function value. Three case studies are examined—considering the average, the worst and the best results for the obtained model parameters, computation time and objective function value. Moreover, a schema of ICrA of ICrA results to examine the 14 differently tuned GAs performances is proposed. Applying ICrA, some relations and dependencies between the defined criteria are established. Based on the used scale for defining the consonance and dissonance between each pair of criteria, a discussion which criteria are in consonance and dissonance as well as the degree of their dependence are presented. Some conclusions about the preferred tuned GAs for the considered model parameter identification in terms of achieved accuracy for a given computation time are given.

Acknowledgments The work presented here is partially supported by the Bulgarian National Scientific Fund under Grant DFNI-I02/5 and by the Polish-Bulgarian collaborative Grant "Parallel and Distributed Computing Practices".

References

1. Angelova, M., Roeva, O., Pencheva, T.: InterCriteria analysis of crossover and mutation rates relations in simple genetic algorithm. In: Proceedings of the 2015 Federated Conference on Computer Science and Information Systems, pp. 419–424 (2015)
2. Atanassov, K.: Generalized index matrices. Comptes Rendus de l'Academie Bulgare des Sciences **40**(11), 15–18 (1987)
3. Atanassov, K.: On index matrices, part 1: standard cases. Adv. Stud. Contemp. Math. **20**(2), 291–302 (2010)
4. Atanassov, K.: On index matrices, part 2: intuitionistic fuzzy case. Proc. Jangjeon Math. Soc. **13**(2), 121–126 (2010)
5. Atanassov, K.: On Intuitionistic Fuzzy Sets Theory. Springer, Berlin (2012)
6. Atanassov, K., Mavrov, D., Atanassova, V.: Intercriteria decision making: a new approach for multicriteria decision making, based on index matrices and intuitionistic fuzzy sets. Iss. Intuitionistic Fuzzy Sets Gen. Nets **11**, 1–8 (2014)
7. Atanassov, K., Atanassova, V., Gluhchev, G.: InterCriteria analysis: ideas and problems. Not Intuitionistic Fuzzy Sets **21**(1), 81–88 (2015)
8. Bastin, G., Dochain, D.: On-line Estimation and Adaptive Control of Bioreactors. Elsevier Scientific Publications, Amsterdam (1991)
9. Boussaïd, I., Lepagnot, J., Siarry, P.: A survey on optimization metaheuristics. Inf. Sci. **237**, 82–117 (2013)
10. Doughabadi, M.H., Bahrami, H., Kolahan, F.: Evaluating the effects of parameters setting on the performance of genetic algorithm using regression modeling and statistical analysis. J. Ind. Eng. Spec. Iss. 61–68 (2011)
11. Goldberg, D.E.: Genetic algorithms in search, optimization and machine learning. Addison Wesley Longman, London (2006)
12. Ilkova, T., Petrov, M.: Intercriteria analysis for identification of *Escherichia coli* fed-batch mathematical model. J. Int. Sci. Publ.: Mater., Meth. Technol **9**, 598–608 (2015)
13. Pencheva, T., Angelova, M., Atanassova, V., Roeva, O.: InterCriteria analysis of genetic algorithm parameters in parameter identification. Notes Intuitionistic Fuzzy Sets **21**(2), 99–110 (2015)
14. Pencheva, T., Angelova, M., Vassilev, P., Roeva, O.: InterCriteria analysis approach to parameter identification of a fermentation process model. Adv Intell Syst Comput **401**, 385–397 (2016)
15. Picek, S., Golub, M., Jakobovic, D.: Evaluation of crossover operator performance in genetic algorithms with binary representation. Bio-Inspired Computing and Applications. Lecture Notes in Computer Science, vol. 6840, pp. 223–230. Springer, Berlin (2011)
16. Razali, N.M., Geraghty, J.: Genetic algorithm performance with different selection strategies in solving TSP. In: Proceedings of the World Congress on Engineering 2011 – WCE 2011, vol. II (2011)
17. Roeva, O.: Sensitivity analysis of E. coli fed-batch cultivation local models. Mathematica Balkanica. New Series **25**(4), 395–411 (2011)

18. Roeva, O., Vassilev, P.: InterCriteria analysis of generation gap influence on geneticalgorithms performance. Adv. Intell. Syst. Comput. **401**, 301–313 (2016)
19. Roeva, O., Pencheva, T., Hitzmann, B., Tzonkov, St.: A genetic algorithms based approach for identification of *Escherichia coli* fed-batch fermentation. Int. J. Bioautom. **1**, 30–41 (2004)
20. Roeva, O., Fidanova, S., Paprzycki, M.: Influence of the population size on the genetic algorithm performance in case of cultivation process modelling. In: Proceedings of the 2013 Federated Conference on Computer Science and Information Systems, pp. 371–376 (2013)
21. Roeva, O., Pencheva, T., Tzonkov, S., Hitzmann, B.: Functional state modelling of cultivation processes: dissolved oxygen limitation state. Int. J. Bioautom. **19**(1 Suppl.1), S93–S112 (2015)
22. Roeva, O., Fidanova, S., Paprzycki, M.: InterCriteria analysis of ACO and GA hybrid algorithms. Stud Comput Intell **610**, 107–126 (2016)

Exploring Sparse Covariance Estimation Techniques in Evolution Strategies

Silja Meyer-Nieberg and Erik Kropat

Abstract When considering continuous search spaces, evolution strategies are among the well-performing metaheuristics. In contrast to other evolutionary algorithms, their main search operator is mutation which necessitates its adaptation during the run. Here, the covariance matrix plays an important role. Modern Evolution Strategies apply forms of covariance matrix adaptation. However, the quality of the common estimate of the covariance is known to be questionable for high search space dimensions. This paper presents a new approach by considering sparse covariance matrix techniques together with a space transformation.

1 Introduction

Evolutionary computation has a long research tradition. The field comprises today the main classes genetic algorithms, genetic programming, evolution strategies, evolutionary programming, and differential evolution. Evolution strategies (ESs), on which the research presented in this paper focuses, are primarily used for optimizing continuous functions. The function is not required to be analytical.

Evolution strategies rely on mutation, i.e., on the random perturbation of candidate solutions to navigate the search space. The process must be controlled in order to achieve good performance. For this, modern ESs apply covariance matrix adaptation in several variants [1]. Nearly all approaches take the sample covariance matrix into account. This estimator is known to be problematic in the case of small sample sizes compared to the search space dimensionality. Since the population size in evolution strategies is typically considerably smaller, this paper argues that the adaptation process may profit from the introduction of different estimators.

S. Meyer-Nieberg (✉) · E. Kropat
Department of Computer Science, Universität der Bundeswehr München,
Werner-Heisenberg Weg 37, 85577 Neubiberg, Germany
e-mail: silja.meyer-nieberg@unibw.de

E. Kropat
e-mail: erik.kropat@unibw.de

© Springer International Publishing Switzerland 2016
S. Fidanova (ed.), *Recent Advances in Computational Optimization*,
Studies in Computational Intelligence 655, DOI 10.1007/978-3-319-40132-4_15

So far, evolutionary algorithms or related approaches have only seldom considered statistical estimation methods targeted at high-dimensional spaces. The reason may be twofold: The improved quality of the estimators induces increased computational costs which may lower the convergence velocity of the algorithm. In addition, the estimators are developed and analyzed for samples of independently, identically distributed random variables. Since evolutionary algorithms deploy selection based on rank or fitness, the assumption of the same distribution is not valid. This may be the reason as to why the literature research has resulted in only one previous approach [7]. There, the authors considered Gaussian based estimation of distribution algorithms. The problem they were faced with concerned a non-positive definiteness of the estimated covariance matrix. Therefore, Dong and Yao augmented the algorithm with a shrinkage procedure to guarantee positive definiteness. Shrinkage is one of the common methods to improve the quality of the sample covariance, see e.g. [19]. While the approach in [7] resembles the Ledoit–Wolf estimator [19], it adapted the shrinkage intensity during the run.

This paper extends the work presented in [22, 23], where Ledoit–Wolf shrinkage estimators were analyzed, combined with a maximum entropy approach, and integrated into evolution strategies. While the results were promising, the question remained how to adapt the parameter of the estimator. Therefore, in this paper, another computational simple estimation method is investigated: thresholding.

The paper is structured as follows. First, modern evolution strategies with covariance adaptation are introduced. Afterwards, a short motivation as to why we think that the covariance computation in ESs may profit from estimation theory for high-dimensional spaces is provided. The next section describes the new approach developed and is followed by the experimental section which compares the new approach against the original ES. Conclusions and a discussion of potential future research constitute the last part of the paper.

1.1 Modern Evolution Strategies

This section provides a short introduction into evolutionary algorithms focussing on evolution strategies and covariance matrix adaptation. Evolutionary algorithms (EAs) [8] in general are population-based stochastic search and optimization algorithms used when only direct function measurements are possible.

Their iterative search process requires the definition of termination criteria and stops if these are fulfilled. In each generation, a series of operations is performed: selection for reproduction, followed by offspring creation, i.e. recombination and mutation processes, and finally survivor selection. The initial population of candidate solutions is either drawn randomly from the permissable search space or is initialized based on information already obtained. First of all, the offspring population has to be created. For this, a subset of the parents is determined during *parent selection*. The creation of the offspring is based on recombination and mutation. Recombination combines traits from two or more parents resulting in one or more intermediate offspring. In contrast, mutation is an unary operator changing the components of an individual randomly. After the offspring have been created, survivor selection is performed to determine the next parent population. The different variants of evolutionary algorithms adhere to the same principles in general, but they may differ in the representation of the solutions and how the selection, recombination, and mutation processes are realized.

Evolution Strategies

Evolution strategies (ESs) [25, 28] are used for continuous optimization $f : \mathbb{R}^N \to \mathbb{R}$. Several variants have been introduced see e.g. [1, 3]. In many cases, a population of μ parents is used to create a set of λ offspring, with $\mu \leq \lambda$. For recombination, ρ parents are chosen uniformly at random without replacement and are then recombined. Recombination usually consists of determining the (weighted) mean or centroid of the parents [3]. The result is then mutated by adding a normally distributed random variable with zero mean and covariance matrix $\sigma^2 \mathbf{C}$. While there are ESs that operate without recombination, the mutation process is seen as the essential process. It is often interpreted as the main search operator. After the offspring have been created, the individuals are evaluated using the function to be optimized or a derived function which allows an easy ranking of the population. Only the rank of an individual is important for the selection. In the case of continuous optimization, the old parent population is typically discarded with the selection considering only the λ offspring of which the μ best are chosen.

The covariance matrix which is central to the mutation must be adapted during the run: Evolution strategies with ill-adapted parameters converge only slowly or may even fail in the optimization. Therefore, research on methods for adapting the scale factor σ or the full covariance matrix has a long research tradition in ESs dating back to their origins [25]. The next section describes one of the current approaches.

Updating the Covariance Matrix

To our knowledge, covariance matrix adaptation comprises two main classes: one applied in the *covariance matrix adaptation evolution strategy* (CMA-ES) [17] and an alternative used in the *covariance matrix self-adaptation evolution strategy* (CMSA-ES) [4]. Both consider information from the present population combining it with information from the search process so far. The CMA-ES is one of the most powerful evolution strategies and often referred to as the standard in ESs. However, as pointed out in [4], its scaling behavior with the population size may not be good. Beyer and Sendhoff [4] showed that the CMSA-ES performs comparably to the CMA-ES for smaller populations but that is less computational expensive for larger population sizes.

Therefore, the present paper focuses on the CMSA-ES leaving the CMA-ES for future research. The CMSA-ES uses weighted intermediate recombination, in other words, the weighted centroid $\mathbf{m}^{(g)}$ of the μ best individuals of the population is computed. To create the offspring, random vectors are drawn from the multivariate normal distribution $\mathcal{N}(\mathbf{m}^{(g)}, (\sigma^{(g)})^2 \mathbf{C}^{(g)})$. The notation of covariance matrix as $(\sigma^{(g)})^2 \mathbf{C}^{(g)}$ illustrates that the actual covariance matrix is interpreted as the combination of a general scaling factor (or step-size or mutation strength) with a rotation matrix. Following the usual practice in literature on evolution strategies the latter matrix $\mathbf{C}^{(g)}$ is referred to as *covariance matrix* in the remainder of the paper.

The covariance matrix update is based upon the common estimate of the covariance using the newly created population. Instead of considering all offspring for deriving the estimates, though, it introduces a bias towards good search regions by taking only the μ best individuals into account. Furthermore, it does not estimate the mean anew but uses the weighted mean $\mathbf{m}^{(g)}$. Following [17],

$$\mathbf{z}_{m:\lambda}^{(g+1)} := \frac{1}{\sigma^{(g)}} \left(\mathbf{x}_{m:\lambda}^{(g+1)} - \mathbf{m}^{(g)} \right) \tag{1}$$

are determined with $\mathbf{x}_{m:\lambda}$ denoting the mth best of the λ particle according to the fitness ranking. The rank-μ update then obtains the covariance matrix as

$$\mathbf{C}_{\mu}^{(g+1)} := \sum_{m=1}^{\mu} w_m \mathbf{z}_{m:\lambda}^{(g+1)} (\mathbf{z}_{m:\lambda}^{(g+1)})^{\mathrm{T}} \tag{2}$$

which is usually a positive semi-definite matrix since $\mu \ll N$. The weights w_m should fulfill $w_1 \geq w_2 \geq \cdots \geq w_\mu$ with $\sum_{m=1}^{\mu} w_i = 1$. To derive reliable estimates larger population sizes are required which would lower the algorithm's speed. Therefore, past covariance matrices are taken into account via the convex combination of (2) with the sample covariance being shrunk towards the old covariance

$$\mathbf{C}^{(g+1)} := (1 - \frac{1}{c_\tau})\mathbf{C}^{(g)} + \frac{1}{c_\tau}\mathbf{C}_{\mu}^{(g+1)} \tag{3}$$

with the weights usually set to $w_m = 1/\mu$ and

$$c_\tau = 1 + \frac{N(N+1)}{2\mu}, \tag{4}$$

see [4]. As long as $\mathbf{C}^{(g)}$ is positive semi-definite, (3) will result in a positive definite matrix.

Step-Size Adaptation

The CMSA implements the step-size using *self-adaptation* first introduced in [25] and developed further in [28]. Here, evolution is used for fitting the strategy parameters of the mutation process. In other words, the scaling parameter or in its full form, the complete covariance matrix, undergoes recombination, mutation, and indirect selection processes. The working principle is based on an indirect stochastic linkage between good individuals and appropriate parameters: Well adapted parameters should result more often in better offspring than too large or too small values or misleading directions. Although self-adaptation has been developed to adapt the whole covariance matrix, it is applied today mainly to adapt the step-size or a diagonal covariance matrix. In the case of the mutation strength, usually a log-normal distribution

$$\sigma_l^{(g)} = \sigma_{\text{base}}\exp(\tau \mathscr{N}(0, 1)) \tag{5}$$

is used for mutation. The parameter τ is called the *learning rate* and is usually chosen to scale with $1/\sqrt{2N}$. The baseline σ_{base} is either the mutation strength of the parent or if recombination is used the recombination result. For the step-size, it is possible to apply the same type of recombination as for the positions although different forms—for instance a multiplicative combination—could be used instead. The self-adaptation of the step-size is referred to as σ-*self-adaptation* (σSA) in the remainder of this paper.

The newly created mutation strength is then directly used in the mutation of the offspring. If the resulting offspring is sufficiently good, the scale factor is passed to the next generation.

Self-adaptation with recombination has been shown to be "robust" against noise [2] and is used in the CMSA-ES as the update rule for the scaling factor.

1.2 Concerning the Covariance Matrix Adaptation …

In the case of $\lambda > 1$, the sample covariance (2) appears in nearly any adaptation process. Disregarding the distortion due to selection, the sample covariance as the maximum likelihood estimator of the true covariance matrix is known as a good and reliable estimate if $\mu \gg N$. Evolution strategies typically operate with $\mu < N$, however. For example, following [15] the sizes of the parent and offspring populations in the standard CMA-ES should be chosen as $\lambda = \lfloor\log(3N)\rfloor + 4$ and $\mu = \lfloor\lambda/2\rfloor$.

Unfortunately, $\mu < N$ leads to problems with respect to the covariance estimation. This is a well-known problem in statistics [29, 30], giving raise to a broad range on literature on alternative estimators e.g. [5, 6, 10, 12, 13, 20, 21, 27, 30]. The quality of a maximum likelihood estimate may be insufficient—especially for high-dimensional spaces, see e.g. [27]. For example, Marčenko and Pastur showed that if $N/\mu \nrightarrow 0$ but $N/\mu \in (0, 1)$, instead, the eigenvalues of the covariance matrix are distributed in the interval $((1 - \sqrt{N/\mu})^2, (1 + \sqrt{N/\mu})^2)$ in the case of the standard normal distribution [21].

Equation (3) actually attempts to counteract the singularity of the population covariance matrix by using the well-known concept of shrinking. However, some distinctive differences are present. First of all, the target is a full covariance matrix whereas shrinkage typically considers simpler regulation forms as e.g. a diagonal matrix. Secondly, the parameter is usually determined via optimizing a performance measure.

Seeing that evolution strategies already apply some kind of shrinkage, some questions arise: Can we improve the estimator further by not only "shrinking" the population or sample covariance matrix but by applying further concepts stemming from the

estimation of high-dimensional covariance matrices? And considering that (3) is one
regulation technique among several, is it possible to find another well-performing
substitute? Or did research in evolution strategies already happen upon the best technique possible?

2   A Sparse Covariance Matrix Adaptation

This section introduces the new covariance adaptation technique which uses thresholding to transform the population covariance matrix. The decision for thresholding
is based upon the comparatively computational efficiency of the approach.

2.1   Space Transformation

The ideal covariance matrix for the search depends on the function landscape which
is unknown in practical applications. Considering the smooth test functions of typical
black-box optimization suites, shows that the Hessians of several functions, as e.g.
the separable functions, can be classified as sparse or approximately sparse matrices
following the definitions introduced later.

Therefore, sparse structures of the covariance matrix suffice which is exemplified
by the separable CMA-ES [26] which restricts the covariance to a diagonal matrix
in case of separability to allow fast progress to the optimal solution. For the general
case, a spare structure may not be suitable, however.

For this reason, the paper does not require sparseness of the original covariance
matrix, although it would be interesting to see how such a variant would perform on
the test suites. Instead, it considers a transformation. As argued in [18], an change
of the coordinate system may improve the performance of an evolution strategy.
Therefore, an adaptive encoding was introduced. In each iteration, the covariance
matrix is adapted following the rules of the CMA-ES. Its spectral decomposition
is used to change the basis. The creation of new search points is carried out in the
eigenspace of the current covariance matrix and the main search parameters of the
CMA-ES are updated there. After selection, the covariance matrix is adapted and
utilized for a renewed decoding and encoding.

This paper also addresses a change of the coordination system. However, we
address the covariance matrix adaptation and estimation itself which in [18] occurs
in the original space. Here, we argue that a switch to the eigenspace of the old
covariance matrix $\mathbf{C}^{(g)}$ may be beneficial for the estimation of the covariance matrix
itself.

Let the covariance matrix $\mathbf{C}^{(g)}$ be a symmetric, positive definite $N \times N$ matrix.
The condition holds for the original adaptation since (3) combines a positive definite
with a positive semi-definite matrix. As we will see below, in the case of thresholding
the condition may not always be fulfilled. Assuming a positive definite matrix allows

carrying out a spectral decomposition: Let $\mathbf{v}_1, \ldots, \mathbf{v}_N$ denote the N eigenvectors with the eigenvalues $\lambda_1, \ldots, \lambda_N, \lambda_j > 0$. Note, the eigenvectors form a orthonormal basis of $\mathbb{R}^N$, i.e., $\mathbf{v}_i^T \mathbf{v}_i = 1$ and $\mathbf{v}_i^T \mathbf{v}_j = 0$, if $i \neq j$. We define $\mathbf{V} := (\mathbf{v}_1, \ldots, \mathbf{v}_N)$ as the modal matrix. It then holds that $\mathbf{V}^{-1} = \mathbf{V}^T$. Switching to the eigenspace of $\mathbf{C}^{(g)}$ results in the representation of the covariance matrix

$$\Lambda^{(g)} = \mathbf{V} \mathbf{C}^{(g)} \mathbf{V}^T \tag{6}$$

as a diagonal matrix with the eigenvalues as the diagonal entries. Diagonal matrices are sparse matrices, thus for the estimation of the covariance matrix the more efficient procedures for sparse structures could be used. However, it is not the goal to re-estimate $\mathbf{C}^{(g)}$ but to estimate the true covariance matrix of the distribution indicated by the sample $\mathbf{z}_{1;\lambda}, \ldots, \mathbf{z}_{\mu;\lambda}$.

Before continuing, it should be noted that several definitions of sparseness exist. Usually, it is demanded that the number of non-zero elements in a row may not exceed a predefined limit $s_0(N) > 0$, i.e.,

$$\max_i \sum_{j=1}^{N} \delta(|a_{ij}| > 0) \leq s_0(N), \tag{7}$$

which should grow only slowly with N. The indicator function δ fulfills $\delta(\cdot) = 1$ if the condition is met and is zero otherwise. This definition can, however, be relaxed to a more general definition of sparseness, also referred to as approximate sparseness. Cai and Liu [5] consider the following uniformity class of sparse matrices

Definition 1 Let $s_0(N) > 0$ and let $\cdot \succ 0$ denote positive definiteness. Then a class of sparse covariance matrices is defined as

$$\begin{aligned} \mathcal{U}_q^* &:= \mathcal{U}_q^*(s_0(N)) \\ &= \left\{ \Sigma : \Sigma \succ 0, \max_i \sum_{j=1}^{p} (\sigma_{ii}\sigma_{jj})^{\frac{(1-q)}{2}} |\sigma_{ij}|^q \leq s_0(N) \right\} \end{aligned} \tag{8}$$

for some $0 \leq q < 1$.

Definition 1 requires the entries of the covariance matrix to lie within a weighted l_q ball. The weight is given by the variances. Cai and Liu [5] introduce a thresholding estimator that requires the assumption above. Its convergence rate towards the true covariance depends on $s_0(N)(\log(N)/\mu)^{(1-q)/2}$. Therefore, the number $s_0(N) > 0$ should again grow only "slowly" for $N \to \infty$.

Definition 1 leads to the main assumption of the paper. Consider an evolution strategy in the search space. The new sample that is the offspring population has been created with the help of the old covariance matrix. The covariance matrix of the selected sample differs from the previous. The deviations of from its structure stem from finite sampling characteristics and rank-based selection. Assuming that the form of the covariance matrix will not change considerably in one iteration, the new underlying covariance matrix should be sparse in the eigenspace of the old covariance, however.

Assumption 1 Let $\Sigma^{(g+1)}$ denote the true covariance matrix of the selected offspring. Consider the old covariance $\mathbf{C}^{(g)}$ with its modal matrix $\mathbf{V}$. Then $\hat{\Lambda} = \mathbf{V}\Sigma^{(g+1)}\mathbf{V}^{\mathrm{T}}$ is approximately sparse, i.e. $\hat{\Lambda} \in \mathcal{U}_q^*$ for some $0 \leq q < 1$.

Assuming the validity of the assumption, we change the coordinate system in order to perform the covariance matrix estimate. Reconsider the normalized (apart from the covariance matrix) mutation vectors $\mathbf{z}_{1;\lambda}, \ldots, \mathbf{z}_{\mu;\lambda}$ that were associated with the μ best offspring. Their representation in the eigenspace reads

$$\hat{\mathbf{z}}_{m;\lambda} = \mathbf{V}^{\mathrm{T}}\mathbf{z}_{m;\lambda} \text{ for } m = 1, \ldots, \mu. \tag{9}$$

The transformed population covariance is then estimated as

$$\hat{\mathbf{C}}_\mu = \sum_{m=1}^{\mu} w_m \hat{\mathbf{z}}_{m;\lambda} \hat{\mathbf{z}}_{m;\lambda}^{\mathrm{T}}. \tag{10}$$

The estimate (10) will be used to compute the final estimator. In the next section, we discuss potential estimators for sparse covariance matrices.

2.2 Sparse Covariance Matrix Estimation

In recent years, covariance matrix estimation in high-dimensional spaces has received a lot of attention. In the case of sparse covariance matrices, banding, tapering, and thresholding can be applied, see e.g. [24] All three make use of the fact that many entries of the matrix that shall be estimated are actually zero or at least very small. Banding and tapering differ from thresholding in that they assume a specific matrix structure in other words they assume an ordering of the variables which is for instance often the case in time-series analysis. Banding and tapering approaches typically lead to consistent estimators if $\log(N)/\mu \to 0$.

Thresholding does not assume a natural order of the variables. Instead, it discards entries which are smaller than a given threshold $\varepsilon > 0$. For a matrix $\mathbf{A}$, the thresholding operator $T_\varepsilon(\mathbf{A})$ is defined as

$$T_\varepsilon(\mathbf{A}) := (a_{ij}\delta(|a_{ij}| \geq \varepsilon))_{N \times N}. \tag{11}$$

The choice of the threshold is critical for the quality of the resulting estimate.

Equation (11) represents a example of universal thresholding with a hard thresholding function. Equation (11) can be extended in several ways. On the one hand, the threshold may depend on the entry itself, and on the other hand, instead of the hard threshold applied, a generalized thresholding function $s_\lambda(\cdot)$ can be used. Following [5], the function $s_\lambda(\cdot)$ should have the following properties

(i) $\exists c > 0$: $s_\lambda(x) \leq c|y| \; \forall x, y$ which satisfy $|x - y| \leq \lambda$,
(ii) $s_\lambda(x) = 0 \; \forall x \leq \lambda$,
(iii) $|s_\lambda(x) - x| \leq \lambda \; \forall x \in \mathbb{R}$.

Several functions have been introduced that fulfill i)-iii), as e.g. the soft-thresholding

$$s_\lambda(x) = \text{sign}(x)(|x| - \lambda)_+ \tag{12}$$

or the Lasso

$$s_\lambda(x) = |x| \left(1 - |\frac{\lambda}{x}|^\eta\right)_+ \tag{13}$$

with $(x)_+ := \max(0, x)$. In this paper, the threshold λ_{ij} is defined component-wise and not universal. Since its correct choice is difficult to decide a priori, adaptive thresholding is applied as in [5], setting

$$\lambda_{ij} := \lambda_{ij}(\delta) = \delta \sqrt{\frac{\hat{\theta}_{ij} \log N}{\mu}} \tag{14}$$

with $\delta > 0$ can be either chosen as a constant or adapted data driven. The variable $\hat{\theta}_{ij}$ that appears in (14) is obtained as

$$\hat{\theta}_{ij} = \frac{1}{\mu} \sum_{m=1}^{\mu} [(\hat{z}_{mi} - \overline{Z^i})((\hat{z}_{mj} - \overline{Z^j}) - \hat{c}_{ij}^\mu]^2 \tag{15}$$

with $\hat{c}_{ij}^\mu$ denoting the (i,j)-entry of $\hat{\mathbf{C}}_\mu^{(g+1)}$, $\hat{z}_{mi}$ the ith component of $\hat{\mathbf{z}}_{m:\lambda}$, and $\overline{Z^i} := (1/\mu) \sum_{m=1}^{\mu} \hat{z}_{mi}$. Other thresholds have been introduced, see e.g. [9] and will be considered in future work.

While thresholding respects symmetry and non-negativeness properties, it results only in asymptotically positive definite matrices. Thus, for finite sample sizes, it does neither preserve nor induce positive definiteness in general. This holds for hard thresholding as well as for most cases of potential thresholding functions. As shown in [14], a positive semi-definiteness can only be guaranteed for a small class of functions for general matrices. In the case that the condition number of the matrix is sufficiently small, the group of functions that preserve positive definiteness can be widened to include also polynomials. In [9], procedures are discussed that result in positive definite matrices. As this paper aims for a proof of concept, it does not consider repair mechanisms.

2.3   Evolution Strategies with Sparse Covariance Adaptation

Component-wise adaptive thresholding can be integrated readily into evolution strategies. Figure 1 illustrates the main points of the algorithm. There are several ways to design the operator $T_{S_{\lambda_{ij}}}$. The first choice concerns the thresholding function $s_{\lambda_{ij}}(\cdot)$. The second question concerns whether thresholding should be applied to all entries of the covariance matrix (11) or only to the off-diagonal elements. This question is difficult to decide beforehand in the application context considered. Therefore, two variants are investigated

1. CMSA-Thres-ES (abbreviated to Thres): An evolution strategy with CMSA which applies thresholding in the eigenspace of the covariance, using the operator

$$T_{S_{\lambda_{ij}}}(\mathbf{A})_{ij} = s_{\lambda_{ij}}(a_{ij}) \tag{16}$$

and
2. CMSA-Diag-ES (abbreviated to Diag): An ES with covariance matrix adaptation which uses thresholding in the eigenspace of the covariance and excepts the diagonal elements with

$$T_{S_{\lambda_{ij}}}(\mathbf{A})_{ij} = \begin{cases} a_{ij} & \text{if } i = j \\ s_{\lambda_{ij}}(a_{ij}) & \text{if } i \neq j \end{cases}. \tag{17}$$

Fig. 1 The CMSA-ES with thresholding. The generation counter g is sometimes left out in order to simplify the notation. The symbol *spectral* stands for the spectral decomposition of the matrix into the modal matrix **V** and the diagonal matrix containing the eigenvalues **D**. Rank-based deterministic selection of the μ best offspring is performed in line 8 based on the fitness f

Require: $\lambda, \mu, \mathbf{C}^{(0)}, \mathbf{m}^{(0)}, \sigma^{(0)}, \tau, c_\tau$
1: $g = 0$
2: **while** termination criteria not met **do**
3:   **for** $l = 1$ **to** λ **do**
4:     $\sigma_l = \sigma^{(g)} \exp(\tau \mathcal{N}(0,1))$
5:     $\mathbf{x}_l = \mathbf{m}^{(g)} + \sigma_l \mathcal{N}(0, \mathbf{C}^{(g)})$
6:     $f_l = f(\mathbf{x}_l)$
7:   **end for**
8:   Select $(\mathbf{x}_{1:\lambda}, \sigma_{1:\lambda}), \ldots, (\mathbf{x}_{\mu:\lambda}, \sigma_{\mu:\lambda})$
9:   $\mathbf{m}^{(g+1)} = \sum_{m=1}^{\mu} w_m \mathbf{x}_{m:\lambda}$
10:   $\sigma^{(g+1)} = \sum_{m=1}^{\mu} w_m \sigma_{m:\lambda}$
11:   $\mathbf{z}_{m:\lambda} = \frac{\mathbf{x}_{m:\lambda} - \mathbf{m}^{(g)}}{\sigma^{(g)}}$ for $m = 1, \ldots, \mu$
12:   $\mathbf{V}, \mathbf{D} \leftarrow \text{spectral}(\mathbf{C}^{(g)})$
13:   $\hat{\mathbf{z}}_{m:\lambda} = \mathbf{V}^{\mathsf{T}} \mathbf{z}_{m:\lambda}$ for $m = 1, \ldots, \mu$
14:   $\hat{\mathbf{C}}_\mu = \sum_{m=1}^{\mu} w_m \hat{\mathbf{z}}_{m:\lambda} \hat{\mathbf{z}}_{m:\lambda}^{\mathsf{T}}$
15:   $\hat{\mathbf{C}}_{\text{thres}} = T_{S_{\lambda_{ij}}}(\hat{\mathbf{C}}_\mu)$
16:   $\mathbf{C}_\mu = \mathbf{V}^{\mathsf{T}} \hat{\mathbf{C}}_{\text{thres}} \mathbf{V}$
17:   $\mathbf{C}^{(g+1)} = (1 - \frac{1}{c_\tau}) \mathbf{C}^{(g)} + \frac{1}{c_\tau} \mathbf{C}_\mu$
18:   $g = g + 1$
19: **end while**

In statistics, thresholding is often applied only to the off-diagonal entries. Keeping the diagonal unchanged may however result in a too strong reliance on the structure of the old covariance matrix in our case. This may make a change of the search directions difficult. Therefore, both variants are taken into account.

3 Experiments

The experiments are performed for the search space dimensions $N = 10$ and 20. Since we aim for a general approach, the performance of the new techniques should also be analyzed for lower dimensional spaces. The maximal number of fitness evaluations is set to $FE_{max} = 2 \times 10^5 N$. The start position of the algorithms is randomly chosen from $[-4, 4]^N$. The population size were chosen as $\lambda = \lfloor \log(3N) + 8 \rfloor$ and $\mu = \lceil \lambda/2 \rceil$. The weights w_m were set to $w_m = 1/\mu$.

A run terminates before reaching the maximal number of evaluations, if the difference between the best value obtained so far and the optimal fitness value $|f_{best} - f_{opt}|$ is below a predefined target precision set to 10^{-8}. For each fitness function and dimension, 15 runs are used in accordance to the practice of the black box optimization workshops, see below. If the search stagnates, indicated by changes of the best values being below 10^{-8} for $10 + \lceil 30N/\lambda \rceil$ generations, the ES is restarted. The Lasso thresholding function (13) with $\eta = 4$ was chosen as the thresholding function.

An open point concerns the question of choosing the parameter δ in (15). Therefore, a preliminary series of experiments was conducted.

3.1 Test Suite

For the experiments, the algorithms were implemented in MATLAB. The paper uses black box optimization benchmarking (BBOB) software framework and the test suite introduced for the black box optimization workshops, see [16]. The goal of the workshop is to benchmark and to compare metaheuristics and other direct search methods for continuous optimization. The framework[1] allows the plug-in of algorithms adhering to a common interface and provides a comfortable way of generating the results in form of tables and figures.

The test suite contains noisy and noise-less functions with the position of the optimum changing randomly from run to run. This paper focuses on the noise-less test suite which contains 24 functions [11]. They can be divided into four classes: separable functions (function ids 1–5), functions with low/moderate conditioning (ids 6–9), functions with high conditioning (ids 10–14), and two groups of multimodal functions (ids 15–24). Among the unimodal functions with only one optimal point,

---

[1]Latest version under http://coco.gforge.inria.fr.

Table 1 Some of the test functions used for the comparison of the algorithms

Sphere	$f(\mathbf{x}) = \|\mathbf{z}\|^2$
Rosenbrock	$f(\mathbf{x}) = \sum_{i=1}^{N-1} 200(z_i^2 - z_{i+1})^2 + (z_i - 1)^2$
Ellipsoidal	$f(\mathbf{x}) = \sum_{i=1}^{N} 10^{6 \frac{i-1}{N-1}} z_i^2$
Discus	$f(\mathbf{x}) = 10^6 z_1^2 + \sum_{i=2}^{N} z_i^2$
Rastrigin	$f(\mathbf{x}) = 10\big(N - \sum_{i=1}^{N} \cos(2\pi z_i)\big) + \|\mathbf{z}\|^2$

there are separable functions given by the general formula

$$f(\mathbf{x}) = \sum_{i=1}^{N} f_i(x_i) \qquad (18)$$

which can be solved by optimizing each component separately. The simplest member of this class is the (quadratic) sphere with $f(\mathbf{x}) = \|\mathbf{x}\|^2$. Other functions include ill-conditioned functions, like for instance the elliposoidal function, and multimodal functions (Rastrigin) which represent particular challenges for the optimization (Table 1). The variable $\mathbf{z}$ denotes a transformation of $\mathbf{x}$ in order to keep the algorithm from exploiting certain particularities of the function, see [11].

3.2 Performance Measure

The following performance measure is used in accordance to [16]. The expected running time (ERT) gives the expected value of the function evaluations (f-evaluations) the algorithm needs to reach the target value with the required precision for the first time, see [16]. In this paper, we use

$$\text{ERT} = \frac{\#(FEs(f_{\text{best}} \geq f_{\text{target}}))}{\#succ} \qquad (19)$$

as an estimate by summing up the fitness evaluations $FEs(f_{\text{best}} \geq f_{\text{target}})$ of each run until the fitness of the best individual is smaller than the target value, divided by all successfull runs.

3.3 Choosing the Threshold: The Influence of the Scaling Factor

In order to gain more insights into the choice the scaling factor δ in (15) we start from the expression

$$\delta = \rho \max(\hat{\mathbf{C}}_\mu). \qquad (20)$$

The maximal entry of the population covariance was chosen to make the parameter data dependent and not too small. In future research, we will take a closer look at the choice above. We investigated seven values for the factor ρ with $\rho \in \{0.001, 0.01, 0.1, 1, 2, 5, 10\}$. The goal of this paper is to gain more insights into the question whether larger or smaller scalings are preferable. The experiments are conducted for $N = 20$. Of course, the search space dimensionality may additionally influence the parameter settings and will be investigated in future research. Finally, the decision of using the maximal entry of the population covariance matrix is reconsidered by investigating additionally a data-independent choice setting δ equal to ρ

$$\delta = \rho. \tag{21}$$

the first variant (20). The Tables 2, 3 and 4 show the results on selected functions of the test suite. For each function, 30 runs were conducted. The subset includes two separable functions, the sphere with id 1 and the ellipsoidal with id 2. followed by the attractive sector with id 6 and the rotated Rosenbrock (id 9) as representatives

Table 2 The results for different settings of the parameter ρ in (15) on selected functions for the CMSA-Thres-ES

Δf_{opt}	1e1	1e0	1e-1	1e-2	1e-3	1e-5	1e-7	#succ	Δf_{opt}	1e1	1e0	1e-1	1e-2	1e-3	1e-5	1e-7	#succ
f1	43	43	43	43	43	43	43	15/15	**f9**	1716	3102	3277	3379	3455	3594	3727	15/15
$\rho=0.001$	4.5(1.0)	8.9(1.0)	14(2)	19(2)	24(3)	34(5)	44(3)	30/30	$\rho=0.001$	14(9)	34(99)	35(7)	35(61)	35(94)	35(32)	34(34)	27/30
$\rho=0.01$	4.8(1)	8.9(2)	14(2)	19(2)	24(3)	34(2)	44(2)	30/30	$\rho=0.01$	16(5)	40(66)	41(63)	41(34)	41(59)	40(29)	40(29)	26/30
$\rho=0.1$	5.2(1)	10(2)	15(2)	20(2)	25(3)	35(4)	46(7)	30/30	$\rho=0.1$	15(8)	24(37)	26(7)	26(60)	26(6)	26(32)	26(7)	29/30
$\rho=1$	4.8(3)	9.1(5)	14(3)	19(3)	23(3)	33(2)	42(4)	30/30	$\rho=1$	16(8)	46(67)	46(35)	46(60)	46(3)	45(57)	44(56)	25/30
$\rho=2$	4.9(2)	9.3(3)	14(4)	18(4)	22(2)	32(2)	41(4)	30/30	$\rho=2$	16(7)	40(36)	41(7)	41(64)	41(60)	40(89)	39(32)	26/30
$\rho=5$	5.1(1)	9.2(3)	14(2)	19(4)	23(3)	32(3)	41(4)	30/30	$\rho=5$	17(7)	30(4)	31(3)	31(36)	32(33)	31(32)	31(31)	28/30
$\rho=10$	5.0(1)	9.1(2)	13(3)	18(3)	23(3)	32(3)	41(2)	30/30	$\rho=10$	18(8)	41(65)	41(62)	42(62)	41(5)	41(7)	40(29)	26/30

Δf_{opt}	1e1	1e0	1e-1	1e-2	1e-3	1e-5	1e-7	#succ	Δf_{opt}	1e1	1e0	1e-1	1e-2	1e-3	1e-5	1e-7	#succ
f2	385	386	387	388	390	391	393	15/15	**f10**	7413	8661	10735	13641	14920	17073	17476	15/15
$\rho=0.001$	165(44)	222(38)	255(39)	271(24)	283(24)	291(26)		30/30	$\rho=0.001$	8.3(3)	10(2)	9.3(1.0)	7.7(0.7)	7.3(0.4)	6.6(1)	6.7(1)	30/30
$\rho=0.01$	161(44)	217(32)	246(39)	261(14)	267(38)	277(31)	286(25)	30/30	$\rho=0.01$	8.0(2)	9.4(2)	8.8(1)	7.5(0.7)	7.0(1)	6.4(0.7)	6.5(0.4)	30/30
$\rho=0.1$	150(46)	206(58)	240(40)	258(24)	266(38)	278(28)	286(46)	30/30	$\rho=0.1$	7.7(2)	9.2(2)	8.5(2)	7.0(1.0)	6.6(0.9)	6.1(0.7)	6.2(1)	30/30
$\rho=1$	164(41)	215(61)	245(44)	262(30)	269(24)	279(18)	287(22)	30/30	$\rho=1$	8.8(2)	10(2)	9.0(0.9)	7.5(0.8)	7.0(0.5)	6.5(0.6)	6.5(0.6)	30/30
$\rho=2$	177(48)	225(35)	260(31)	274(30)	280(24)	290(24)	297(55)	30/30	$\rho=2$	8.7(0.9)	10(3)	9.2(1)	7.6(0.9)	7.2(0.8)	6.5(0.9)	6.6(0.7)	30/30
$\rho=5$	166(61)	225(79)	252(76)	269(55)	284(28)	296(29)	304(22)	30/30	$\rho=5$	8.5(2)	10(3)	9.2(2)	7.6(0.7)	7.2(1)	6.6(0.5)	6.7(0.7)	30/30
$\rho=10$	169(47)	227(64)	260(48)	273(34)	283(23)	293(28)	299(23)	30/30	$\rho=10$	8.7(3)	10(2)	9.4(1)	7.8(1)	7.4(0.4)	6.7(0.7)	6.7(0.6)	30/30

Δf_{opt}	1e1	1e0	1e-1	1e-2	1e-3	1e-5	1e-7	#succ	Δf_{opt}	1e1	1e0	1e-1	1e-2	1e-3	1e-5	1e-7	#succ
f6	1296	2343	3413	4255	5220	6728	8409	15/15	**f12**	1042	1938	2740	3156	4140	12407	13827	15/15
$\rho=0.001$	1.7(0.9)	2.5(1)	4.9(3)	14(50)	32(69)	113(112)	$\infty 4e5$	0/30	$\rho=0.001$	6.4(5)	12(14)	16(10)	18(12)	16(7)*	6.6(2)*	6.8(2)*	30/30
$\rho=0.01$	2.0(3)	3.4(1)	8.9(4)	23(4)	41(79)	261(279)	$\infty 4e5$	0/30	$\rho=0.01$	10(25)	19(14)	21(9)	22(7)	20(7)	8.0(3)	8.2(2)	30/30
$\rho=0.1$	1.7(0.8)	2.7(0.4)	4.6(4)	12(14)	28(75)	98(87)	332(329)	4/30	$\rho=0.1$	20(25)	24(19)	25(9)	26(10)	22(7)	8.9(2)	8.8(1)	30/30
$\rho=1$	2.0(0.5)	2.8(2)	3.7(4)	5.9(7)	10(15)	48(72)	176(285)	7/30	$\rho=1$	12(18)	23(22)	24(12)	25(11)	22(9)	8.9(2)	8.9(2)	30/30
$\rho=2$	4.0(6)	4.0(6)	5.0(3)	11(3)	17(35)	48(71)	455(506)	3/30	$\rho=2$	19(28)	27(20)	30(7)	30(8)	25(10)	10(4)	10(3)	30/30
$\rho=5$	2.3(4)	3.8(5)	4.2(3)	7.2(20)	15(42)	44(46)	315(286)	4/30	$\rho=5$	10(0.1)	20(22)	25(14)	27(6)	23(6)	9.2(2)	9.1(2)	30/30
$\rho=10$	2.3(3)	4.2(2)	4.8(2)	11(20)	26(29)	71(147)	257(173)	5/30	$\rho=10$	11(24)	26(20)	27(15)	27(18)	24(10)	10(2)	9.5(1)	30/30

Expected running time (ERT in number of function evaluations) divided by the respective best ERT measured during BBOB-2009 in dimension 20. The ERT and in braces, as dispersion measure, the half difference between 90 and 10 %-tile of bootstrapped run lengths appear for each algorithm and target, the corresponding best ERT in the first row. The different target Δf_{best}-values are shown in the top row. #succ is the number of trials that reached the (final) target $f_{opt} + 10^{-8}$. The median number of conducted function evaluations is additionally given in *italics*, if the target in the last column was never reached. Entries, succeeded by a star, are statistically significantly better (according to the rank-sum test) when compared to all other algorithms of the table, with $p = 0.05$ or $p = 10^{-k}$ when the number k following the star is larger than 1, with Bonferroni correction by the number of instances

Table 3 The results for different settings of the parameter ρ in (15) on selected functions for the CMSA-Diag-ES

Δf_{opt}	1e1	1e0	1e-1	1e-2	1e-3	1e-5	1e-7	#succ	Δf_{opt}	1e1	1e0	1e-1	1e-2	1e-3	1e-5	1e-7	#succ
f1	43	43	43	43	43	43	43	15/15	f9	1716	3102	3277	3379	3455	3594	3727	15/15
ρ=0.001	4.5(0.7)	8.9(1)	14(2)	19(2)	24(3)	34(4)	44(6)	30/30	ρ=0.001	14(5)	34(40)	35(66)	35(7)	35(33)	35(7)	34(55)	27/30
ρ=0.01	4.8(2)	8.9(3)	14(1)	19(2)	24(4)	34(3)	44(4)	30/30	ρ=0.01	16(4)	40(65)	41(33)	41(31)	41(32)	40(57)	40(5)	26/30
ρ=0.1	5.2(1)	10(2)	15(3)	20(3)	25(4)	35(3)	46(4)	30/30	ρ=0.1	15(4)	24(6)	26(6)	26(60)	26(5)	26(8)	26(5)	29/30
ρ=1	4.8(3)	9.1(3)	14(3)	19(3)	23(4)	33(5)	42(2)	30/30	ρ=1	16(5)	46(66)	46(62)	46(92)	46(63)	45(56)	44(56)	25/30
ρ=2	4.9(2)	9.3(2)	14(3)	18(4)	22(4)	32(2)	41(4)	30/30	ρ=2	16(10)	40(36)	41(5)	41(4)	41(31)	40(56)	39(56)	26/30
ρ=5	5.1(2)	9.2(1)	14(2)	19(3)	23(5)	32(4)	41(3)	30/30	ρ=5	17(8)	30(6)	31(8)	31(4)	32(3)	31(33)	31(55)	28/30
ρ=10	5.0(2)	9.1(2)	13(4)	18(3)	23(2)	32(3)	41(3)	30/30	ρ=10	18(6)	41(37)	41(34)	42(34)	41(7)	41(34)	40(56)	26/30
Δf_{opt}	1e1	1e0	1e-1	1e-2	1e-3	1e-5	1e-7	#succ	Δf_{opt}	1e1	1e0	1e-1	1e-2	1e-3	1e-5	1e-7	#succ
f2	385	386	387	388	390	391	393	15/15	f10	7413	8661	10735	13641	14920	17073	17476	15/15
ρ=0.001	165(58)	222(41)	255(32)	265(35)	271(33)	283(25)	291(32)	30/30	ρ=0.001	8.3(3)	10(2)	9.3(2)	7.7(1)	7.3(0.7)	6.6(0.7)	6.7(0.4)	30/30
ρ=0.01	161(53)	217(41)	246(42)	261(25)	267(42)	277(32)	286(38)	30/30	ρ=0.01	8.0(2)	9.4(2)	8.8(2)	7.5(1)	7.0(0.9)	6.4(0.8)	6.5(0.8)	30/30
ρ=0.1	150(70)	206(75)	240(26)	258(51)	266(22)	278(50)	286(30)	30/30	ρ=0.1	7.7(1)	9.2(3)	8.5(2)	7.0(1)	6.6(0.9)	6.1(0.8)	6.2(1)	30/30
ρ=1	164(62)	215(43)	245(40)	262(18)	269(25)	279(20)	287(18)	30/30	ρ=1	8.8(1)	10(2)	9.0(0.8)	7.5(1.0)	7.0(0.7)	6.5(0.6)	6.5(0.7)	30/30
ρ=2	177(38)	225(45)	260(32)	274(32)	280(21)	290(31)	297(51)	30/30	ρ=2	8.7(3)	10(2)	9.2(1)	7.6(0.8)	7.2(0.9)	6.5(1)	6.6(0.7)	30/30
ρ=5	166(60)	225(66)	252(64)	269(54)	284(33)	296(43)	304(34)	30/30	ρ=5	8.5(3)	10(3)	9.2(1)	7.6(2)	7.2(1)	6.6(0.6)	6.7(1)	30/30
ρ=10	169(48)	227(53)	260(38)	273(22)	283(25)	293(16)	299(15)	30/30	ρ=10	8.7(2)	10(2)	9.4(1)	7.8(0.8)	7.4(0.7)	6.7(0.7)	6.7(0.9)	30/30
Δf_{opt}	1e1	1e0	1e-1	1e-2	1e-3	1e-5	1e-7	#succ	Δf_{opt}	1e1	1e0	1e-1	1e-2	1e-3	1e-5	1e-7	#succ
f6	1296	2343	3413	4255	5220	6728	8409	15/15	f12	1042	1938	2740	3156	4140	12407	13827	15/15
ρ=0.001	1.7(0.9)	2.5(2)	4.9(6)	14(31)	32(40)	113(76)	∞4e5	0/30	ρ=0.001	6.4(13)	12(13)	16(11)	18(10)	16(8)*	6.6(3)*	6.8(2)*	30/30
ρ=0.01	2.0(0.5)	3.4(7)	8.9(29)	23(11)	41(102)	261(418)	∞4e5	0/30	ρ=0.01	10(23)	19(14)	21(9)	22(11)	20(7)	8.0(4)	8.2(2)	30/30
ρ=0.1	1.7(1)	2.7(4)	4.6(5)	12(4)	28(64)	98(171)	332(449)	4/30	ρ=0.1	20(29)	24(20)	25(15)	26(9)	22(6)	8.9(3)	8.8(2)	30/30
ρ=1	2.0(2)	2.8(3)	3.7(0.8)	5.9(8)	10(40)	48(100)	176(231)	7/30	ρ=1	12(10)	23(17)	24(15)	25(14)	22(7)	8.9(2)	8.9(2)	30/30
ρ=2	4.0(2)	4.0(2)	5.0(2)	11(33)	17(12)	48(101)	455(714)	3/30	ρ=2	19(29)	27(22)	30(7)	30(11)	25(3)	10(2)	10(3)	30/30
ρ=5	2.3(1.0)	3.8(5)	4.2(5)	7.2(25)	15(41)	44(61)	315(202)	4/30	ρ=5	10(14)	20(26)	25(6)	27(10)	23(7)	9.2(2)	9.1(3)	30/30
ρ=10	2.3(0.7)	4.2(6)	4.8(4)	11(19)	26(50)	71(136)	257(367)	5/30	ρ=10	11(0.1)	26(20)	27(13)	27(12)	24(5)	10(4)	9.5(3)	30/30

Please refer to Table 2 for more information

Table 4 The results for different settings of the parameter ρ in (21) on selected functions for the CMSA-Diag-ES

Δf_{opt}	1e1	1e0	1e-1	1e-2	1e-3	1e-5	1e-7	#succ	Δf_{opt}	1e1	1e0	1e-1	1e-2	1e-3	1e-5	1e-7	#succ
f1	43	43	43	43	43	43	43	15/15	f9	1716	3102	3277	3379	3455	3594	3727	15/15
ρ=0.001	5.0(2)	9.4(2)	15(3)	19(2)	24(3)	35(4)	45(5)	30/30	ρ=0.001	16(4)	67(70)	67(155)	66(61)	65(33)	64(90)	62(140)	22/30
ρ=0.01	4.5(1)	9.2(2)	14(2)	19(2)	24(4)	34(5)	44(3)	30/30	ρ=0.01	14(8)	33(98)	34(7)	35(6)	35(35)	34(32)	34(6)	27/30
ρ=0.1	4.5(0.9)	9.5(2)	14(2)	19(2)	24(4)	34(4)	44(5)	30/30	ρ=0.1	16(7)	41(4)	42(63)	42(34)	42(32)	41(3)	40(3)	26/30
ρ=1	5.5(1)	10(2)	15(2)	20(2)	25(2)	34(4)	44(4)	30/30	ρ=1	22(7)	76(66)	77(72)	77(96)	77(87)	75(86)	73(81)	22/30
ρ=2	5.0(1)	9.4(1)	14(2)	19(2)	23(3)	32(3)	42(4)	30/30	ρ=2	46(39)	97(18)	148(92)	3449(4143)	3373(2489)	∞	∞4e5	0/30
ρ=5	4.5(2)	9.0(1)	14(2)	19(1)	23(2)	33(3)	42(4)	30/30	ρ=5	54(26)	126(75)	183(149)	3509(4202)	∞	∞	∞4e5	0/30
ρ=10	4.6(0.9)	9.0(3)	14(2)	19(2)	23(2)	33(2)	43(3)	30/30	ρ=10	42(12)	81(7)	135(125)	3451(1923)	3456(6223)	∞	∞4e5	0/30
Δf_{opt}	1e1	1e0	1e-1	1e-2	1e-3	1e-5	1e-7	#succ	Δf_{opt}	1e1	1e0	1e-1	1e-2	1e-3	1e-5	1e-7	#succ
f2	385	386	387	388	390	391	393	15/15	f10	7413	8661	10735	13641	14920	17073	17476	15/15
ρ=0.001	149(43)	219(59)	246(64)	261(26)	269(34)	282(53)	291(44)	30/30	ρ=0.001	8.3(2)	9.3(2)	8.7(2)	7.3(1)	7.0(0.8)	6.3(0.9)	6.4(0.9)	30/30
ρ=0.01	156(32)	209(70)	249(48)	260(26)	269(30)	279(21)	288(32)	30/30	ρ=0.01	8.5(4)	10(3)	8.8(2)	7.6(1)	7.2(1)	6.5(0.6)	6.6(0.8)	30/30
ρ=0.1	182(53)	247(65)	269(26)	276(28)	281(23)	291(31)	300(41)	30/30	ρ=0.1	9.1(4)	10(2)	9.1(2)	7.5(1)	7.1(0.9)	6.5(0.6)	6.5(0.7)	30/30
ρ=1	239(92)	349(70)	399(56)	419(41)	430(47)	449(37)	459(33)	30/30	ρ=1	12(4)	15(4)	14(3)	12(1)	11(2)	10(1)	10(1)	30/30
ρ=2	95(11)	113(20)	121(16)	124(10)	125(11)	127(14)	128(7)	30/30	ρ=2	∞	∞	∞	∞	∞	∞	∞4e5	0/30
ρ=5	96(16)	115(15)	122(10)	124(11)	126(9)	129(7)	129(6)	30/30	ρ=5	∞	∞	∞	∞	∞	∞	∞4e5	0/30
ρ=10	93(16)	110(17)	121(17)	125(17)	126(15)	129(13)	130(14)	30/30	ρ=10	∞	∞	∞	∞	∞	∞	∞4e5	0/30
Δf_{opt}	1e1	1e0	1e-1	1e-2	1e-3	1e-5	1e-7	#succ	Δf_{opt}	1e1	1e0	1e-1	1e-2	1e-3	1e-5	1e-7	#succ
f6	1296	2343	3413	4255	5220	6728	8409	15/15	f12	1042	1938	2740	3156	4140	12407	13827	15/15
ρ=0.001	1.6(0.5)	3.1(13)	7.8(25)	15(8)	34(47)	114(88)	1393(880)	1/30	ρ=0.001	6.4(7)	12(12)	15(11)	17(10)	16(7)	6.7(2)	6.9(2)	30/30
ρ=0.01	1.2(0.5)	2.0(1)	4.2(4)	15(18)	35(41)	143(105)	∞4e5	0/30	ρ=0.01	4.7(7)	11(5)	14(12)	16(7)	15(10)	6.2(2)	6.5(2)	30/30
ρ=0.1	1.8(0.9)	3.1(5)	6.0(8)	16(13)	30(59)	105(94)	261(440)	5/30	ρ=0.1	6.5(16)	12(19)	15(14)	16(10)	15(8)	6.5(4)	6.8(3)	30/30
ρ=1	1.5(0.7)	2.0(0.8)	2.5(0.7)	3.5(5)	5.3(7)	10(14)*4	19(32)*	28/30	ρ=1	14(52)	28(23)	31(27)	33(23)	30(19)	12(6)	13(5)	30/30
ρ=2	3.4(16)	36(91)	59(117)	77(79)	120(165)	124(117)	207(159)	6/30	ρ=2	118(288)	271(361)	950(1971)	775(2979)	∞	∞	∞4e5	0/30
ρ=5	16(54)	50(144)	98(69)	135(194)	146(312)	213(183)	326(209)	4/30	ρ=5	78(0.2)	182(413)	480(766)	635(602)	2802(2947)	935(935)	∞4e5	0/30
ρ=10	8.1(2)	29(18)	70(147)	92(128)	156(122)	164(157)	184(150)	7/30	ρ=10	97(288)	237(206)	950(1569)	775(1553)	∞	∞	∞4e5	0/30

Please refer to Table 2 for more information

for functions with moderate conditioning. The group of ill-condition functions is represented by another ellipsoidal with id 10 and the bent cigar (id 12). The parameter investigation does not consider the multimodal functions since our experiments showed that these pose challenges for the algorithms especially for larger search space dimensionalities. The first series of experiments considers CMSA-ESs which apply (15). Table 2 provides the findings for the variant which also subjects the diagonal entries of the covariance matrix to thresholding. It reveals different experimental findings for the functions. In the case of the sphere, the choice of the scaling factor does not affect the outcome considerably. This changes when the other functions are considered. In the case of the elliposoidal with id 2, the size of the parameter does not influence whether the optimization is successfull or not. However, it influences the number of fitness functions necessary to reach the intermediate target precisions. In the case of the elliposoidal, choices between 0.01 and 1 lead to lower values. Larger and smaller values lead to a gradual loss of performance. The case of the attractive sector (id 6) shows that the choice of the scaling factor may strongly influences the optimization success. Before continuing,it should be taken into account, however, that in all cases the number of successfull outcomes is less that 30% of all runs. Therefore, it cannot be excluded that initialization effects may have played a role. For this reason, experiments with more repeats will be conducted in future work. Here, we can state that too small choices do not allow the ES to achieve the final target precision. Otherwise, successfull runs are recorded. However, no clear dependency of the performance on the factor ρ emerges. Regarding the experiment series present, the factor $\rho = 1$ should be preferred. The remaining functions of the subset do not pose the same challenge for the ES as the attractive sector. The majority of the runs leads to successfull outcomes. In the case of the Rosenbrock function, setting $\rho = 0.1$ is beneficial for the number of successes as well as for the number of fitness evaluations required of reaching the various target precisions. The findings roughly transfer to the next function, the ill-conditioned elliposoidal. Here, however, the outcomes do not vary as much as previously. The bent cigar represents a outlier of the function set chosen. In contrast to the other functions, the ESs with low scaling factors ρ perform better. Since the factor is coupled with the maximal entry of the covariance matrix, two potential explanations can be given: First, in some cases, thresholding should be conducted only for few values. Or, second, the covariance matrix estimated contains extreme elements which require a reconsideration of choosing (15). To summarize the findings so far: Relatively robust choices of the scaling factor are presented by medium-sized values which lie between 0.1 and 1 for the experiments conducted so far. This, however, may not be the optimal choice for all functions of the test suite. Future research will therefore investigate adaptive techniques. The findings transfer to the ES variant that does not apply thresholding to the diagonal entries of the covariance matrix as Table 3 shows.

Table 4 reports the results for the data-independent scaling factor setting $\delta = \rho$, see (21). Here, experiments were only conducted for the CMSA-Diag-ES. The results vary strongly over the subset of fitness functions. In the case of the separable functions, the performance can be influenced by the choice of scaling factor. Again, this is more pronounced for the elliposoidal than for the sphere. Interestingly, in the case of the remaining functions, a correct choice of the factor is decisive leading to successfull runs for all 30 trials or to a complete miss of the final target precision. In general, larger values are detrimental and smaller values achieve better outcomes. The exception is the attractive sector (id 6) where a scaling factor of one leads to successes in nearly all trials. This setting cannot be transferred to the other functions. As we have seen, choosing the scaling factor represents an important point in thresholding. Therefore, future research will focus on potential adaptation approaches.

For the remainder of the paper, the data-dependend version (15) is used. The first series of experiments indicated that the values of $\rho = 0.1$ and $\rho = 1$ lead to comparably good results, therefore the parameter was set to 0.5 for the evaluation experiments with the black box optimization benchmarking test suite. The results are provided in the next subsection.

3.4 Results and Discussion

The findings are interesting—indicating advantages for thresholding in many but not in all cases. The result of the comparison depends on the function class. In the case of the separable functions with ids 1–5, the strategies behave on the whole very similar in the case of both dimensionalities 10D and 20D. This can be seen in the empirical cumulative distribution functions plots in Figs. 2 and 3 for example.

Concerning the particular functions, differences are revealed as Tables 5 and 6 show for the expected running time (ERT) which is provided for several precision targets. The expected running time is provided relative to the best results achieved during the black-box optimization workshop in 2009. The first line of the outcomes for each function reports the ERT of the best algorithm of 2009. However, not only the ERT values but also the number of successes is important. The ERT can only be measured if the algorithm achieved the respective target in the run. If the number of trials where is the full optimization objective has been reached is low then the remaining targets should be discussed with care. If only a few runs contribute to the result, the findings may be strongly influenced by initialization effects. To summarize, only a few cases end with differences that are statistically significant. To achieve this, the algorithm has to perform significantly better than both competing methods—the other thresholding variant and the original CMSA-ES.

In the case of the sphere (function with id 1), slight advantages for the thresholding variants are revealed. A similar observation can be made for the second function, the separable ellipsoid. Here, both thresholded ESs are faster, with the one that only shrinks the off-diagonal elements more strongly (Table 6). This is probably due to the enforced more regular structure.

No strategy is able to reach the required target precision in the case of the separable Rastrigin (id 3) and the separable Rastrigin-Bueche (id 4). Since all strategies only achieve the lowest target precision of 10^1, a comparison is not performed. The linear slope is solved fast by all, with the original CMSA-ES the best strategy.

In the case of the function class containing test functions with low to moderate conditioning (ids 6–9), different findings can be made for the two search space dimensionalities. This is also shown by the empirical cumulative distribution functions plots in Figs. 2 and 3, especially for $N = 10$. Also in the case of $N = 10$, Table 5 shows that the strategies with thresholding achieve a better performance in a majority of cases. In addition, thresholding that is not applied to the diagonal appears to lead

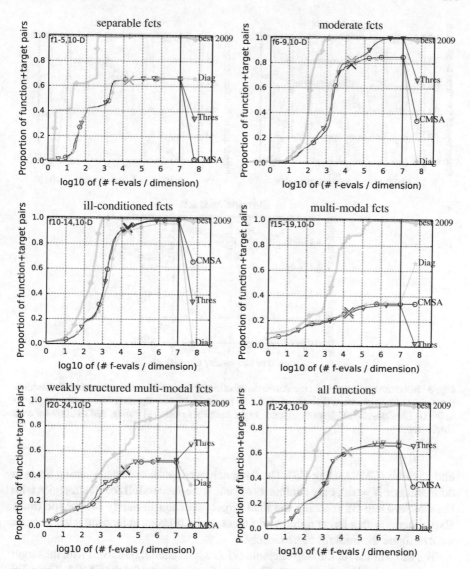

Fig. 2 Bootstrapped empirical cumulative distribution of the number of objective function evaluations divided by dimension (FEvals/DIM) for 50 targets in $10^{[-8,2]}$ for all functions and subgroups in 10-D. The "best 2009" *line* corresponds to the best ERT observed during BBOB 2009 for each single target

to a well-performing strategy with the exception of f6, where the CMSA-Diag-ES appears highly successfull.

The results for f6, the so-called attractive sector, in 10D are astonishing. While the original CMSA-ES could only reach the required target precision in six of the 15 runs, the thresholding variants were able to succeed 13 times (CMSA-Thres-ES)

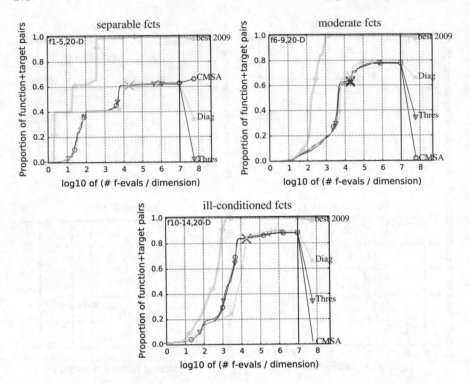

Fig. 3 Bootstrapped empirical cumulative distribution of the number of objective function evaluations divided by dimension (FEvals/DIM) for 50 targets in $10^{[-8.2]}$ for all functions and subgroups in 20-D. The "best 2009" *line* corresponds to the best ERT observed during BBOB 2009 for each single target

and 15 times (CMSA-Diag-ES). The latter achieved lower expected running times in addition. For $N = 20$, only a minority of runs were successfull for all strategies with the CMSA-Diag-ES reaching the final target precision in nearly half of the trials. Experiments with a larger number of fitness evaluations must be conducted in order to investigate the findings more closely.

The same holds for the step ellipsoid (id 7) which cannot be solved with the target precision required by any strategy. The exception is one run for the CMSA-Thres-ES which is able to reach the target precision of 10^{-8} in one case for $N = 10$. Since this may be due to the initialization, it is not taken into further consideration. Concerning the lower precision targets, sometimes the CMSA-ES and sometimes the CMSA-Diag-ES appears superior. However, more research is required, since the number of runs entering the data for some of the target precisions is low and initial positions may be influential.

On the original Rosenbrock function (id 8), the CMSA-ES and the strategies with thresholding show a similar behavior with the CMSA-ES performing better for the first intermediate target precision whereas the CMSA-Diag-ES shows slightly lower

Table 5 Expected running time (ERT in number of function evaluations) divided by the respective best ERT measured during BBOB-2009 in 10

Δf_opt	1e1	1e0	1e-1	1e-2	1e-3	1e-5	1e-7	#succ
f1	22	23	23	23	23	23	23	15/15
CMSA	4.0(1)	8.6(3)	14(3)	19(4)	26(5)	38(5)	50(6)	15/15
Thres	3.7(2)	8.6(2)	14(2)	20(4)	25(3)	37(4)	49(4)	15/15
Diag	4.3(2)	8.6(3)	14(4)	19(4)	25(3)	37(6)	48(5)	15/15

Δf_opt	1e1	1e0	1e-1	1e-2	1e-3	1e-5	1e-7	#succ
f13	387	596	797	1014	4587	6208	7779	15/15
CMSA	15(24)	19(42)	31(34)	58(65)	28(25)	89(131)	185(224)	2/15
Thres	5.5(5)	19(22)	34(25)	61(33)	35(47)	108(106)	∞2e5	0/15
Diag	8.2(17)	20(31)	59(37)	138(196)	87(81)	461(693)	∞2e5	0/15

Δf_opt	1e1	1e0	1e-1	1e-2	1e-3	1e-5	1e-7	#succ
f2	187	190	191	191	193	194	195	15/15
CMSA	65(27)	85(31)	96(28)	105(31)	109(27)	113(22)	129(53)	15/15
Thres	68(48)	88(26)	97(26)	105(26)	109(11)	116(39)	122(39)	15/15
Diag	83(37)	115(29)	135(27)	142(57)	149(23)	158(35)	166(60)	15/15

Δf_opt	1e1	1e0	1e-1	1e-2	1e-3	1e-5	1e-7	#succ
f14	37	98	133	205	392	687	4305	15/15
CMSA	1.1(0.6)	2.2(0.9)	2.8(0.7)	3.5(0.7)	4.3(0.9)	8.7(2)	4.8(3)	15/15
Thres	1.2(0.5)	2.4(2)	3.2(0.6)	3.7(0.8)	4.9(2)	8.8(4)	4.6(3)	15/15
Diag	1.0(0.6)	2.0(0.7)	2.4(0.6)	3.0(0.7)	3.9(2)	10(2)	5.7(3)	15/15

Δf_opt	1e1	1e0	1e-1	1e-2	1e-3	1e-5	1e-7	#succ
f3	1739	3600	3609	3636	3642	3646	3651	15/15
CMSA	20(12)	∞	∞	∞	∞	∞	∞2e5	0/15
Thres	27(29)	∞	∞	∞	∞	∞	∞2e5	0/15
Diag	21(40)	∞	∞	∞	∞	∞	∞2e5	0/15

Δf_opt	1e1	1e0	1e-1	1e-2	1e-3	1e-5	1e-7	#succ
f15	4774	39246	73643	74669	75790	77814	79834	12/15
CMSA	7.0(4)	∞	∞	∞	∞	∞	∞2e5	0/15
Thres	6.7(6)	∞	∞	∞	∞	∞	∞2e5	0/15
Diag	4.8(5)	∞	∞	∞	∞	∞	∞2e5	0/15

Δf_opt	1e1	1e0	1e-1	1e-2	1e-3	1e-5	1e-7	#succ
f4	2234	3626	3660	3695	3707	3744	28767	12/15
CMSA	60(62)	∞	∞	∞	∞	∞	∞2e5	0/15
Thres	123(104)	∞	∞	∞	∞	∞	∞2e5	0/15
Diag	69(45)	∞	∞	∞	∞	∞	∞2e5	0/15

Δf_opt	1e1	1e0	1e-1	1e-2	1e-3	1e-5	1e-7	#succ
f16	425	7029	15779	45669	51151	65798	71570	15/15
CMSA	1.0(0.7)	1.8(2)	13(17)	31(32)	∞	∞	∞2e5	0/15
Thres	1.4(1)	2.1(3)	19(31)	65(94)	∞	∞	∞2e5	0/15
Diag	2.3(5)	2.1(1)	19(16)	63(76)	∞	∞	∞2e5	0/15

Δf_opt	1e1	1e0	1e-1	1e-2	1e-3	1e-5	1e-7	#succ
f5	20	20	20	20	20	20	20	15/15
CMSA	12(7)	17(10)	17(5)	17(10)	17(7)	17(8)	17(9)	15/15
Thres	15(6)	19(5)	20(9)	20(10)	20(7)	20(10)	20(10)	15/15
Diag	14(7)	19(13)	20(12)	20(13)	20(11)	20(13)	20(12)	15/15

Δf_opt	1e1	1e0	1e-1	1e-2	1e-3	1e-5	1e-7	#succ
f17	26	429	2203	6329	9851	20190	26503	15/15
CMSA	0.71(0.5)	18(46)	23(38)	30(46)	140(221)	∞	∞2e5	0/15
Thres	1.3(0.9)	30(65)	29(41)	76(77)	∞	∞	∞2e5	0/15
Diag	1.1(2)	7.2(23)	29(29)	58(30)	∞	∞	∞2e5	0/15

Δf_opt	1e1	1e0	1e-1	1e-2	1e-3	1e-5	1e-7	#succ
f6	412	623	826	1039	1292	1841	2370	15/15
CMSA	1.4(0.2)	3.3(4)	11(3)	14(16)	19(17)	25(19)	163(188)	6/15
Thres	1.6(0.8)	2.2(0.6)	3.0(2)	3.9(4)	8.0(7)	12(13)	43(63)	13/15
Diag	1.2(0.5)	1.7(0.6)	2.3(0.5)	2.6(0.2)	3.4(1)	9.5(9)	13(10)	15/15

Δf_opt	1e1	1e0	1e-1	1e-2	1e-3	1e-5	1e-7	#succ
f18	238	836	7012	15928	27536	37234	42708	15/15
CMSA	68(0.9)	129(220)	124(115)	∞	∞	∞	∞2e5	0/15
Thres	162(409)	286(491)	∞	∞	∞	∞	∞2e5	0/15
Diag	1.1(0.7)	119(155)	400(628)	∞	∞	∞	∞2e5	0/15

Δf_opt	1e1	1e0	1e-1	1e-2	1e-3	1e-5	1e-7	#succ
f7	172	1611	5099	5141	5389			15/15
CMSA	4.0(4)	26(40)	85(142)	∞	∞	∞	∞2e5	0/15
Thres	2.1(2)	22(16)	678(1001)	558(696)	553(292)	553(924)	528(650)	1/15
Diag	3.3(2)	18(20)	208(203)	∞	∞	∞	∞2e5	0/15

Δf_opt	1e1	1e0	1e-1	1e-2	1e-3	1e-5	1e-7	#succ
f19	1	1	10609	9.8e5	1.4e6	1.4e6	1.4e6	15/15
CMSA	18(2)	1.5e5(1e5)	∞	∞	∞	∞	∞2e5	0/15
Thres	19(15)	8.2e4(7e4)	∞	∞	∞	∞	∞2e5	0/15
Diag	16(4)	8.7e4(1e5)	∞	∞	∞	∞	∞2e5	0/15

Δf_opt	1e1	1e0	1e-1	1e-2	1e-3	1e-5	1e-7	#succ
f8	326	921	1114	1217	1267	1315	1343	15/15
CMSA	3.3(5)	17(8)	18(8)	18(10)	18(3)	19(6)	19(5)	15/15
Thres	7.1(6)	20(12)	20(8)	19(5)	19(9)	19(13)	20(18)	15/15
Diag	9.3(15)	18(7)	18(7)	18(4)	18(5)	18(6)	18(5)	15/15

Δf_opt	1e1	1e0	1e-1	1e-2	1e-3	1e-5	1e-7	#succ
f20	32	15426	5.5e5	5.7e5	5.7e5	5.8e5	5.9e5	15/15
CMSA	1.9(2)	25(25)	∞	∞	∞	∞	∞2e5	0/15
Thres	1.7(0.9)	23(22)	∞	∞	∞	∞	∞2e5	0/15
Diag	2.1(1)	20(17)	∞	∞	∞	∞	∞2e5	0/15

Δf_opt	1e1	1e0	1e-1	1e-2	1e-3	1e-5	1e-7	#succ
f9	200	648	857	993	1065	1138	1185	15/15
CMSA	2.3(2)	25(13)	24(12)	22(7)	22(11)	21(7)	21(9)	15/15
Thres	2.5(0.5)	21(5)	20(5)	19(10)	19(5)	18(10)	18(8)	13/15
Diag	4.1(5)	26(14)	24(11)	22(8)	22(4)	21(11)	21(7)	15/15

Δf_opt	1e1	1e0	1e-1	1e-2	1e-3	1e-5	1e-7	#succ
f21	130	2236	4392	4487	4618	5074	11329	8/15
CMSA	9.5(21)	23(62)	20(14)	20(19)	19(61)	17(18)	7.8(11)	12/15
Thres	14(11)	15(10)	19(35)	19(34)	18(22)	17(25)	7.6(16)	12/15
Diag	2.9(7)	15(46)	13(23)	13(13)	13(11)	11(17)	5.2(5)	14/15

Δf_opt	1e1	1e0	1e-1	1e-2	1e-3	1e-5	1e-7	#succ
f10	1835	2172	2455	2728	2802	4543	4739	15/15
CMSA	6.5(4)	7.6(2)	7.7(3)	7.5(2)	7.6(3)	4.9(2)	4.9(2)	15/15
Thres	7.1(3)	8.6(3)	8.6(3)	8.3(3)	8.3(3)	5.5(3)	5.5(2)	15/15
Diag	10(3)	11(3)	10(3)	10(2)	10(2)	6.6(2)	6.6(2)	15/15

Δf_opt	1e1	1e0	1e-1	1e-2	1e-3	1e-5	1e-7	#succ
f22	98	2839	6353	6620	6798	8296	10351	6/15
CMSA	25(21)	6.3(7)	13(13)	12(13)	12(17)	10(12)	8.1(8)	13/15
Thres	30(91)	4.0(6)	8.7(17)	8.5(3)	8.4(14)	7.0(11)	5.8(7)	14/15
Diag	26(65)	8.9(12)	13(16)	12(16)	12(16)	10(24)	8.3(15)	12/15

Δf_opt	1e1	1e0	1e-1	1e-2	1e-3	1e-5	1e-7	#succ
f11	266	1041	2602	2954	3338	4092	4843	15/15
CMSA	14(5)	6.2(3)	3.2(1.0)	3.4(2)	3.5(2)	3.3(1)	3.0(1)	15/15
Thres	16(3)	6.0(2)	3.1(0.5)	3.1(0.9)	3.0(1)	2.9(1)	2.6(0.9)	15/15
Diag	28(15)	10(5)	4.5(3)	4.5(2)	4.3(2)	3.8(2)	3.4(1)	15/15

Δf_opt	1e1	1e0	1e-1	1e-2	1e-3	1e-5	1e-7	#succ
f23	2.8	915	16425	1.8e5	2.0e5	2.1e5	2.1e5	15/15
CMSA	1.5(2)	391(291)	∞	∞	∞	∞	∞2e5	0/15
Thres	3.1(2)	256(205)	∞	∞	∞	∞	∞2e5	0/15
Diag	1.5(1.0)	154(201)	∞	∞	∞	∞	∞2e5	0/15

Δf_opt	1e1	1e0	1e-1	1e-2	1e-3	1e-5	1e-7	#succ
f12	515	896	1240	1390	1569	3660	5154	15/15
CMSA	4.2(0.2)	10(10)	13(17)	15(9)	16(13)	8.8(5)	7.8(3)	15/15
Thres	4.3(0.2)	13(11)	14(12)	18(9)	19(10)	10(7)	8.2(5)	15/15
Diag	13(12)	18(23)	24(18)	28(14)	28(10)	14(5)	13(5)	15/15

Δf_opt	1e1	1e0	1e-1	1e-2	1e-3	1e-5	1e-7	#succ
f24	98761	1.0e6	7.5e7	7.5e7	7.5e7	7.5e7	7.5e7	1/15
CMSA	4.8(4)	∞	∞	∞	∞	∞	∞2e5	0/15
Thres	6.4(8)	∞	∞	∞	∞	∞	∞2e5	0/15
Diag	14(15)	∞	∞	∞	∞	∞	∞2e5	0/15

The ERT and in braces, as dispersion measure, the half difference between 90 and 10 %-tile of bootstrapped run lengths appear for each algorithm and target, the corresponding best ERT in the first row. The different target Δf-values are shown in the top row. #succ is the number of trials that reached the (final) target $f_{opt} + 10^{-8}$. The median number of conducted function evaluations is additionally given in *italics*, if the target in the last column was never reached. Entries, succeeded by a star, are statistically significantly better (according to the rank-sum test) when compared to all other algorithms of the table, with $p = 0.05$ or $p = 10^{-k}$ when the number k following the star is larger than 1, with Bonferroni correction by the number of instances

Table 6 Expected running time (ERT in number of function evaluations) divided by the respective best ERT measured during BBOB-2009 in 20

Δf_{opt}	1e1	1e0	1e-1	1e-2	1e-3	1e-5	1e-7	#succ
f1	43	43	43	43	43	43	43	15/15
CMSA	4.9(2)	10(1)	15(2)	19(2)	25(2)	34(3)	45(3)	15/15
Thres	**4.3**(0.8)	**8.8**(2)	14(2)	19(3)	23(2)	33(3)	44(2)	15/15
Diag	4.6(2)	9.0(1.0)	**13**(2)	**18**(3)	23(3)	33(3)	42(2)	15/15

Δf_{opt}	1e1	1e0	1e-1	1e-2	1e-3	1e-5	1e-7	#succ
f8	2039	3871	4040	4148	4219	4371	4484	15/15
CMSA	11(8)	**30**(56)	**31**(32)	**31**(54)	**31**(25)	**31**(28)	**31**(89)	13/15
Thres	18(3)	56(9)	56(53)	56(77)	55(74)	54(25)	53(29)	11/15
Diag	20(10)	58(76)	59(80)	60(27)	60(72)	59(48)	59(67)	11/15

Δf_{opt}	1e1	1e0	1e-1	1e-2	1e-3	1e-5	1e-7	#succ
f2	385	386	387	388	390	391	393	15/15
CMSA	173(33)	240(42)	265(37)	273(34)	277(45)	**285**(54)	**293**(49)	15/15
Thres	**155**(38)	**224**(41)	**254**(39)	**263**(37)	**271**(38)	286(20)	295(22)	15/15
Diag	442(48)	511(75)	540(73)	555(52)	557(77)	573(72)	580(60)	15/15

Δf_{opt}	1e1	1e0	1e-1	1e-2	1e-3	1e-5	1e-7	#succ
f9	1716	3102	3277	3379	3455	3594	3727	15/15
CMSA	17(6)	40(37)	41(95)	41(33)	41(5)	40(3)	40(30)	13/15
Thres	15(2)	40(2)	41(32)	41(90)	41(2)	40(59)	40(110)	13/15
Diag	22(9)	37(4)	**39**(62)	**39**(2)	**39**(2)	**39**(30)	**39**(3)	14/15

Δf_{opt}	1e1	1e0	1e-1	1e-2	1e-3	1e-5	1e-7	#succ
f3	5066	7626	7635	7637	7643	7646	7651	15/15
CMSA	∞	∞	∞	∞	∞	∞	∞4e5	0/15
Thres	∞	∞	∞	∞	∞	∞	∞4e5	0/15
Diag	∞	∞	∞	∞	∞	∞	∞4e5	0/15

Δf_{opt}	1e1	1e0	1e-1	1e-2	1e-3	1e-5	1e-7	#succ
f10	7413	8661	10735	13641	14920	17073	17476	15/15
CMSA	10(5)	11(4)	9.2(2)	7.8(2)	7.3(1)	6.7(0.6)	6.8(0.8)	15/15
Thres	**8.9**(3)	10(3)	**9.1**(1)	**7.7**(0.6)	**7.2**(0.7)	**6.6**(0.9)	**6.7**(0.6)	15/15
Diag	23(7)	23(3)	19(3)	16(3)	15(3)	13(2)	13(2)	15/15

Δf_{opt}	1e1	1e0	1e-1	1e-2	1e-3	1e-5	1e-7	#succ
f4	4722	7628	7666	7686	7700	7758	1.4e5	9/15
CMSA	∞	∞	∞	∞	∞	∞	∞4e5	0/15
Thres	∞	∞	∞	∞	∞	∞	∞4e5	0/15
Diag	∞	∞	∞	∞	∞	∞	∞4e5	0/15

Δf_{opt}	1e1	1e0	1e-1	1e-2	1e-3	1e-5	1e-7	#succ
f11	1002	2228	6278	8586	9762	12285	14831	15/15
CMSA	12(1.0)	7.5(1)	3.1(0.5)	2.6(0.5)	2.6(0.3)	2.5(0.6)	2.5(0.5)	15/15
Thres	13(2)	7.3(0.9)	2.9(0.4)	2.4(0.3)	2.3(0.3)	2.2(0.3)	2.1(0.3)	15/15
Diag	60(25)	38(12)	14(5)	11(4)	10(3)	8.2(3)	7.0(1)	15/15

Δf_{opt}	1e1	1e0	1e-1	1e-2	1e-3	1e-5	1e-7	#succ
f5	41	41	41	41	41	41	41	15/15
CMSA	12(4)	15(4)	15(4)	15(6)	15(5)	15(4)	15(5)	15/15
Thres	14(2)	19(10)	20(10)	20(9)	20(9)	20(10)	20(10)	15/15
Diag	13(3)	17(6)	17(5)	18(7)	18(7)	18(5)	18(6)	15/15

Δf_{opt}	1e1	1e0	1e-1	1e-2	1e-3	1e-5	1e-7	#succ
f12	1042	1938	2740	3156	4140	12407	13827	15/15
CMSA	2.5(4)	10(9)	13(5)	16(6)	14(5)	5.9(2)	6.2(2)	15/15
Thres	**1.4**(0.1)	**8.7**(13)	16(16)	18(13)	18(3)	7.2(2)	7.5(1)	15/15
Diag	47(59)	111(40)	101(11)	99(10)	80(5)	28(2)	26(1)	15/15

Δf_{opt}	1e1	1e0	1e-1	1e-2	1e-3	1e-5	1e-7	#succ
f6	1296	2343	3413	4255	5220	6728	8409	15/15
CMSA	**1.5**(1)	**2.5**(2)	4.6(7)	12(23)	34(40)	80(126)	331(292)	2/15
Thres	2.7(2)	3.1(3)	**4.3**(5)	**7.7**(12)	24(24)	123(107)	146(304)	4/15
Diag	2.5(1)	5.6(11)	7.8(14)	14(10)	18(26)	**34**(74)	**63**(119)	7/15

Δf_{opt}	1e1	1e0	1e-1	1e-2	1e-3	1e-5	1e-7	#succ
f13	652	2021	2751	3507	18749	24455	30201	15/15
CMSA	156(0.7)	545(841)	2037(2072)	∞	∞	∞	∞4e5	0/15
Thres	225(307)	397(841)	583(473)	1598(2196)	∞	∞	∞4e5	0/15
Diag	97(154)	227(445)	292(473)	742(342)	∞	∞	∞4e5	0/15

Δf_{opt}	1e1	1e0	1e-1	1e-2	1e-3	1e-5	1e-7	#succ
f7	1351	4274	9503	16523	16524	16524	16969	15/15
CMSA	622(1184)	∞	∞	∞	∞	∞	∞4e5	0/15
Thres	1285(1760)	∞	∞	∞	∞	∞	∞4e5	0/15
Diag	593(592)	∞	∞	∞	∞	∞	∞4e5	0/15

Δf_{opt}	1e1	1e0	1e-1	1e-2	1e-3	1e-5	1e-7	#succ
f14	75	239	304	451	932	1648	15661	15/15
CMSA	1.8(0.8)	1.9(0.5)	2.5(0.5)	3.2(0.5)	**5.2**(0.6)	11(2)	4.2(1)	15/15
Thres	1.9(1)	2.0(0.5)	2.5(1)	**3.1**(0.6)	6.0(1)	13(2)	**3.6**(0.9)	15/15
Diag	**1.8**(0.9)	**1.8**(0.6)	2.2(0.3)	3.2(0.7)	7.6(2)	46(12)	11(1)	15/15

The ERT and in braces, as dispersion measure, the half difference between 90 and 10 %-tile of bootstrapped run lengths appear for each algorithm and target, the corresponding best ERT in the first row. The different target Δf-values are shown in the top row. #succ is the number of trials that reached the (final) target $f_{opt} + 10^{-8}$. The median number of conducted function evaluations is additionally given in *italics*, if the target in the last column was never reached. Entries, succeeded by a star, are statistically significantly better (according to the rank-sum test) when compared to all other algorithms of the table, with $p = 0.05$ or $p = 10^{-k}$ when the number k following the star is larger than 1, with Bonferroni correction by the number of instances

running times for the later. The version which subjects the complete covariance to thresholding performs always slightly worse. The roles of the CMSA-Diag-ES and the CMSA-Thres-ES change for the rotated Rosenbrock (id 9). Here, the best results can be observed for the complete thresholding variant.

In the case of ill-conditioned functions (ids 10–14), the findings are mixed. In general, thresholding without including the diagonal does not appear to improve the performance. The strategy performs worst of all—an indicator that keeping the diagonal unchanged may be sometimes inappropriate due to the space transformation. However, since there are interactions with the choice of the thresholding parameters which may have resulted in comparatively too large diagonal elements, we need to address this issue further before coming to a conclusion. First of all for $N = 10$, all strategies are successfull in all cases for the ellipsoid (id 10), the discus (id 11), the bent cigar (id 12), and the sum of different powers (id 14). Only the CMSA-ES reaches the optimization target in the case of the sharp ridge (id 13). This, however, only twice. The reasons for this require further analysis. Either the findings may be due to a violation of the sparseness assumption or considering that this is only a weak assumption the choice of the thresholding parameters and the function should be reconsidered.

All strategies exhibit problems in the case of the group of multi-modal functions, Rastrigin (id 15), Weierstrass (id 16), Schaffer F7 with condition number 10 (id 17), Schaffer F7 with condition 1000 (id 18), and Griewank-Rosenbrock F8F2 (id 19). Partly, this may be due to the maximal number of fitness evaluations permitted. Even the best performing methods of the 2009 BBOB workshop required more evaluations than we allowed in total. Thus, experiments with larger values for the maximal function evaluations should be conducted in future research. Concerning the preliminary targets with lower precision, the CMSA-ES achieves the best results in a majority of cases. However, the same argumentation as for the step ellipsoid applies.

In the case of $N = 20$, the number of function evaluations that were necessary in the case of the best algorithms of 2009 to reach even the lower precision target of 10^{-1} exceeds the budget chosen here. Therefore, the function group is excluded from the analysis for $N = 20$ and not shown in Fig. 3 and Table 6.

The remaining group consists of multi-modal functions with weak global structures. Here, especially the functions with numbers 20 (Schwefel $x \sin(x)$), 23 (Kaatsuuras), and 24 (Lunacek bi-Rastrigin) represent challenges for the algorithms. In the case of $N = 10$, they can only reach the first targets of 10^1 and 10^0. Again, the maximal number of function evaluations should be increased to allow a more detailed analysis on these functions. For the case of the remaining functions, function 21, Gallagher 101 peaks, and function 22, Gallagher 21 peaks, the results indicate a better performance for the CMSA-ES versions with thresholding compared with the original algorithm. Again due to similar reasons as for the first group of multi-modal functions, the results are only shown for $N = 10$.

4   Conclusions and Outlook

This paper presents covariance matrix adaptation techniques for evolution strategies. The original versions are based on the sample covariance—an estimator known to be problematic. Especially in high-dimensional search spaces, where the population size does not exceed the search space dimensionality, the agreement of the estimator and the true covariance may be low. Therefore, thresholding, a comparably computationally simple estimation technique, has been integrated into the covariance adaptation process. Thresholding stems from estimation theory for high-dimensional spaces and assumes an approximately sparse structure of the covariance matrix. The matrix entries are therefore thresholded, meaning a thresholding function is applied. The paper considered adaptive entry-wise thresholding. Since the covariance matrix cannot be assumed to be sparse in general, a basis transformation was carried out and the thresholding process was performed in the transformed space. The performance of the resulting new covariance matrix adapting evolution strategies was compared to the original variant on the black-box optimization benchmarking test suite. Two main variants were considered: A CMSA-ES which subjected the complete covariance to thresholding and a variant which left the diagonal elements unchanged. While the latter is more common in statistics, it is not easy to justify its preforation in optimization. The first findings were interesting with the new variants performing better for several function classes. While this is promising, more experiments and analyses are required and will be conducted in future research. The choice of the thresholding function and the scaling parameter for the threshold represent important research questions. In this paper, the scaling factor was analyzed by a small series of experiments. The findings were then used in the benchmarking experiments. They represent, however, only the first steps of the research. Techniques to make the parameter adaptive are currently under investigation.

References

1. Bäck, T., Foussette, C., Krause, Peter: Contemporary Evolution Strategies. Natural Computing. Springer, Berlin (2013)
2. Beyer, H.-G., Meyer-Nieberg, S.: Self-adaptation of evolution strategies under noisy fitness evaluations. Genet. Program. Evolv. Mach. **7**(4), 295–328 (2006)
3. Beyer, H.-G., Schwefel, H.-P.: Evolution strategies: a comprehensive introduction. Nat. Comput. **1**(1), 3–52 (2002)
4. Beyer, H.-G., Sendhoff, B.: Lecture Notes in Computer Science. In: Rudolph, G., et al. (eds.) PPSN. Covariance matrix adaptation revisited - the CMSA evolution strategy, vol. 5199, pp. 123–132. Springer, Berlin (2008)
5. Cai, Tony, Liu, Weidong: Adaptive thresholding for sparse covariance matrix estimation. J. Am. Stat. Assoc. **106**(494), 672–684 (2011)
6. Chen, X., Wang, Z.J., McKeown, M.J.: Shrinkage-to-tapering estimation of large covariance matrices. IEEE Trans. Signal Process. **60**(11), 5640–5656 (2012)
7. Dong, W., Yao, X.: Covariance matrix repairing in gaussian based EDAs. In: 2007 IEEE Congress on Evolutionary Computation, 2007. CEC, pp. 415–422 (2007)
8. Eiben, A.E., Smith, J.E.: Introduction to Evolutionary Computing. Natural Computing Series. Springer, Berlin (2003)
9. Fan, J., Liao, Y., Liu, H.: An overview on the estimation of large covariance and precision matrices. arXiv:1504.02995
10. Fan, J., Liao, Y., Mincheva, Martina: Large covariance estimation by thresholding principal orthogonal complements. J. R. Stat. Soc.: Ser. B (Stat. Methodol.) **75**(4), 603–680 (2013)
11. Finck, S., Hansen, N., Ros, R., Auger, A.: Real-parameter black-box optimization benchmarking 2010: presentation of the noiseless functions. Technical report, Institute National de Recherche en Informatique et Automatique (2010) 2009/22
12. Fisher, T.J., Sun, Xiaoqian: Improved Stein-type shrinkage estimators for the high-dimensional multivariate normal covariance matrix. Comput. Stat. Data Anal. **55**(5), 1909–1918 (2011)
13. Friedman, J., Hastie, T., Tibshirani, R.: Sparse inverse covariance estimation with the graphical lasso. Biostatistics **9**(3), 432–441 (2008)
14. Guillot, D., Rajaratnam, B.: Functions preserving positive definiteness for sparse matrices. Trans. Am. Math. Soc. **367**(1), 627–649 (2015)
15. Hansen, N.: The CMA evolution strategy: a comparing review. In: Lozano, J.A. et al., (ed.) Towards a new evolutionary computation. Advances in estimation of distribution algorithms, pp. 75–102. Springer (2006)

16. Hansen, N., Auger, A., Finck, S., Ros, R.: Real-parameter black-box optimization benchmarking 2012: Experimental setup. Technical report, INRIA (2012)
17. Hansen, N., Ostermeier, A.: Completely derandomized self-adaptation in evolution strategies. Evolut. Comput. **9**(2), 159–195 (2001)
18. Hansen, Nikolaus: Adaptive encoding: How to render search coordinate system invariant. In: Rudolph, G., Jansen, T., Beume, N., Lucas, Simon, Poloni, Carlo (eds.) Parallel Problem Solving from Nature PPSN X. Lecture Notes in Computer Science, vol. 5199, pp. 205–214. Springer, Berlin (2008)
19. Ledoit, O., Wolf, Michael: A well-conditioned estimator for large dimensional covariance matrices. J. Multivar. Anal. Arch. **88**(2), 265–411 (2004)
20. Levina, E., Rothman, A., Zhu, J.: Sparse estimation of large covariance matrices via a nested lasso penalty. Ann. Appl. Stat. **2**(1), 245–263 (2008)
21. Marčenko, V.A., Pastur, L.A.: Distribution of eigenvalues for some sets of random matrices. Sbornik: Math. **1**(4), 457–483 (1967)
22. Meyer-Nieberg, S., Kropat, E: Adapting the covariance in evolution strategies. In: Proceedings of ICORES 2014, pp. 89–99. SCITEPRESS (2014)
23. Meyer-Nieberg, S., Kropat, E.: Communications in Computer and Information Science. In: Pinson, E., Valente, F., Vitoriano, B. (eds.) Operations Research and Enterprise System. A new look at the covariance matrix estimation in evolution strategies, vol. 509, pp. 157–172. Springer International Publishing, Berlin (2015)
24. Pourahmadi, M.: High-Dimensional Covariance Estimation: With High-Dimensional Data. Wiley, New York (2013)
25. Rechenberg, I.: Evolutionsstrategie: Optimierung technischer Systeme nach Prinzipien der biologischen Evolution. Frommann-Holzboog, Stuttgart (1973)
26. Ros, R., Hansen, N.: Parallel Problem Solving from Nature – PPSN X: 10th International Conference, Dortmund, Germany, Sept 13-17, 2008. Proceedings, chapter A Simple Modification in CMA-ES Achieving Linear Time and Space Complexity, pp. 296–305. Springer, Heidelberg (2008)
27. Schäffer, J., Strimmer, K.: A shrinkage approach to large-scale covariance matrix estimation and implications for functional genomics. Stat. Appl. Genet. Mol. Biol. **4**(1), 32 (2005)
28. Schwefel, H.-P.: Numerical Optimization of Computer Models. Wiley, Chichester (1981)
29. Stein, C.: Inadmissibility of the usual estimator for the mean of a multivariate distribution. In: Proceedings of 3rd Berkeley Symposium on Mathematical Statistics Probability, vol.1, pp. 197–206. Berkeley, CA (1956)
30. Stein, C.: Estimation of a covariance matrix. In: Rietz Lecture, 39th Annual Meeting. IMS, Atlanta, GA (1975)

Parallel Metaheuristics for Robust Graph Coloring Problem

Z. Kokosiński, Ł. Ochał and G. Chrząszcz

Abstract In this chapter a new formulation of the robust graph coloring problem (RGCP) is proposed. In opposition to classical GCP defined for the given graph $G(V, E)$ not only elements of E but also $\bar{E}$ can be subject of color conflicts in edge vertices. Conflicts in $\bar{E}$ are assigned penalties $0 < P(e) < 1$. In addition to satisfying constraints related to the number of colors and/or a threshold of the acceptable sum of penalties for color conflicts in graph complementary edges (rigidity level), a new bound called the relative robustness threshold (RRT) is proposed. Then three metaheuristics—SA, TS, EA and their parallel analogues PSA, PTS, PEA—for that version of RGCP are presented and experimentally tested. For comparison we use DIMACS graph coloring instances in which a selected percentage of graph edges E is randomly moved to $\bar{E}$. Since graph densities and chromatic numbers of DIMACS GCP instances are known in advance, the RGCP instances generated on their basis are more suitable for testing algorithms than totally random instances used so far. The results of the conducted experiments are presented and discussed.

1 Introduction

The classical graph k-colorability problem belongs to the class of NP-hard combinatorial problems [12]. This decision problem is defined for an undirected graph $G = (V, E)$ and positive integer $k \leq |V|$: is there an assignment of available k colors to graph vertices, providing that adjacent vertices receive different colors? In optimization version of the basic problem called GCP, a conflict-free coloring with minimum number of colors k is searched.

Many particular colorings of graph vertices and/or edges represent solutions of variety of practical problems that can be modeled by graphs with specific constraints put on the elements of the sets V and E. With additional assumptions many variants of the coloring problem can be defined such as equitable coloring, sum coloring,

Z. Kokosiński (✉) · Ł. Ochał · G. Chrząszcz
Faculty of Electrical and Computer Engineering, Cracow University of Technology,
Warszawska 24, 31-155 Cracow, Poland
e-mail: zk@pk.edu.pl

© Springer International Publishing Switzerland 2016
S. Fidanova (ed.), *Recent Advances in Computational Optimization*,
Studies in Computational Intelligence 655, DOI 10.1007/978-3-319-40132-4_16

contrast coloring, harmonious coloring, circular coloring, consecutive coloring, list coloring, total coloring etc. [17, 24].

One of the most interesting variants of GCP is the robust graph coloring problem (RGCP) [33]. It models a class of vertex coloring problems in which adjacency relation between graph vertices is not "stable". In certain circumstances nonadjacent vertices u and v can become adjacent and the edge (u, v) is assigned a penalty $0 < P(u, v) < 1$ when there is a color conflict: $c(u) = c(v)$. If all penalties $P(u, v)$ are known it is possible to define requirements for solution feasibility. A conflict-free coloring of all vertices in E is required, while a number of penalized color conflicts in the set of complementary edges $\bar{E}$, not exceeding certain threshold (f.i. rigidity level) can be tolerated. Given a number of colors k, the coloring with a lower rigidity level is more robust. In general, feasibility conditions can be expressed in terms of the maximum number of colors used and an upper bound on a cost function.

Similarly as GCP also RGCP is known to be NP-hard [33]. Therefore, application of approximate algorithms and metaheuristics for solving this problem is reasonable [1, 13, 14]. In literature no r-approximation algorithms for RGCP were reported so far. Research conducted in this area contains a number of algorithms and metaheuristics for RGCP [2, 25, 26, 31, 32] a new formulation of specific robust coloring problems [4] and a combination of system robustness and fuzziness [10, 15]. Research results were gathered and summarized in [31]. Other recent papers on system robustness are [9, 30].

In research papers [25, 26] there appears a problem in experimental verification of the investigated metaheuristic methods. How to measure quality of the solution, when nothing is known about chromatic properties of the given graph? How to validate the assumed penalty threshold for a feasible solution? In our approach RGCP is redefined in order to allow the system designer to use a new cost function—the relative robustness of the solution, which can be expressed by a percentage—the relative robustness $RR = 100\%$ means a conflict-free vertex coloring in both edges E and complementary edges $\bar{E}$ with $P(e) > 0$. This view is very natural and meets common expectations of system designers. For experimental verification three metaheuristics—Tabu Search (TS), Simulated Annealing (SA), Evolutionary Algorithm (EA) and their parallel versions—PTS, PTA and PEA—are used. In standard and parallel versions they were applied earlier in similar research [3, 7, 11, 19–21, 23–29]. As input graphs we used DIMACS graph coloring instances [6, 8] in which a constant percentage of graph edges E, denoted by E' is assigned penalties $0 < P(e) < 1$. Since graph densities and chromatic numbers of DIMACS GCP instances are known in advance [18], the RGCP instances generated on their basis are more suitable for testing algorithms than totally random instances used so far. The obtained results justify both theoretical assumptions and application of parallel metaheuristics for solving RGCP problem. The presented work is an extended version of the conference paper [22].

In the next section RGCP problem is defined together with its new formulation. TS/PTS, SA/PSA, EA/PEA algorithms and their parameters are presented in Sect. 3. Then, in Sect. 4, computer experiments are described and their results discussed. In conclusion some general suggestions related to the obtained results and future research in this area are derived.

2   Robust Graph Coloring Problem

GCP is defined for an undirected graph $G(V, E)$ as an assignment of available colors $\{1, \ldots, k\}$ to graph vertices providing that adjacent vertices receive different colors and the number of colors k is minimal. The resulting coloring c is called conflict-free and k is called graph chromatic number $\chi(G)$.

RGCP—A Simple Formulation

RGCP is defined for undirected weighted graph $G(V, E)$ with function $w(e) = p_{uv} \in [0, 1]$, as an assignment of available colors $\{1, \ldots, k\}$ to graph vertices, providing that

$$\forall (u, v) \in E : (p_{uv} = 1) \Rightarrow \quad c(u) \neq c(v) \tag{1}$$

and rigidity level for $\bar{E}$ is minimum

$$RL(c) = \sum_{((u,v)\in\overline{E})\wedge(c(u)=c(v))\wedge(p(u,v)>0)} p_{uv} \tag{2}$$

In some cases weight (penalty) p_{uv} may be considered as a probability of edge existence; in the classical vertex coloring $p_{uv} \in \{0, 1\}$.

For most practical problems it suffices, that

$$RL(c) \leq T \tag{3}$$

where:

T—is an assumed threshold.

The question is what value of T is reasonable for the modeled problem? How it reflects system robustness? What level of T should be guaranteed for the given system? In order to answer such questions an alternative formulation of RGCP problem is proposed. The alternative formulation of RGCP does not change the nature of the problem, but allows the system designer to apply the relative robustness level instead of the absolute robustness level which is not known in advance.

RGCP—An Alternative Formulation

Let us characterize system robustness more precisely. The robustness threshold is given by the following formula

$$RT(c) = \frac{RRT(c)}{100\%} \sum_{(e\in\overline{E})\wedge(p((u,v))>0)} p_{uv}, \tag{4}$$

where:

$RRT(c)$—is a relative robustness threshold set up by the designer and expressed in (%).

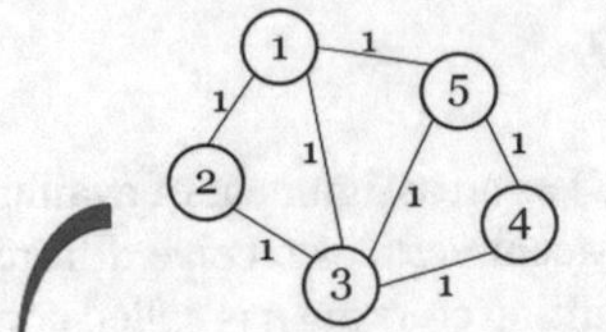

$$A = \begin{bmatrix} 0 & 1 & 1 & 0 & 1 \\ 1 & 0 & 1 & 0 & 0 \\ 1 & 1 & 0 & 1 & 1 \\ 0 & 0 & 1 & 0 & 1 \\ 1 & 0 & 1 & 1 & 0 \end{bmatrix}$$

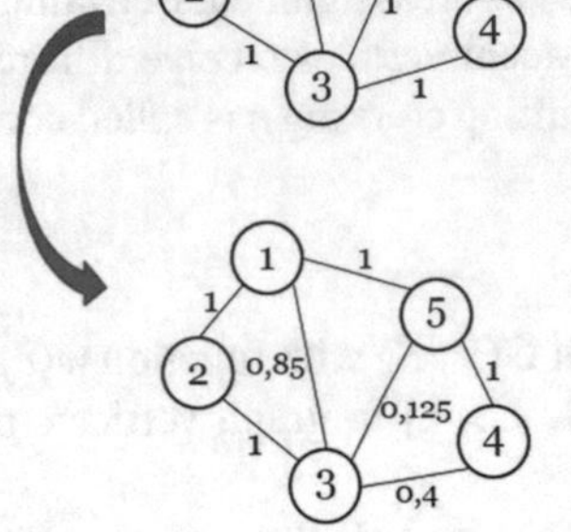

$$A = \begin{bmatrix} 0 & 1 & 0{,}85 & 0 & 1 \\ 1 & 0 & 1 & 0 & 0 \\ 0{,}85 & 1 & 0 & 0{,}4 & 0{,}125 \\ 0 & 0 & 0{,}4 & 0 & 1 \\ 1 & 0 & 0{,}125 & 1 & 0 \end{bmatrix}$$

Fig. 1 Generation of RGCP graph instances from GCP instances (42.86% of E moved to $\bar{E}$; $\bar{E}$ percentage increased from 30 to 60%)

Thus, our optimization goal is to find a coloring c satisfying Eq. (1) and inequalities (5) and (6):

$$\sum_{((u,v)\in\overline{E})\wedge(c(u)\neq c(v))\wedge(p(u,v)>0)} p_{uv} \geq RT(c) \qquad (5)$$

$$\frac{\sum_{((u,v)\in\overline{E})\wedge(c(u)\neq c(v))\wedge(p(u,v)>0)} p_{uv}}{\sum_{((u,v)\in\overline{E})\wedge(p(u,v)>0)} p_{uv}} \cdot 100\% \geq RRT(c) \qquad (6)$$

Generation of RGCP Instances

Parametrized RGCP instances can be generated by a random modification of GCP instances: a percentage of $E' \epsilon E$ is moved to $\bar{E}$ with weights $0 < p(u, v) < 1$. The E' size is denoted in the graph name: *name_size*, f.i. *games120_40* denotes 40% of E' in E of the graph *games120*. In Fig. 1 three out of seven edges are selected at random and assigned new values $p(u, v)$.

Cost Functions for RGCP

In computing cost function for the given coloring $cf(c)$ the priority is given to conflict-free coloring of edges with $p(u, v) = 1$. Otherwise, the solution is not feasible. Feasible solutions in TS/PTS and SA/PSA have the cost function:

$$cf(c) = \sum_{(u,v)\in E\cup\overline{E}} q_{uv} \qquad (7)$$

where:

$$q_{uv} = \begin{cases} 1, & if\,(c(u) = c(v))\ and\ p_{uv} = 1 \\ p_{uv} & if\,(c(u) = c(v))\ and\ p_{uv} < 1 \\ 0 & if\,(c(u) \neq c(v)) \end{cases} \qquad (8)$$

In EA/PEA algorithms the experimentally found cost function is more complex. It is defined as follows:

$$cf(c) = \sum_{(u,v)\in E\cup \bar{E}} q_{uv} + e + d + k \tag{9}$$

where:

$$q_{uv} = \begin{cases} 3 \Leftrightarrow c(u) = c(v) \wedge p_{uv} = 1 \\ p_{uv} \Leftrightarrow c(u) = c(v) \wedge p_{uv} < 1 \\ 0 \Leftrightarrow c(u) \neq c(v) \end{cases} \tag{10}$$

and

$$e = \begin{cases} 1.5 \Leftrightarrow \displaystyle\sum_{((u,v)\in\bar{E})\wedge(c(u)=c(v))\wedge(p_{uv}>0)} p_{uv} > T \\ 0 \Leftrightarrow \displaystyle\sum_{((u,v)\in\bar{E})\wedge(c(u)\neq c(v))\wedge(p_{uv}>0)} p_{uv} \leq T \end{cases} \tag{11}$$

and

$$d = \begin{cases} 3 \Leftrightarrow \exists(u,v) \in F : (p_{uv} = 1) \Rightarrow c(u) = c(v) \\ 0 \Leftrightarrow \forall(u,v) \in E : (p_{uv} = 1) \Rightarrow c(u) \neq c(v) \end{cases} \tag{12}$$

and

k—number of colors used; $k \geq \chi(G)$.

3 PTS, PSA and PEA Metaheuristics

The applications of basic metaheuristics for RGCP was reported in [25]. The first parallel metaheuristic for RGCP—Parallel Evolutionary Algorithm—was presented in [5]. In the present paper we deal also with two other popular parallel metaheuristics PTS and PSA. The details of their implementation are skipped here for the sake of brevity. In order to determine their parameters at first we investigate algorithms TS, SA and EA.

Parameters for Tabu Search Algorithm

Tabu Search metaheuristic presented in [25] is adapted for parallelization. PTS algorithm includes three TS processes that periodically exchange information when 1/3 and 2/3 of the required RR is obtained. There are at least two key parameters of TS/PTS algorithms that have to be set [7]: tMAX and MaxTenure. This parameters were found experimentally. The results of conducted experiments are shown in Tables 1 and 2.

The values of parameters recommended for TS and PTS algorithms are as follows: tMAX = 10 and MaxTenure = 15. As a selection criterion majority of optimum solutions with respect to relative robustness RRT was used.

Table 1 Efficiency of TSA with tMax (MaxTenure = 10)

Graph G(V, E)	tMax: 5			tMax: 10			tMax: 15		
	c.f.	Time (s)	RR (%)	c.f.	Time (s)	RR (%)	c.f.	Time (s)	RR (%)
queen5.5_40 $\chi(G)$ = 5 dens. = 53.3 %	6.3	0.4	91.0	**4.9**	**0.3**	**93.0**	6.2	0.5	91.2
games120_40 $\chi(G)$ = 9 dens. = 8.9 %	0	260	100	**0**	**247**	**100**	0	253	100
myciel7_40 $\chi(G)$ = 8 dens. = 13 %	1.8	795	99.8	**0**	**766**	**100**	0	745	100

Table 2 Efficiency of TSA with MaxTenure (tMAX = 10)

Graph G(V, E)	MaxTenure: 5			MaxTenure: 10			**MaxTenure: 15**		
	c.f.	Time (s)	RR (%)	c.f.	Time (s)	RR (%)	c.f.	Time (s)	RR (%)
queen5.5_40 $\chi(G)$ = 5 dens. = 53.3 %	5.2	0.5	92.6	6.5	0.4	90.7	**3.8**	**0.3**	**94.6**
games120_40 $\chi(G)$ = 9 dens. = 8.9 %	0	252	100	0	260	100	**0**	**250**	**100**
myciel7_40 $\chi(G)$ = 8 dens. = 13 %	0.2	879	100	0.7	**776**	100	0.4	833	**100**

Parameters for Simulated Annealing Algorithm

A Simulated Annealing metaheuristic for RGCP presented in [25] is adapted for parallelization. There are three important parameters of SA and PSA algorithms that have to be set [27]: MinIteration, ControlFactor (speed of convergence) and Tmax. These parameters were also found experimentally. The results of conducted experiments are shown in Tables 3, 4 and 5. We can assume Tmin = 0.25.

PSA algorithm includes also three SA processes that periodically exchange information when 1/3 and 2/3 of the required RRT is obtained. All processes resume computations with new best solution. The values of parameters recommended for PSA are the following: MinIteration = 5, ControlFactor = 0.9 and Tmax = 10.

Similarly, as a selection criterion for a given parameter the majority of optimum solutions with respect to relative robustness RR was used.

SA/PSA algorithm has SA2/PSA2 version with automatic computation of Tmax (the initial temperature), cf. [27].

Parameters for Evolutionary Algorithm

A parallel evolutionary metaheuristic for GCP presented in [19, 20] is used. For RGCP the classical GCP crossover and mutation operators may be applied. From the set of best operators proposed so far we selected the following: Conflict Elimination Crossover (CEX) [19, 20] Greedy Partition Crossover (GPX) [11], Sum-Product Par-

Table 3 Efficiency of SA algorithm with MinIteration (Tmin = 0.25; Tmax = 10; Control Factor = 0.9)

Graph G(V, E)	MinIteration: 5			MinIteration: 10			MinIteration: 15		
	c.f.	Time (s)	RR (%)	c.f.	Time (s)	RR (%)	c.f.	Time (s)	RR (%)
queen5.5_40 $\chi(G)$ = 5 dens. = 53.3 %	7.7	**0.3**	89.1	**6.5**	0.6	**90.8**	7.9	0.9	88.8
games120_40 $\chi(G)$ = 9 dens. = 8.9 %	**3.7**	29.9	**98.6**	15.9	**14.2**	94.0	9.3	18.3	96.5
myciel7_40 $\chi(G)$ = 8 dens. = 13 %	**30**	121	**97.2**	11k	58.8	–	31k	**45.7**	–

Table 4 Efficiency of SA algorithm with ControlFactor (Tmin = 0.25; Tmax = 10; Min Iteration = 5)

Graph G(V, E)	ControlFactor: 0.85			ControlFactor: 0.9			ControlFactor: 0.95		
	c.f.	Time (s)	RR (%)	c.f.	Time (s)	RR (%)	c.f.	Time (s)	RR (%)
queen5.5_40 $\chi(G)$ = 5 dens. = 53.3 %	14	**0.2**	79.3	**6.4**	0.3	**90.9**	7.9	0.5	88.9
games120_40 $\chi(g)$ = 9 dens. = 8.9 %	104	**18.3**	96.1	**4.1**	28.7	**98.5**	6.4	57.3	97.6
myciel7_40 $\chi(g)$ = 8 dens. = 13 %	2k	**75.2**	–	13.8	119	98.8	**5.4**	240	**99.5**

Table 5 Efficiency of SA algorithm with Tmax (Tmin = 0.25; MinIteration = 5; Control Factor = 0.9)

Graph G(V, E)	Tmax: 5			Tmax: 10			Tmax: 15		
	c.f.	Time (s)	RR (%)	c.f.	Time (s)	RR (%)	c.f.	Time (s)	RR (%)
queen5.5_40 $\chi(G)$ = 5 dens. = 54 %	10	**0.2**	85.8	**9.5**	0.3	**86.6**	12.0	0.3	83.1
games120_40 $\chi(G)$ = 9 dens. = 9 %	**4.0**	**24.2**	**98.5**	5.4	29.8	98.0	6.9	30.7	97.4
myciel7_40 $\chi(G)$ = 8 dens. = 13 %	19	**94.1**	98.3	**14.5**	122	**98.7**	18.1	137	98.3

tition Crossover (SPPX) [19, 20] Best Coloring Crossover (BCX) [28] and mutation First Fit (FF).

In conflict-based crossovers for GCP the assignment representation of colorings is used [16] and the offspring tries to copy conflict-free colors from the parents.

In CEX each parental chromosome $p1$ and $p2$ is partitioned into two blocks. The first block consists of conflict-free nodes while the second one is built of the remaining nodes that participate in conflicts. The latest block in both chromosomes is replaced by corresponding block of colors taken from the other parent. As a result two offspring chromosomes $o1$ and $o2$ are obtained. In many cases a significant reduction of color conflicts is noticed.

In BCX conflict numbers for all vertices in any colorings is computed and consecutive colors of single offspring are selected on this basis. In some cases the tournament mechanism is adopted in BCX for resolving color selection.

For definitions of GPX, SPPX and FF operators the reader may refer the bibliography.

If not stated otherwise the EA parameters are as follows: mutation First Fit, crossover probability 0.8, mutation probability 0.8, population size $102 = 3 \times 34$, roulette selection, max number of iterations 5000, relative robustness threshold: 70 %. PEA parameters: island migration scheme (3 islands, subpopulation size 34, all-to-all migration, best individuals replace the worst), migration size: 15 % of subpopulations, migration rate 7.

The above parameters were used for all experiments except of those where particular settings are given.

4   Experimental Results

The application for testing of TS/PTS and SA/PSA methods for RGCP was written in C++, while GUI in C#, accordingly. Microsoft Visual Studio 2008 v.9.0 was used. Computer experiments we performed on a machine with Intel Core 2 Duo, CPU P8400, 2.27 GHz, 4 GB RAM. Experiments with EA/PEA metaheuristics were conducted in order to confirm expected PEA behavior, compare evolutionary operators and check efficiency of PEA for solving RGCP. The application was created in C++ using Visual Studio and run on a machine with Pentium 4, 1.8 GHz, 1 GB RAM.

Goals of Optimization and Algorithms

The programs solve RGCP problem providing value of cost function, relative robustness RR and the number of colors. There is a pool of algorithm's variants to choose from, including sequential and parallel versions.

The main purpose of optimization was obtaining the best available robustness with minimum number of colors used. It is possible to:

a. compute maximal relative robustness (RR) for the given number of colors (algorithms: TS, SA, SA2)
b. compute the above in parallel (algorithms: PTS, PSA2)
c. find a robust coloring with minimal number of colors for the given RRT (algorithms TS, SA^C, SA2^C)
d. compute the above in parallel (algorithms PTS^C, PSA2^C)

In addition it is possible to:

e. compare serial and parallel version of EA (algorithms EA, PEA)
f. verify efficiency of four crossover operators (algorithms EA, PEA)
g. compare solutions of RGCP for 6 random graph instances with different E' size (algorithms EA, PEA)

In next subsections a number of experiments performed with the help of both programs is reported.

TS Versus SA

The first experiment was devoted to efficiency comparison of sequential versions of the two basic metaheuristics. For comparison 9 DIMACS graphs were selected the number of colors was set up to $k = \chi(G)$. The results are shown in Fig. 2. For most combinations of test graphs and the size of the set E' the TS outperforms SA in terms of relative robustness RR of the modeled system. Typically, TS was able to achieve 100 % RR and never less than 95 %. SA issued a bit worse results: for only three graphs maximum RR = 100 % was obtained. In majority of cases RR was within the range 91–99 %. In a single case when SA algorithm failed to achieve a conflict-free coloring for a graph with density 46 %, the value k was incremented. Basically, more dense graph are more difficult to color. SA is simpler than TS, much faster for bigger graphs and its power relies on randomization in a higher degree than TS which is more precise in searching for a good solution, checking all color combinations for all vertices in each iteration. Regardless of E' size both algorithms delivered solutions with similar values of cost function and RR. However, when E' size is bigger, the number of iterations required to obtain a conflict-free coloring decreases in both methods and the speed of TS decreases. The graph density is more essential than the graph size.

TS^C Versus SA^C and SA2^C

Three subsequent experiments were based on eight graphs instances with the percentage of E' equal 60 %. The number of colors was computed that allows to achieve the given level of system relative reliability RRT on the levels 70, 85, and 95 % respectively.

In Figs. 3, 4 and 5 the order of bars characterizing experiments for the given input graph is as follows: $\chi(G)$, TS^C, SA^C and SA2^C. The results depicted in Fig. 3 present the number of colors used by the corresponding methods for the set of all 8 graphs with $RRT = 70$ %. The average number of colors used is as follows: TS$^C = 4.5$, SA$^C = 4.625$ and SA2$^C = 4.375$, with the average sum of $\chi(G)$ equal 8.5.

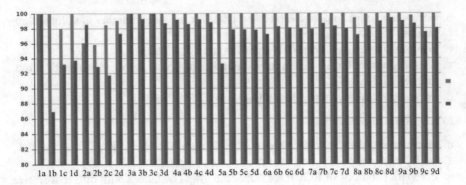

Fig. 2 Relative robustness *RR* [TS—*blue*, SA—*red*]. Graphs: *1*—queen5.5, *2*—queen6.6, *3*—myciel, *4*—huck, *5*—david, *6*—games120, *7*—anna, *8*—mulsol.i.4, *9*—myciel7. E' : a = 10%, b = 20%, c = 40%, d = 60%. Number of colors $k = \chi(G)$

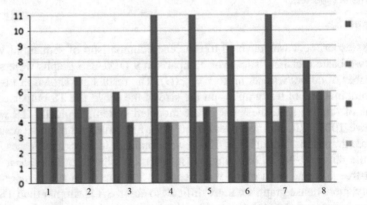

Fig. 3 Number of colors required for $RRT = 70\%$, [$\chi(G)$, TS^C, SA^C, $SA2^C$]. Basic graphs: *1*—queen5.5, *2*—queen6.6, *3*—myciel, *4*—huck, *5*—david, *6*—games120, *7*—anna, *8*—myciel7; $E' - 60\%$

Similarly, the results depicted in Fig. 4 can be characterized in short by average number of colors used by the corresponding methods for the set of all 8 graphs $RRT = 85\%$: $TS^C = 5.125$, $SA^C = 5.125$ and $SA2^C = 5.0$ with the same sum of $\chi(G)$.

Finally, the general results depicted in Fig. 5 can be summarized by average number of colors used by the corresponding methods for the set of all 8 graphs with $RRT = 95\%$: $TS^C = 5.625$, $SA^C = 7.0$ and $SA2^C = 6.75$ with respect to the sum of $\chi(G)$ as above.

PTS^C Versus $PSA2^C$

The experiment reported in previous subsection was then repeated for parallel meta-heuristics PTS^C and $PSA2^C$ (with an automatic computing of initial temperature Tmax). Three subsequent experiments were based on eight graphs instances with

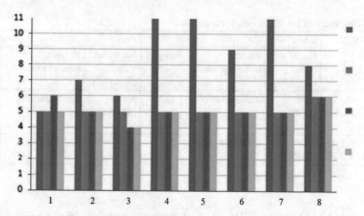

Fig. 4 Number of colors required for $RRT = 85\%$, $[\chi(G), TS^C, SA^C, SA2^C]$. Graphs: *1*—queen5.5, *2*—queen6.6, *3*—myciel, *4*—huck, *5*—david, *6*—games120, *7*—anna, *8*—myciel7; $E' - 60\%$

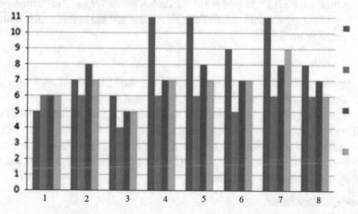

Fig. 5 Number of colors required for $RRT = 95\%$, $[\chi(G), TS^C, SA^C, SA2^C]$. Graphs: *1*—queen5.5, *2*—queen6.6, *3*—myciel, *4*—huck, *5*—david, *6*—games120, *7*—anna, *8*—myciel7; $E' - 60\%$

the percentage of E' equal 60%. The number of colors was computed that allows to achieve the given level of system relative reliability RRT on the levels 70, 85, and 95% respectively.

Results of the research concerning minimization of colors in a conflict free robust graph coloring with fixed RR level can be summarized by the average number of colors used by the corresponding sequential and parallel methods for the set of all eight graphs from previous subsection: $PTS^C = 5.5$, $TS^C = 5.625$, $PSA2^C = 6.625$ and $SA2^C = 6.75$ when the average $\chi(G)$ is 8.5. As expected, the results obtained by parallel metaheuristics are slightly improved in comparison to classical metaheuristics.

Table 6 Efficiency of EA algorithm versus PEA

Graph G(V, E)	EA			PEA		
	No. of sol. with cf. $\leq \chi(G)$	No. iter. min./avg./max.	Avg. t (s)	No. of sol. with cf. $\leq \chi(G)$	No. iter. min./avg./max	Avg. t (s)
queen7.7_40 $\chi(G) = 7$ dens. = 40%	5/30 (17%)	22/1362/4897	12.8	9/30 (30%)	103/**1030**/3071	**9.8**
queen8.8.40 $\chi(G)$ = 9 dens. = 36%	3/30 (10%)	1948/2222/2636	23	8/30 (26.6%)	129/**2051**/4548	**21.7**

In addition total computation time of sequential and parallel versions of both metaheuristics was compared for the set of all eight graphs from previous subsection. Average processing time of PTS^C is 666.8 (s) while TS^C 693.2 (s). The average processing time of $PSA2^C$ is 674.8 (s) while $SA2^C$ 435.9 (s). Solutions generated by PTS^C are often repeatable while $PSA2^C$ results are less stable and with similar quality as those from $SA2^C$.

EA Versus PEA

In this computer experiment simulated migration based PEA was favorable compared with EA. Test graphs: queen7.7 with $\chi(G) = 7$ and queen8.8 with $\chi(G) = 9$ were used for generation of graphs queen7.7.r and queen8.8.r, in which 40% of edges in E was moved at random to $\bar{E}$ and assigned random costs $0 < P(e) < 1$. RRT was set at 70%. Number of trials 30. The obtained results show that PEA is able to find in average more good solutions with cf. $\leq \chi(G)$, with less iterations and faster than EA with analogous parameters (see Table 6).

Crossover Quality in EA

The next experiment was devoted to time and iteration efficiency comparison of four crossover operators known from literature: CEX [20], BCX [28], GPX [11] and SPPX [20]. The BCX is taken for comparison for the first time. In the experiment First Fit mutation and roulette wheel selection were used. The population size was 70, crossover probability 0.8, mutation probability 0.2, $RRT = 90\%$. In SPPX probabilities of SUM and PRODUCT operations were both equal 0.8.

The results of the experiment are presented in Table 7. CEX operator provides a conflict-free solution in minimal time, while BCX requires the least number of iterations. This is because of relative complexity of BCX which is most elaborated operator from the set.

Efficiency of EA/PEA With Input Graphs With Various Size of E′

In the final experiment the percentage of E' edges, which are randomly selected from E and then moved to $\bar{E}$, varies from 10 to 60% for 6 graph coloring instances. This leads to a relaxation of constrains put on the coloring function since less color conflicts in E is possible. The penalties for color conflicts in E' are generated at

Table 7 Time and iteration efficiency of crossover operators in EA

Graph G(V, E)	CEX			BCX			GPX			SPPX		
	Av. no. iter.	Av. t(s)	t(s)/100 iter.	Av. no. iter	Av. t(s)	t(s)/100 iter.	Av. no. iter.	Av. t(s)	t(s)/100 iter.	Av. no. iter.	Av. t(s)	t(s)/100 iter.
myciel7.20 $\chi(G) = 8$ dens. = 13%	60.8	**0.55**	**0.9**	**24.2**	0.97	4	46.6	0.66	1.42	32.5	0.98	1.73
games120.20 $\chi(G) = 9$ dens. = 8.9%	38.8	**0.14**	**0.36**	**15.7**	0.18	1.2	41.2	0.22	0.53	31.2	0.29	0.93
le450.15b.20 $\chi(G) = 15$ dens. = 8.01%	150	**4.44**	**2.95**	**86.3**	14.8	17.1	113	4.53	4	121	4.87	4.01

Table 8 Efficiency of EA/PEA with input graphs with various sizes of E'

Graph G(V, E)	Percentage of E' edges of E with $0 < p(e) < 1$ (%)	Cost of the coloring			Number of colors			Number of conflicts		
		Avg.	Max.	Min.	Avg.	Max	Min	Avg.	Max	Min
david.r $\chi(G) \leq 11$ dens. = 10.8 %	10	10.9	13	10.6	10.2	11	10	0.8	1	0
	20	10.2	10.4	9.5	9.5	10	9	1.9	3	1
	40	10.1	10.3	9.7	9	10	8	2.4	4	1
	60	9.2	9.5	8.6	8.7	9	8	3.3	6	2
anna.r $\chi(G) \leq 11$ dens. = 5.2 %	10	10.6	12.4	10.3	9.6	10	9	1.3	2	1
	20	8.7	9.3	8.0	7.7	8	7	3.8	4	3
	40	8.7	9.4	7.9	7.8	8	7	3.8	5	2
	60	8.8	9.4	8.3	7.5	8	7	4.1	5	3
games120.r $\chi(G) \leq 9$ dens. = 8.9 %	10	9.1	10.3	8.5	8.9	9	8	0.4	2	0
	20	9.5	12.0	8.3	8.2	9	8	1.8	2	0
	40	8.8	10.1	8.5	8	8	8	2.1	4	1
	60	8.6	9.7	8.1	8	8	8	2.1	3	2
huck.r $\chi(G) \leq 11$ dens. = 11.4 %	10	9.8	10.3	9.3	9.2	10	9	2.4	3	1
	20	10.5	12.4	9.6	9.8	10	9	2.4	3	1
	40	10.1	10.6	9.8	9.1	10	9	2.9	4	1
	60	9.5	8.9	10.1	8.8	10	8	3.4	6	1

(continued)

Table 8 (continued)

Graph G(V, E)	Percentage of E' edges of E with 0 < p(e) < 1 (%)	Cost of the coloring			Number of colors			Number of conflicts		
		Avg.	Max.	Min.	Avg.	Max	Min	Avg.	Max	Min
myciel7.r χ(G) ≤ 8 dens. = 13 %	10	7.3	7.6	7.1	7.1	8	7	0.9	1	0
	20	7.3	7.6	7.1	7	7	7	1	1	1
	40	7.2	7.9	7.1	7	7	7	1	1	1
	60	7.1	7.3	7.1	7	7	7	1	1	1
queen5.5.r χ(G) ≤ 5 dens. = 53.3 %	10	5.7	9.1	5	5.2	6	5	0.5	2	0
	20	5.6	6.9	5	5.3	6	5	0.9	4	0
	40	5.2	6.9	5	5.1	6	5	0.4	3	0
	60	5.4	7.5	5	5.2	6	5	0.3	2	0

random. Three measures are taken into account for efficiency comparison: cost of the coloring, the number of colors used and the number of conflicts. One can notice several regularities in the obtained output data (see Table 8). At first, the average cost of the coloring decreases when the percentage of E' edges increases. The average number of colors also decreases with the size of E'. The maximal number of colors is equal or less then $\chi(G)$ of the original DIMACS graph and as a rule the minimal number of colors is much lower. For "easy" graphs the number of conflicts in an accepted solution is higher when the percentage of E' is high. For more "difficult" graphs like queen5.5.r the above observations are not valid.

5 Conclusions

In this paper new formulation of RGCP problem is given that seems to be more appropriate for designers of robust systems. Relative robustness is a versatile measure for characterization of any robust system modeled by a graph. For experimental verification two popular parallel metaheuristics TS/PTS and SA/PSA were used. In addition applicability and efficiency of EA/PEA metaheuristic for RGCP was investigated.

The optimization goal was to satisfy a new measure—the relative reliability threshold. We proposed a new method of test instance generation. Instead of totally random test graphs used so far the experimental verification was performed on a set of benchmark graphs derived from the DIMACS graph coloring instances by random modification of a given percentage E' of graph edges$E$. The results confirm that the proposed approach and the used tools can be efficiently used for practical applications.

The EA/PEA implementation contained two most efficient crossover operators for GCP: CEX and BCX, both having different characteristics. It would be desirable in EA/PEA to use simultaneously both operators or create a new combined randomized operator. This shall be a topic of further work.

An interesting goal of the future research is to apply to RGCP—and verify experimentally—more metaheuristics like Parallel Immune Algorithm (PIA), Ant Colony Optimization (ACO), Particle Swarm Optimization (PSO) and others [13, 14]. For particular applications the robustness measures can be modified to reflect specific properties of the given system.

References

1. Alba, E. (ed.): Parallel Metaheuristic - A New Class of Algorithms. Wiley, Hoboken (2005)
2. Archetti, C., Bianchessi N, Hertz, A.: A branch-and-price algorithm for the robust graph coloring problem. Les Cahiers du Gerad, G-2011–75, Montreal (2011)
3. Bouziri, H., Jouini, M.: A tabu search approach for the sum coloring problem. Electron. Notes Discrete Math. **36**, 915–922 (2010)

4. Bracho, R.L., Rodriguez, J.R., Martinez, F.J.: Algorithms for robust graph coloring on paths. In: Proceedings of 2nd International Conference on Electrical and Electronics Engineering, Mexico, pp. 9–12. IEEE (2005)
5. Chrzaszcz, G.: Parallel evolutionary algorithm for robust scheduling in power systems. M.Sc. thesis, Cracow University of Technology (in Polish) (2009)
6. COLOR web site. http://mat.gsia.cmu.edu/COLOR/instances.html
7. Dabrowski, J.: Parallelization techniques for tabu search. In: Proceediongs of 8th International Conference on Applied Parallel Computing: State of the Art in Scientific Computing (2007)
8. DIMACS ftp site. ftp://dimacs.rutgers.edu/pub/challenge/graph/benchmarks/
9. Deleplanque, S., Derutin, J.-P., Quilliot, A.: Anticipation in the dial-a-ride problem: an introduction to the robustness. In: Proceedings of the 2013 Federated Conference on Computer Science and Information Systems, FedCSIS'2013, pp. 299–305. Kraków, Poland (2013)
10. Dey, A., Pradhan, R., Pal, A., Pal, T.: The fuzzy robust graph coloring problem. In: Satapathy, S.C., et al. (eds.) Proceedings of the 3rd International Conference on Frontiers of Intelligent Computing: Theory and Applications (FICTA) 2014 - 1. Advances in Intelligent Systems and Computing Proceedings, vol. 327, pp. 805–813. Springer, New York (2015)
11. Galinier, P., Hao, J.-P.: Hybrid evolutionary algorithm for graph coloring. J. Comb. Optim. 3(4), 374–397 (1999)
12. Garey, R., Johnson, D.S.: Computers and Intractability: A Guide to the Theory of NP-Completeness. Freeman, New York (1979)
13. Gendreau, M., Potvin, J.Y. (eds.): Handbook of Metaheuristics. International Series in Operations Research & Management Science. Springer, New York (2010)
14. Glover, F., Kochenberger, G.A. (eds.): Handbook of Metaheuristics. Kluwer, Boston (2003)
15. Gładysz, B.: Fuzzy robust courses scheduling problem. Fuzzy Optim. Decis. Mak. 6, 155–161 (2007)
16. Hutchinson, G.: Partitioning algorithms for finite sets. Commun. ACM 6, 613–614 (1963)
17. Jensen, T.R., Toft, B.: Graph Coloring Problems. Wiley Interscience, New York (1995)
18. Johnson, D.S., Trick, M.A.: Cliques, Coloring and Satisfiability: Second DIMACS Implementation Challenge. DIMACS Series in Discrete Mathematics and Theoretical Computer Science, vol. 26 (1996)
19. Kokosiński, Z., Kołodziej, M., Kwarciany, K.: Parallel genetic algorithm for graph coloring problem. In: Proceedings of the International Conference on Computational Science, ICCS'2004, LNCS, vol. 3036, pp. 215–222 (2004)
20. Kokosiński, Z., Kwarciany, K., Kołodziej, M.: Efficient graph coloring with parallel genetic algorithms. Comput. Inf. 24, 123–147 (2005)
21. Kokosiński, Z.: Parallel metaheuristics in graph coloring. Bulletin of the National University "Lviv Politechnic". Series: Computer sciences and information technologies, vol. 744, pp. 209–214 (2012)
22. Kokosiński, Z., Ochał, Ł.: Evelution of metaheuristics for robust graph coloring problem. In: Proceedings of the 2015 Federated Conference on Computer Science and Information Systems, FedCSIS'2015, Łódź, Poland. Annals of Computer Science and Information Systems, vol. 5, pp. 519–524 (2015)
23. Kong, Y., Wang, F., Lim, A., Guo, S.: A new hybrid genetic algorithm for the robust graph coloring problem. AI 2003, LNAI, vol. 2903, pp. 125-136 (2003)
24. Kubale, M. (ed.): Graph Colorings. American Mathematical Society, Providence (2004)
25. Lim, A., Wang, F.: Metaheuristic for robust graph coloring problem. In: Proceedings of the 16th IEEE International Conference on Tools with Artificial Intelligence, ICTAI (2004)
26. Lim, A., Wang, F.: Robust graph coloring for uncertain supply chain management. In: Proceedings of 38th Annual Hawaii International Conference on System Science, HICSS 2005, IEEE, 81b (2005)
27. Łukasik, S., Kokosiński, Z., Świętoń, G.: Parallel simulated annealing algorithm for graph coloring problem. In: Proceedings of International Conference Parallel Processing and Applied Mathematics, PPAM'2007, LNCS, vol. 4967, pp. 229–238 (2008)

28. Myszkowski, P.B.: Solving scheduling problems by evolutionary algorithms for graph coloring problem. In: Xhafa, F., Abraham, A. (eds.): Metaheuristics for Scheduling in Industrial and Manufacturing Applications. Studies in Computational Intelligence, vol. 128, pp. 145–167 (2008)
29. Pahlavani, A., Eshghi, K.: A hybrid algorithm of simulated annealing and tabu search for graph colouring problem. Int. J. Oper. Res. **11**(2), 136–159 (2011)
30. Ruta, D.: Robust method of sparse feature selection for multi-label classification with naive Bayes. In: Proceedings of the 2014 Federated Conference on Computer Science and Information Systems, FedCSIS'2014, Warsaw, Poland, pp. 375–380 (2014)
31. Wang, F., Xu, Z.: Metaheuristics for robust graph coloring. J. Heuristics **19**(4), 529–548 (2013)
32. Xu, M., Wang, Y., Wei, A.: Robust graph coloring based on the matrix semi-tensor product with application to examination timetabling. Control Theory Technol. **12**(2), 187–197 (2014)
33. Yáñez, J., Ramirez, J.: The robust coloring problem. Eur. J. Oper. Res. **148**(3), 546–558 (2003)

Author Index

© Springer International Publishing Switzerland 2016
S. Fidanova (ed.), *Recent Advances in Computational Optimization*,
Studies in Computational Intelligence 655, DOI 10.1007/978-3-319-40132-4

303

Printed in the United States
By Bookmasters